THERMAL ANALYSIS

Techniques and Applications

THERMAL ANALYSIS

Techniques and Applications

Prof. N. K. Kaushik

Former Head

Department of Chemistry

University of Delhi, Delhi

Dr. S. K. Shukla

Associate Professor

Department of Polymer Science

Bhaskaracharya College of Applied Sciences

University of Delhi, Delhi

I.K. International Pvt. Ltd.

NEW DELHI

Published by

I.K. International Pvt. Ltd.
4435-36/7, Ansari Road, Daryaganj
New Delhi–110 002 (India)
E-mail: tech_info@techsarworld.com
Website: www.techsarworld.com

ISBN: 978-93-86768-73-5

Published/Printed by Krishan Makhijani for I.K. International Pvt. Ltd., 4435-36/7, Ansari Road, Daryaganj, New Delhi–110 002.

Foreword

The present book is a commendable effort which gives an exhaustive account of various techniques of thermal analysis depending on changes in chemical, physical, mechanical, electrical and optical properties of materials, along with theoretical concepts and applications. It provides a comprehensive treatment with clarity of concepts and principles of thermal analytical techniques.

The book is divided into nine chapters. Chapter 1 provides basic information and historical developments, basic concepts involved in different techniques of thermal analysis. Chapter 2 deals with detailed discussion of different thermal properties as involved in associated techniques. Thermogravimetric analysis discussed in Chapter 3, while differential thermal analysis, differential scanning calorimetry and thermometric titrations have been discussed in chapter 4. Chapter 5 deals with oxidation-induction time/temperature and a few other techniques while Chapter 6 deals with several techniques involved in thermomechanometry. A critical account of thermophotometry has been presented in Chapter 7. Integration of two techniques such as TG-DTA sometimes has interesting applications and these have been discussed in Chapter 8. Chapter 9 deals with details of experimental procedure to be followed in teaching, industrial and research laboratories.

Thus, the book gives a detailed account of various sub-techniques of thermal analysis with emphasis on concepts and applications. It would be very useful for postgraduate students, research scientists and technologists (engaged in materials research and development). It can even be used as a reference book in general and also for courses in Analytical Chemistry.

Prof. R. P. Rastogi,
(INSA Hon. Scientist)
Ex. Vice Chancellor, B.H.U.
Former Head, Chemistry Dept.
Gorakhpur University, India

Preface

Overview of Book

Authors acknowledge the platform provided by Delhi University to gather valuable information and experience to compile this book on Thermal Analysis. The objective of this book is to be a self-contained, practical, entry-level text on Thermal Analysis, which connects basic science to technology. The book is intended to be accessible to a wide range of audience from graduate students to a researcher in chemistry, physics, pharmacy, engineering and electronics. Thus, may be useful for industry personnel, laboratory technicians and chemists. The exposition is based upon integrated basic principles of thermal changes applied to material characterization, with applied computer science and instrumentation. The chapters intend to bridge the gap between the case studies and research problems, dealing with new applications and often these new applications also require chemical models beyond those already existing. So, it is important to have a reasonable understanding of the literature on the topic with proofs and results available elsewhere, however the important ones were complied in the books.

The book is based upon the author's teaching, research and working experience with the instruments. In addition, the results from research papers on chemical synthesis, characterization and observation during instrument operation were incorporated. Every attempt is made to keep the book self-contained through an integrated approach, without depending heavily on prerequisites, especially with a diverse readership and interdisciplinary topics.

Organisation of the Book

Chapters 1 and 2 cover the basics for the entire techniques with objective to generate interest in the readers. The salient topic of the first chapter: history of techniques, milestone discoveries in the field, nomenclatures, etc. The second chapter deals with the different types of thermal events of a material such as thermal conductivity, thermal resistivity, inertia, etc. which is the basis for development of thermal techniques and also suitable for technical development of the thermal techniques.

Advanced and special techniques are covered in Chapters 3 to 7 and can be selected according to the instructor's or reader's interests.

Chapter 3 covers more advanced and general topics for mass change upon heat treatment of materials. The derivatization of the information is able to enhance information gathering capacity such as DTG explained with suitable examples.

Chapter 4 introduces the theory, basic instrumentation and applications of the DTA and DSC techniques.

Chapter 5 introduces the thermal analysis in control oxidative atmosphere suitable for industrial estimations. The least explored techniques are also discussed in brief.

Chapter 6 deals with the techniques suitable for estimation of mechanical properties in different conditions such as free, dynamic, static force with suitable illustrations.

Chapter 7 discusses techniques that are important for surface scientists. These techniques are introduced for microscopic, luminescence measurement of a material under heat treatment.

The blending of techniques is developed as a complimentary platform to enhance an application range for a scientist. However, the integration of two techniques adds some advantage, however also restricts its sensitivity.

Chapter 8 continues the discussion of the techniques from the previous chapter with suitable technical applications. Some complimentary techniques such as TG-DTA, TG-DSC, TG-MS, TG-IR and TG-GC are explained keeping in view their wide applications.

Chapter 9 deals with the use of techniques in laboratory for introduction of an experiment. These experiments may be introduced in an educational, research or an industrial laboratory.

The book concludes with appendices containing a list of worldwide manufacturers of instruments, books, national and international events along with a glossary of technical terms.

Distinct Features of the Book

- Both theoretical and practical methods are emphasized based on the utility, with respect to the scientific problems.
- It clearly shows that thermal treatment on discontinuous as well as non-smooth properties (the random properties) need to be estimated for scientific development. These ideas are elaborated keeping in the mind the interest of the audience.
- Basic principles of thermal techniques are used to set up the practical foundations through clear and systematic explanation, making the book accessible as a reference to many who work with applications.
- Insightful and useful materials are used so that the book can be readily used for realistic applications.
- Clarity has been the utmost priority for entry level students. The notations are clearly identified with the quantity it symbolizes, rather than arbitrarily selected. However, some compromise has been made for some standard notations.

The authors will be grateful to the readers for valuable suggestions towards improvement of the book.

N. K. Kaushik
S. K. Shukla

The Authors

Professor N. K. Kaushik retired from Department of Chemistry, University of Delhi. He has more than 40 years of teaching experience. He taught Inorganic and Analytical Chemistry including Thermo-analytical methods of analysis to MSc, MPhil and PhD students. He has also served as Head of the Department from 2004-2007. He served as President Section of Chemical Sciences of 95th National Science Conference held in Visakhapatnam in January 2008. His major research interests are microanalytical techniques in inorganic analysis, synthetic and structural chemistry, coordination behavior of organometallics with biologically active ligands, Thermoanalytical behavior of organometallic complexes and calculation of kinetic parameters under non-isothermal conditions. Professor Kaushik has supervised research work of more than 35 PhD and 21 MPhil students. He has published more than 200 research papers in national and international journals of repute. He has delivered many invited talks at various national and international conferences. He is the life member of Indian Science Congress Association, National Academy of Sciences, India; member of International Confederation of Thermal Analysis and Calorimetry (ICTAC), Indian Thermal Analysis Society, BARC, India for several years. He was a member of the committee appointed by the Ministry of Health and Family Welfare (GOI) for setting of Tobacco Testing Laboratories; member of National Writing Team of the NCERT for writing class XI and XII chemistry textbooks in Hindi medium.

Dr. S. K. Shukla is faculty in the Polymer Science Department, Bhaskaracharya College of Applied Sciences, University of Delhi. He has published more than 100 papers and abstracts in different international, national journals and conferences proceedings along with ten contributory chapters in books of international repute. Edited a book entitled" **Advanced Carbon Materials and Technology**, for John Wiley, Scrivener and special issue for Material Today Proceedings. Member of editorial boards and reviewer of many internationals journal like *American Journal of Biosensors & Bioelectronic, Sensors and Actuators and members of various scientific societies.*

Dr. Shukla has also been awarded some of national and international awards for his scientific contributions like, Advance Materials Letter Award-2012, International Association Advance Material Scientists Award- 2011, Indian Association Solid State and Allied Chemists (Bronze Medal-2011), Best Paper award in International Conference on Advance Polymeric Materials- 2013 at Central Institute of Plastics Engineering and Technology, Lucknow, Dr. N.G. Patel Award in NPTS-16 Lucknow University, Advance Material World Congress-2013, Izmir Turkey and Best Teacher for Innovation Award by University of Delhi. His major research interests are Advanced Materials, Microanalytical Techniques, Nanocomposite, Bio-conducting Polymers and Optochemical Devices.

Contents

1

Introduction and Historical Development

1.1 INTRODUCTION

The analytical methods based on the determination of changes in chemical, physical, mechanical, electrical and optical properties of materials on heating at constant rate in a controlled atmosphere are known as **Thermal Methods of Analysis [1].** The definition is usually also extended to include isothermal analysis, in which the property of interest is measured as function of time. The methods now have more than eleven decades of history. During this period, various techniques were invented and used in many fields. Considering the long history of the steady development, it is often thought that thermal analysis is a mature technique and has only little room for further progress as in the IR spectroscopy. However, it is not the case with this technique. Many frontiers explored such as magnetic resonance imaging, nanocalorimetry, atomic force microscopy and new trends are in progress. These methods have been developed out of the scientific studies of the changes in the properties of a sample which occur on heating such as, when ice is heated, it melts at 0°C and then boils at 100°C; when sugar is heated it first melts, and then forms brown caramel. In fact, every material behaves in a characteristic way when heated. While the application of the technique began in research, it has been diffusing into control of processes and qualities. Today, thermal analysis is an essential tool for material research and development, while quality assurance is now one of the hot topics in thermal analysis [2-3].

The more frequently used thermal methods of thermal analysis are shown in Table 1.1, with their commonly used names.

Thermal analysis techniques may also be classified into three groups; depending on the way in which the physical properties are recorded:

1. The absolute value of property itself can be measured; for example, mass.
2. The difference between the same property of the sample and that of standard material, such as temperature difference.

TABLE 1.1 Thermal methods

Technique	Abbreviation	Property	Uses
Thermogravimetry	TG	Mass	Decompositions, oxidations
Differential thermal analysis	DTA	Temperature difference	Phase changes, reactions
Differential scanning calorimetry	DSC	Power difference or heat flow	Heat capacity, phase changes, reactions
Evolved gas analysis	EGA	Evolved gas	Composition of volatile component
Thermomechanical analysis	TMA	Deformations, dimensional change	Mechanical changes
Dynamic mechanical analysis	DMA	Moduli	Expansion, phase changes
Dielectric thermal analysis	DETA	Electrical	Decomposition
Thermo-optometry		Optical	Colour change
Less frequently used techniques			
Thermosonimetry	TS	Sound	Materials change
Thermoluminescence	TL	Light emission	Transition
Thermomagnetometry	TM	Magnetic property	Curie point
Thermoacoustometric Analysis	TAA	Acoustic properties	
Thermomicroscopy	TM	Image	Surface structure
Dilatometry	DIL	Length	Dimensional change
Dynamic reflectance spectroscopy	DRS	Reflectance	Structural transition
Emanation thermal analysis	ETA	Radioactive	Radioactive disintegration

3. The rate at which the property is changing can be measured; it is called derivative measurement.

The various organizations, national and international are working for nomenclature, abbreviations, definitions and formulation of standards, which have formulated a particular system of nomenclature which is opted in the book but is not always used in journals and some fields of science.

1.2 HISTORICAL DEVELOPMENT

These methods of analysis are in use since ancient times, with the efforts to learn the changes in mass of a heated substance. The objective was to follow the course of dehydration, oxidation or especially to discover the temperature at which a substance begins to decompose, a datum of importance for metallurgical operations. That is why throughout the 19th and early 20th centuries, a lot of researchers started constructing weight loss curves as a function of the temperature, weighing being made with an ordinary balance after the specimen was heated. Truchot heated pyrites and pyrrhotite as early as 1907 using a method resembling the present thermogravimetry. In 1912, an electromagnetic compensation balance was constructed by Urbain and Boulanger to follow the efflorescence of dehydrated salt. While the word thermobalance was introduced in science by K. Honda in 1915, he designed and used the first TG apparatus. Later, Shibata rendered the Honda balance into self recording by converting it into a null type balance. Afterwards, Somiya modified it suitably to use it with the furnace at 1300°C. In 1923, Guichards made the realization of linear elevation temperature with respect to time. Further, C. Duval recorded the thermograms by using Chevnard Thermorecording balance of many inorganic compounds known up to 1963 and presented his observations in his book, "Inorganic Thermogravimetry" published by Elsevier. However, the modern thermal instruments are developed with series of milestone discoveries. An effort has been made to compile the milestone discoveries in the field (Table 1.2).

All the methods are based on measurement of relative temperatures except the method which uses a phase change, in which case the measurement consists of determining the amount of phase change that has occurred and multiplying by the enthalpy change for the phase change. All calorimetric measurements thus require at least two separate experiments, one for the measurement and one for calibration, and may require a third for a baseline (zero) determination.

The advent of different scientific discoveries helped the scientists and technologists to update the technique for timely requirements. Most of them are discussed below.

TABLE 1.2 Some milestones in the development of thermal analysis

Year	Scientists	Contributions
1714	Fahrenheit	Mercury thermometer
1742	Celsius	Temperature scale
1760	Black	Ice calorimeter
1784	Lavoisier and Laplace	-do-
1782	Wedgwood	Pyrometer
1822	Seebeck	Thermoelectric effect
1826	Becquerel	Thermocouple
1848	Kelvin	Absolute temperature scale
1871	Siemens	Resistance thermometer
1892	Le Chatelier	Optical pyrometer
1899	Austen-Roberts	DTA
1907	Henning	Dilatometer
1915	Honda	Thermobalance
1950	Ozawa and Suga	Automated instrument
1960	—	Fully automated DSC(Pc-DSC)
1971	Paulik	SCTA
1986	Binnig	Thermal Imaging (AFM)
1992	Reading	Mt-DSC
1999	Price	Micro calorimeter
2005	Toda	Full denconvolution

1.2.1 Sample Controlled Thermal Analysis (SCTA) or Control Rate Thermal Analysis (CRTA)

In this technique, response of the sample to the initial heating programme is used as feedback to the control system to influence continuation of the temperature programme in same way. The simplest method is to set a rate of mass loss and use the difference between the set value and the observed rate of mass loss as the input signal to the influence controller. The method holds the rate of reaction at a constant value by using the measured loss or pressure of evolved gas as feedback to the furnace temperature controller. The temperature time profiles that result are important in understanding what has occurred. Often they correspond to almost isothermal conditions during much reactions, so the technique has also been called quasi isothermal.

Another closely related technique, particularly useful in obtaining kinetic information, is that involving either temperature jump or rate jump. In stepwise isothermal analysis upper and lower limits are set for the reaction rate and when the rate falls below the minimum set value, typically 0.08% of the maximum rate of loss at a normal linear heating rate, the temperature is increased at a generally fast linear rate, until the reaction rate exceeds the maximum set value. The temperature is then held constant until the reaction rate falls below the minimum.

1.2.2 High Resolution or Dynamic Thermogravimetry (Hi-Res TG or Dynamic Rate TG)

Another method, known as high resolution TG or dynamic rate TG produces a marked decrease in the heating rate when the system crosses the upper threshold set for the rate. The changes in the heating rates are controlled by adjustment of the values of two numerical parameters, referred as resolution and sensitivity and example of Hi-res TG is the dehydration of $CuSO_4.5H_2O$. For maximum resolution of multistage reactions the dynamic rate method may be suitable because the slowest heating rate should be applied during the transition from one stage to another, in contrast to the usual method where the heating rate decreases and then increases during the passage from one stage to another. The instrument control parameters are varied as a function of the sample's rate of weight change. This novel feature improves resolution and also reduces analysis time. The maximum heating rate and resolution index are user selectable. Data reported demonstrates the improved resolution of TG curves and very high resolution for difficult-to-resolve complex materials. Results are reported for many types of samples including: polymers, polymer blends, inorganic reference materials, and complex organic mixtures.

1.2.3 Modulated Thermal Analysis

The use of **modulated temperature**, modulated mode of DSC (MT-DSC) applies to a periodic modulation in sample temperature and examines the response in heat flow. The modulated DSC provides additional information of qualitative and quantitative physical and chemical changes than conventional DSC like the separation of reversible and non-reversible thermal events. It also provides additional information particularly for polymers [2]. The effects of baseline slope and curvature are reduced, increasing the sensitivity of the system. Overlapping events such as molecular relaxation and glass transitions can be separated. Heat capacity can be measured directly with modulated DSC in a minimum number of experiments. Both modulated DSC and conventional DSC measure the difference in heat flow to a sample and to an inert reference. The sample and reference cells are identical. However, modulated DSC uses a different heating profile. Whereas DSC measures heat flow as a function of a constant rate of change in temperature, modulated DSC superimposes a sinusoidal temperature modulation on this rate. The sinusoidal change in temperature permits the measurement of heat-capacity effects simultaneously with the kinetic effect. Typical experimental procedure for an initial modulated DSC experiment includes a heating rate from isothermal to 5°C/min and a modulation amplitude from 0.01 to 10°C. The modulation period can vary from 10 to 100 seconds or, expressed as a frequency, from 10 to 100 MHz. The response is represented by a dynamic heat capacity of complex quantity, the plots against angular frequency of modulation, ω, of the dynamic heat capacity obtained by temperature-modulated mode with the DSC used as a single calorimeter.

1.2.4 Pressure Differential Calorimeter

The thermal transitions can be measured in a high pressure DSC cell for novel observations and thermochemistry. For illustration, poly(ethylene terephthalate) / carbon dioxide (PET/CO_2) systems were investigated by using a differential scanning calorimeter accessorized with a high-pressure DSC cell. It was found that the glass transition temperature of PET decreases with an increase in the CO_2 pressure due to the plasticization effect, which is quite noticeable even at rather low CO_2 pressures. The sorbed CO_2 enhances the mobility of the chain segments and depresses the crystallization temperature of the PET. The CO_2 induced crystallization of PET at high pressure is attributed mainly to the plasticization effect. This causes a lower TG than room temperature for PET, and also crystallization of PET can occur at room temperature. The fullerene [4] C60 can be converted into different structures at high pressure and temperature. They are metastable and revert to pristine C60 on reheating to 300°C at ambient pressure for synthesis temperatures between 300° and 400°C and pressures of 5 gigapascal. The same C60, when treated at 500° to 800°C at the same pressure, transforms into a rhombohedral structure with hexagonal lattice structure. The intermolecular distance is small enough that a chemical bond can form, in accordance with the reduced solubility of the pressure-induced phases. The method is also known as oxidation induction technique (OIT).

1.2.5 Thermal Analysis on μ-chip

The electronic advancement results in one single-chip architectural platform that is ideal for BLDC motor control known as the 16-bit digital signal controller (DSC). This type of platform delivers control of a microcontroller, along with the computation and throughout capabilities of a digital signal processor (DSP). In this way, DSC is excellent at executing the complex, high-speed mathematical functions required. This is excellent for real-time control applications that require high robustness and increased reliability. The additional system benefits derived from DSC include:

1. An external crystal, saves critical board space, and reduces system costs
2. High-resolution

3. Improves system performance and reducing software overhead.
4. Pulse width modulation supports the automatic generation of specified waveforms.
5. Accurate and rapid measurement

This calorimeter has proved to be an effective analytical technique for characterizing micronized compounds. It can be used to detect presence of metastable region, kinetics of recrystallisation, making prediction of physical stability, which is not detectable even by 'X-ray Diffraction'.

Another set of developments are associated with the development of **automation** in the sense of analysis of sample without operator intervention and use of program [5]. This mode of instrument is suitable for multistep sample run, either simultaneous or sequentially. The multi-specimen instruments are available, to be weighed and analysed sequentially. The instrument is also available to operate multi-furnace system at a time.

1.2.6 Pulsed Thermal Analysis

This technique was introduced by Maciejenski in 1997 with the objective to sort out confusion with one of the many variations of modulated temperature technique [6-7]. The method also referred to as "Consumed Gas Analysis' is based on the principle of monitoring the composition, mass, enthalpy change resulting from passing a known amount of reactant with a carrier gas in thermochemical chamber.

1.2.7 Future Challenges [8]

Over the past four decades, thermal analysis techniques have been greatly improved and perfected due to the requirements of material characterization. The major developments have been associated with computerizing thermal analysis techniques. However, in the coming decades, thermal analysis techniques will continue to develop in at least two different directions:

(i) More precise measurements can be carried out using traditional thermal analysis techniques. This will lead to a variety of new information regarding a material's structure interface and morphology obtained from the results.
(ii) Thermal analysis techniques should be combined with other in situ temperature-controlled experiments, such as diffraction, scattering, microscopy, and spectroscopy, for investigation of the structure and dynamics of materials. This will generate information of the structural evolution with changing thermal properties and therefore, greatly aid in the understanding of structure-property relationships of the materials.

The most noticeable may be in situ temperature-controlled synchrotron wide angle X-ray diffraction (WAXD) and small angle X-ray scattering (SAXS), temperature-controlled solid state nuclear magnetic resonance spectroscopy NMR and temperature-controlled atomic force microscope (AFM).

1.3 TERMINOLOGY

The International Confederation of Thermal Analysis and Calorimetry (ICTAC) has produced publications and recommendations on the nomenclature to be used, for calibration method and the symbols used to represent the technique utilized. The definitions and the abbreviations are constantly under review. Therefore, new techniques or modifications of the older techniques require to be re-examined. The details can be seen in glossary.

1.4 APPLICATIONS

These methods are good analytical tools to measure one-, two- and presently three-dimensional property changes and are of enormous use. Some are tabulated below.

TABLE 1.3 Applications of thermal methods

Applications field	Applied techniques
Reaction and kinetics	TG, DTA, DSC, EGA, ETA
Structural transitions	DTA, TD, DSC
Stoichiometry	TG, EGA
Sintering and crack formation	TD
Dimensional change	TD
Critical temperature	TE and TM
Phase diagram	DTA, DSC
Vapour pressure and volatility	TG, EGA and EGD
Heat capacity	DSC

The information helps in the characterization of organic and inorganic salts, polymers, metals, semiconductors and other classes of materials. Not only this, thermal methods also help us in understanding the behaviour of chemicals and materials with change of temperature during reactions, such as transport and storage. On the basis of information, one can suggest possible precautions in using particular chemicals. The varieties of materials which can be usefully studied by the thermal methods are: organic and inorganic compounds, pharmaceuticals, polymers and plastics, minerals, soils and clays, ceramics and glasses, cement, catalysts, biological materials, e.g., kidney stone, explosives, fats, oils, soaps and waxes, foodstuffs and additives, liquid crystals, metals, alloys, etc.

However, still the limitations of the technique lie, e.g., studying consecutive reactions. For this research, efforts are still in progress.

EXERCISE

1. Define thermal methods of analysis.
2. Expand the following abbreviations:
   (a) TMA (b) DRS
   (c) TL (d) TS
   (e) DETA
3. Name the discoverers of the following:
   (a) Thermobalance (b) Mercury thermometer
   (c) Thermoelectric effect (d) Pyrometer
   (e) Mt-DSCs
4. List the advantages of modulated thermal analysis.
5. Write a note on thermal analysis on μ-chip.
6. Sketch a brief about historical development of thermal analysis.
7. Compare dilatometer and thermal mechanical analyzer.
8. Discuss the advantageous features of sample controlled thermal analysis and modulated thermal analysis.
9. Define: (a) Optical pyrometer (b) Thermobalance (c) Thermal imaging (d) Denconvolution.
10. Write notes on:
    (a) Dynamic thermogravimetry
    (b) High-pressure DSC cell
    (c) Consumed gas analysis
    (d) Emanation thermal analysis

REFERENCES

1. Mackenzie, R. C. (1993). Nomenclature in thermal analysis. *J. Therm. Anal. Calorim.* 40, 5-28.
2. Lambardi, G. (1980). For Better Thermal Analysis. 2nd Edition, International Confederation for Thermal Analysis.
3. Reading, M. (1993). Modulated differential scanning calorimetry—A new way forward in materials characterization. *Trends Polym. Sci.* 1(8), 248-253.
4. Iwasa, Y., et al. (1994). Newphases of C_{60} synthesized at high pressure. *Science.* 264, 1570-1572.

5. Blaine, R. L. and Hahn, B. K. (1998). Obtaining kinetic parameters by modulated temperature thermogravimetric. *J. Therm. Anal. Calorim.* 54, 695-704.
6. Ozawa, T. (2000). Thermal analysis–Review and prospect. *Thermochimica Acta.* 355, 35-42.
7. Reading, M. and Hourston, D. J., Eds. (2006). Modulated Temperature Differential Scanning Calorimetry: Theoretical and Practical Applications in Polymer Characterisation (Vol. 6). Springer Science & Business Media.
8. Cheng, S. Z. D., Li, C. Y., Calhoun, B. H., Zhu, L. and Zhou, W. W. (2000). Thermal analysis: The next two decades. *Thermochimica Acta.* 355, 59-68.

2

Thermal Effects

2.1 INTRODUCTION

Heating of materials leads to various visible and invisible changes due to their characteristics: physical, chemical, mechanical, electrical, and acoustic. These changes are related to their chemical structure, matrix and phase identity. Thermal effects also lead to many secondary changes associated to their primary changes. The secondary changes are basically, temperature, conductance (resistance), colour, magnetism, etc. A few changes are transitional that highly depend on a particular temperature and play an important role in the use of materials in particular applications. These changes are useful for their characterization and their applications [1-2]. Some basic thermal effects discussed here are for understanding the thermal behaviour of materials:

(a) Thermal conductivity
(b) Thermal capacity
(c) Thermal resistivity
(d) Thermal diffusivity
(e) Thermal inertia

These behaviours, generally investigated in thermal analysis (T.A.), are in solid state. However, the thermal behaviour of liquids is also studied in special techniques, but the gas phase is not normally in the principal reactants in T.A. experiments.

Generally, thermal reactions can be placed in the following types:

(a) Reactions of solids: A (s) (structure 1) $\rightarrow$ A (s) (structure 2)
(b) Decomposition of solids: A (s) $\rightarrow$ B (s) + C (g)
(c) Reaction of solid and gas: A (s) B (g) $\rightarrow$ Ax By (s)
(d) Solid-solid interactions: A (s) B (s) $\rightarrow$ Ax By (s)

Thus, it is important to understand the different thermal effects prior to performing an experiment.

2.2 THERMAL CONDUCTIVITY

2.2.1 Introduction

In physics, thermal conductivity [3], *k*, is the property of a material that indicates its ability to conduct heat. It appears primarily in Fourier's law for heat conduction.

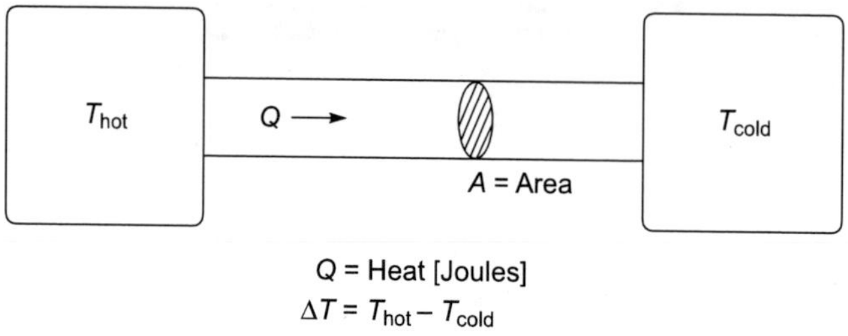

Fig. 2.1 Measuring principle of thermal conductivity

First, we define heat conduction by the formula:

$$H = \frac{\Delta Q}{\Delta t} = k \times A \times \frac{\Delta T}{x} \tag{1}$$

where, $\frac{\Delta Q}{\Delta t}$ is the rate of heat flow, *k* is the thermal conductivity, *A* is the total cross-sectional area of the conducting surface, ΔT is temperature difference and *x* is the thickness of the conducting surface separating the two temperatures, T_{hot} and T_{cold}.

Thus, rearranging equation (1) gives thermal conductivity,

$$k = \frac{\Delta Q}{\Delta t} \times \frac{1}{A} \times \frac{x}{\Delta T} \tag{2}$$

(where, $\frac{\Delta T}{x}$ is the temperature gradient).

In other words, it is defined as the quantity of heat, ΔQ, transmitted during time Δt through a thickness *x*, in a direction normal to a surface of area *A*, due to a temperature difference ΔT, under steady state conditions and when the heat transfer is dependent only on the temperature gradient.

Alternatively, it can be thought of as a flux of heat (energy per unit area per unit time) divided by a temperature gradient (temperature difference per unit length)

$$k = \frac{\Delta Q}{A \times \Delta t} \times \frac{x}{\Delta T} \tag{3}$$

Typical SI units are: W/(m·K)

Thermal conductivity of a material depends on many properties of the material, notably its structure and temperature [4]. For instance, pure crystalline substances exhibit very different thermal conductivities along different crystal axes, due to differences in phonon coupling along a given crystal axis. Sapphire is a notable example of variable thermal conductivity based on orientation and temperature. Thermal conductivity of metals approximately tracks electrical conductivity according to the Wiedemann-Franz law, as freely moving valence electrons transfer not only electric current but also heat energy. However, general correlation between electrical and thermal conductance does not hold for other materials, due to the increased importance of phonon carriers for heat in non-metals. Air and other gases are generally good insulators, in the absence of convection. Therefore, many insulating materials function simply by having a large number of gas filled pockets which prevent large-scale convection. Examples include expanded and extruded polystyrene (popularly referred to as "styrofoam") and silica aerogel. Natural, biological insulators such as fur and feathers achieve similar effects by dramatically inhibiting convection of air or water near an animal's skin.

Light gases, such as hydrogen and helium typically have high thermal conductivity. Dense gases such as xenon and dichlorodifluoromethane have low thermal conductivity. An exception, sulphur hexafluoride, a dense gas, has a relatively high thermal conductivity due to its high heat capacity. Argon, a gas denser than air, is often used in insulated glazing (double paned windows) to improve their insulation characteristics. In Table 2.1, a list of thermal conductivity is compiled for the general reference of a researcher. This is a list of approximate values of thermal conductivity, for some common materials. For more details and accurate values, references can be seen.

TABLE 2.1 Approximate values of thermal conductivity of common materials

Material	Thermal conductivity W/(m·K)
Cement, Portland	0.29
Concrete, stone	1.7
Air	0.025
Wood	0.04 - 0.4
Alcohols and oils	0.1 - 0.21
Hollow fill fibre insulation	0.042
Silica Aerogel	0.004 - 0.04
Soil	1.5
Rubber	0.16
Epoxy (unfilled)	0.59
LPG	0.23 - 0.26
Epoxy (silica-filled)	0.30
Water	0.6

Contd.

Thermal grease	0.7 - 3
Thermal epoxy	1 - 7
Glass	1.1
Ice	2
Sandstone	2.4
Stainless steel	12.11 ~ 45.0
Lead	35.3
Aluminium	237
Gold	318
Copper	401
Mineral oil	0.138
Silver	429
Diamond	900 - 2320

2.2.2 Thermal Conductivity Measurement

There are a number of possibilities to measure thermal conductivity [5], each of them suitable for a limited range of materials, depending on the thermal properties and the medium temperature. Distinction can be made between steady-state and transient techniques.

In general, the steady-state techniques perform a measurement when the temperature of the material that is measured does not change with time. This makes the signal analysis straightforward (steady state implies constant signals). The disadvantage generally is that it takes a well-engineered experimental setup. The 'divided bar' (various types) is the most common device used for consolidated rock samples.

The transient techniques perform a measurement during the process of heating up. The advantage is that measurements can be made relatively quickly. Transient methods are usually carried out by needle probes (inserted into samples or plunged into the ocean floor).

For good conductors of heat, Searle's bar method can be used. For poor conductors of heat, Lee's disc method can be used. An alternative traditional method using real thermometers is described. A brief review of new methods measuring thermal conductivity, thermal diffusivity and specific heat within a single measurement is available in literature. A thermal conductance tester, one of the instruments of gemology, determines if gems are genuine diamonds using diamond's uniquely high thermal conductivity. Another term, thermal transmittance, measured in the same units as thermal conductance is sometimes known as the composite thermal conductance. The term *U*-value is another synonym. Thermal transmittance (*U*-value) $= 1/(\Sigma(L/k))$ + convection + radiation, measured in $W \cdot K^{-1} \cdot m^{-2}$.

2.3 THERMAL DIFFUSIVITY

In heat transfer analysis, thermal diffusivity (symbol: α, but note that the symbols κ, D, and k are also commonly used) is an important parameter. It is the ratio of thermal conductivity to volumetric heat capacity. Its SI unit is m^2/s.

$$\alpha = \frac{k}{\rho C_p} \tag{4}$$

where,

k : thermal conductivity (SI units: W/m·K)

ρ : density (kg/m^3)

C_p : specific heat capacity (J/kg·K)

The term used in the thermal diffusivity expression above, ρc_p, can be identified as the volumetric heat capacity with SI units of $J/m^3 \cdot K$.

Substances with high thermal diffusivity rapidly adjust their temperature to that of their surroundings, because they conduct heat quickly in comparison to their volumetric heat capacity or 'thermal bulk'. A list of thermal diffusivities is given in Table 2.2 for reference.

TABLE 2.2 Some thermal diffusivities of common materials

Material	Thermal diffusivity (m^2/s)
Aluminium	8.418×10^{-5}
Carbon steel (1%)	1.172×10^{-5}
Copper	1.1234×10^{-4}
Pure silver (99.9%)	1.6563×10^{-4}
Aluminium oxide (polycrystalline)	1.20×10^{-5}
Common brick	5.2×10^{-7}
Pyrolytic graphite, parallel to layers	1.22×10^{-3}
Pyrolytic graphite, normal to layers	3.6×10^{-6}
Window glass	3.4×10^{-7}
Nylon	9×10^{-8}
Sandstone	1.12×10^{-6} to 1.19×10^{-6}
Wood (Yellow Pine)	8.2×10^{-8}

Contd.

Air (1 atm, 300 K)	2.2160×10^{-5}
Water vapour (1 atm, 400 K)	2.338×10^{-5}
Engine oil (saturated liquid, 100°C)	7.38×10^{-8}

2.4 THERMAL RESISTIVITY

Thermal resistivity indicates the thermal insulation behaviour of a material. This is always important for understanding the nature of a material and the effect of additive. This is important for engineering design involving the selection of an appropriate sized heat sink for a given heat source. Working in units of thermal resistance greatly simplifies the design calculation. The following formula can be used to estimate the performance:

$$R_{hs} = \frac{\Delta T}{P_{th}} - R_s \qquad (5)$$

where,

R_{hs} is the maximum thermal resistance of the heat sink to ambient, in °C/W
T is the temperature difference (temperature drop), in °C
P_{th} is the thermal power (heat flow), in watts
R_s is the thermal resistance of the heat source, in °C/W

The thermal resistance reported technically as R-value means what is described above as thermal insulance, and thermal conductance means the reciprocal. For materials in series, these thermal resistances (unlike conductances) can simply be added to give a thermal resistance for the whole. The thermal resistance is generally represented in terms of (R value) = L/k, measured in $K \cdot m^2 \cdot W^{-1}$. In some industries, a tog value is quoted as a measure of thermal resistance in place of a measure in SI units.

2.5 THERMAL CAPACITY

2.5.1 Introduction

It is the heat bearing ability of materials [6-7], reported as **specific heat capacity;** also known simply as specific heat, is the measure of the heat energy required to increase the temperature of a unit quantity of a substance by a certain temperature interval. The term originated primarily through the work of Scottish physicist, Joseph Black who conducted various heat measurements and used the phrase –"capacity for heat". More heat energy is required to increase the temperature of a substance with high specific heat capacity than one with low specific heat capacity. For instance, eight times the heat energy is required to increase the temperature of an ingot of magnesium as is required for a lead ingot of the same mass. The specific heat of virtually any substance can be measured, including chemical elements, compounds, alloys, solutions, and composites.

The symbols for specific heat capacity are either C or c depending on how the quantity of a substance is measured. In the measurement of physical properties, the term "specific" means the measure is a bulk property (an intensive property), wherein the quantity of substance must be specified. For example, the heat energy required to raise water's temperature to one kelvin (equal to one Celsius degree) is 4.184 joules per gram—the gram being the specified quantity. Scientifically, this measure would be expressed as $c = 4184$ J kg^{-1} K^{-1}. The specific heat capacity of a material is

$$c = \frac{\partial C}{\partial m} \tag{6}$$

where,

C is the heat capacity of a body made of the material in question
m is the mass of the body.

The equation relating heat energy to specific heat capacity, where the unit quantity is in terms of mass is:

$$\Delta Q = mc\Delta T$$

where ΔQ is the heat energy put into or taken out of the substance, m is the mass of the substance, c is the specific heat capacity, and ΔT is the temperature differential.

Where the unit quantity is in terms of moles, the equation relating heat energy to specific heat capacity (also known as molar heat capacity) is:

$$\Delta Q = nc\Delta T$$

where ΔQ is the heat energy put into or taken out of the substance, n is the number of moles, c is the specific heat capacity, and ΔT is the temperature differential.

A list of specific heat of some materials are shown in Table 2.3.

2.5.2 Factors Affecting Specific Heat Capacity

The specific heat of a material can be affected by many factors [8] as given below:

1. **Degrees of Freedom:** Molecules are quite different from the monatomic gases like helium and argon. With monatomic gases, heat energy comprises only translational motions. Translational motions are ordinary, whole-body movements in 3D space whereby particles move about and exchange energy in collisions—like rubber balls in a vigorously shaking container. These simple movements in the three X, Y, and Z-axes dimensions of space means monatomic atoms have three translational degrees of freedom. Molecules, however, have various internal vibrational and rotational degrees of freedom because they are complex objects; they are a population of atoms that can move about within a molecule in different ways. Heat energy is stored in these internal motions,

but on per-atom basis, the heat capacity of molecules does not exceed the heat capacity of monatomic gases, unless vibrational modes are brought into play. For instance, nitrogen, which is a diatomic molecule, has five active degrees of freedom at room temperature: the three comprising translational motion plus two rotational degrees of freedom internally. Although the constant-volume molar heat capacity of nitrogen (diatomic) at this temperature is five-thirds that of monatomic gases, on a per-mole of atoms basis, it is only five-sixths that of a monatomic gas, because each atom has fewer degrees of freedom at low temperatures, due to the intra-atomic bond. This is expected since two nitrogen atoms would have six degrees of freedom (three for each atom) if free, but when locked together, they lose one degree of freedom, and together possess only five.

2. **Per Mole of Molecules:** When the specific heat capacity, *c*, of a material is measured (*c* means the unit quantity is in terms of mass), different values arise because different substances have different molar masses (essentially, the weight of the individual atoms or molecules). Heat energy arises, in part, due to the number of atoms or molecules that are vibrating. If a substance has a lighter molar mass, then each gram of it has more atoms or molecules available to store heat energy. This is why hydrogen—the lightest substance—has such a high specific heat capacity on a gram basis; one gram of it contains twice the number of molecules as a gram of helium, and the ratio for other substances is even larger. If specific heat capacity is measured on molar basis (C), however, the differences between substances is less pronounced and hydrogen's molar heat capacity is unremarkable.

3. **Per Mole of Atoms:** Conversely, for molecular-based substances (which also absorb heat into their internal degrees of freedom), massive, complex molecules with high atomic count—like gasoline—can store a great deal of energy per mole and yet are quite unremarkable on mass basis, or on per-atom basis. This is because, in fully excited systems, heat is stored independently by each atom in a substance, not primarily by the bulk motion of molecules.

4. **Other Factors:**

   Hydrogen bonds These bonds provide another place where heat energy may be stored as potential energy of vibration, even at comparatively low temperatures. Hydrogen bonds account for the fact that liquid water stores nearly the theoretical limit of 3R per mole of atoms, even at relatively low temperatures (i.e., near the freezing point of water).

   Impurities In the case of alloys, there are several conditions in which small impurity concentrations can greatly affect the specific heat. Alloys may exhibit marked difference in behaviour even in the case of small amounts of impurities being one element of the alloy; for example, impurities in semiconducting ferromagnetic alloys may lead to quite different specific heat properties.

TABLE 2.3 List of specific heats of some common materials

Substance	Phase	C_p kJ kg^{-1} K^{-1}	C_p, m J mol^{-1} K^{-1}	C_v, m J mol^{-1} K^{-1}	Volumetric heat capacity, J cm^{-3} K^{-1}
Air (dry, 0°C)	gas	1.0035	29.07	20.7643	0.001297
Air (typical room conditions A)	gas	1.012	29.19	20.85	–
Aluminium	solid	0.897	24.2	–	2.422
Ammonia	liquid	4.700	80.08	–	3.263
Antimony	solid	0.207	25.2	–	1.386
Argon	gas	0.5203	20.7862	12.4717	–
Arsenic	solid	0.328	24.6	–	1.878
Beryllium	solid	1.82	16.4	–	3.367
Bismuth	solid	0.123	25.7	–	1.20
Copper	solid	0.385	24.47	–	3.45
Carbon dioxide	gas	0.839*	36.94	28.46	–
Diamond	solid	0.5091	6.115	–	1.782
Ethanol	liquid	2.44	112	–	1.925
Gasoline	liquid	2.22	228	–	1.64
Glass	solid	0.84	–	–	–
Gold	solid	0.1291	25.42	–	2.492
Granite	solid	0.790	–	–	2.17
Graphite	solid	0.710	8.53	–	1.534
Helium	gas	5.1932	20.7862	12.4717	–
Hydrogen	gas	14.30	28.82	–	–
Hydrogen sulphide	gas	1.015*	34.60	–	–
Iron	solid	0.450	25.1	–	3.537
Lead	solid	0.127	26.4	–	1.44

Contd.

Lithium	solid	3.58	24.8	–	1.912
Magnesium	solid	1.02	24.9	–	1.773
Mercury	liquid	0.1395	27.98	–	1.888
Nitrogen	gas	1.040	29.12	20.8	–
Neon	gas	1.0301	20.7862	12.4717	–
Oxygen	gas	0.918	29.38	–	–
Paraffin wax	solid	2.5	900	–	2.325
Silica (fused)	solid	0.703	42.2	–	1.547
Silver	solid	0.233	24.9	–	2.44
Tungsten	solid	0.134	24.8	–	2.58
Uranium	solid	0.116	27.7	–	2.216
Water (steam)	gas (100°C)	2.080	37.47	28.03	–
Water	liquid (25°C)	4.1813	75.327	74.53	4.184
Water (ice)	solid (–10°C)	2.050	38.09	–	1.938
Zinc	solid	0.387	25.2	–	2.76

The above data are measured at 25°C.

2.6 HEAT CAPACITY

Heat capacity (symbol: C_p) — as distinct from specific heat capacity [6] is the measure of the heat energy required to increase the temperature of an object by a certain temperature interval. Heat capacity is an extensive property because its value is proportional to the amount of material in the object; for example, a bathtub of water has a greater heat capacity than a cup of water.

Heat capacity is usually expressed in units of J K^{-1} (or J/K). For instance, the gasoline in a 55-gallon drum has an average heat capacity of 347 kJ/K.

Heat capacity is mathematically defined as the ratio of a small amount of heat (δQ) added to the body, to the corresponding small increase in its temperature, dT:

$$C = \left(\frac{\delta Q}{dT}\right)_{cond.} = T\left(\frac{dS}{dT}\right)_{cond.} \tag{7}$$

This information is used to account for different ways that heat can be stored as kinetic energy (energy of motion) and potential energy (energy stored in force fields), as an object expands or contracts. For all real systems, the path through these changes must be explicitly defined, since the value of heat capacity depends on which path from one temperature to another, is chosen. Of particular usefulness in this context are the values of heat capacity for constant volume, C_V, and constant pressure, C_P. These will be defined below.

The state of a simple compressible body with fixed mass is described by two thermodynamic parameters such as temperature (T) and pressure (p). Therefore, as mentioned above, one may distinguish between heat capacity at constant volume, C_V, and heat capacity at constant pressure, C_p:

$$C_V = \left(\frac{\delta Q}{dT}\right)_V = T\left(\frac{\partial S}{\partial T}\right)_V \tag{8}$$

$$C_p = \left(\frac{\delta Q}{dT}\right)_p = T\left(\frac{\partial S}{\partial T}\right)_p \tag{9}$$

where,

δQ is the infinitesimal amount of heat added, and
dT is the subsequent rise in temperature.

Starting from the fundamental thermodynamic relation, one can show a derivation is given here. For an ideal gas this reduces to the simple relation,

$$C_p - C_v = nR \tag{10}$$

where n is the number of moles, and C_p and C_v are extensive (non-molar) properties.

2.7 LATENT HEAT

In thermochemistry, latent heat is the amount of energy in the form of heat released or absorbed by a substance during a change of phase state (i.e., solid, liquid, or gas), also called a phase transition. Two latent heats (or enthalpies) are typically described: latent heat of fusion (melting), and latent heat of vaporization (boiling). The names describe the direction of heat flow from one phase to the next: solid → liquid → gas.

The equation for latent heat is:

$$Q = mL \tag{11}$$

where,

Q is the amount of energy released or absorbed during the change of phase of the substance (in joules),
m is the mass of the substance, and
L is the specific latent heat for a particular substance (J kg^{-1}).

In other words, specific latent heat is found when energy is divided by mass. The latent heats and relevant change of phase temperatures of common fluids and gases are recorded in Table 2.4.

TABLE 2.4 Latent heats and relevant change of phase temperatures

Substance	Latent heat fusion J/g	Melting point °C	Latent heat vaporization J/g	Boiling point °C
Ethyl alcohol	108	–114	855	78.3
Ammonia	339	–75	1369	–33.34
Carbon dioxide	184	–57	574	–78
Helium	–	–	21	–268.93
Hydrogen	58	–259	455	–253
Lead	24.5	372.3	871	1750
Nitrogen	25.7	–210	200	–196
Oxygen	13.9	–219	213	–183
Toluene	–	–93	351	110.6
Turpentine	–	–	293	–
Water	334	0	2500 (at 0°C)	100

2.8 THERMAL EFFUSIVITY

In thermodynamics, thermal effusivity of a material is defined as the square root of the product of the material's thermal conductivity and its volumetric heat capacity. Generally reported as *e*, which is related to heat capacity as per Eq. (12)

$$e = (k\rho C_p)^{1/2} \tag{12}$$

Here, k is the thermal conductivity, ρ is the density and C_p is the specific heat capacity. The product of ρ and C_p is known as the volumetric heat capacity.

Therefore, a material's thermal effusivity is a measure of its ability to exchange thermal energy with its surroundings.

2.9 THERMAL INERTIA

Thermal inertia [9] is a term commonly used by scientists and engineers modelling heat transfers and is a bulk material property related to thermal conductivity and volumetric heat capacity. For example, material has a high thermal inertia, which means that dynamic effects are prevalent in a model, so that a steady-state calculation will yield inaccurate results. Thermal inertia is a measure of the thermal mass and the velocity of the thermal wave which controls the surface temperature of a material. In heat transfer, a higher value of the volumetric heat capacity means a longer time for the system to reach equilibrium.

The thermal inertia (I) of a material is defined as the square root of the product of the material's bulk thermal conductivity (k) and volumetric heat capacity, where the latter is the product of density (ρ) and specific heat capacity (c):

$$I = \sqrt{k\rho c} \tag{13}$$

Thermal inertia cannot be measured directly. Instead, it can be inferred from the diurnal verification of the temperature at the earth's surface depending not only on thermal inertia of the geological material, but also on the many complicated surface heating processes such as latent and sensible heat transfer between the ground and air.

The result indicates that thermal inertia can be a useful tool in separation of geological materials by remote sensing technology. It also forms a valuable set of complimentary data for use with the other types of remote sensing measurements. It includes refinement of surface metrological measurements.

2.10 THERMOCHROMISM

It is a heat dependent reversible change in the colour of materials after heating or cooling of the materials in melted or in solution state. Thermochromism is exhibited by a wide spectrum of materials like organic, inorganic, polymer and hybrid form. The basic mechanisms in thermochromism are temperature induced structural transition, molecular arrangements and phase transitions. A simple example of thermochromism is the structural transition of VO_2 from monoclinic to tetragonal at 68°C. This temperature induced change also exhibits changes from IR transparent to non-transparent as well as transition between metallic to semiconducting phases at particular temperature with transition from reddish brown to greenish blue. The low temperature monoclinic phase has insulating nature while high temperature shows metallic nature with tetragonal structure [10]. The colouring transitions of pigment are used in building with variable temperature dependent changes, a simple example is shown in Fig. 2.2 for the use of thermochromic materials in building as reflective materials for incoming radiation to monitor the amount of heat transmitted as energy efficient smart window.

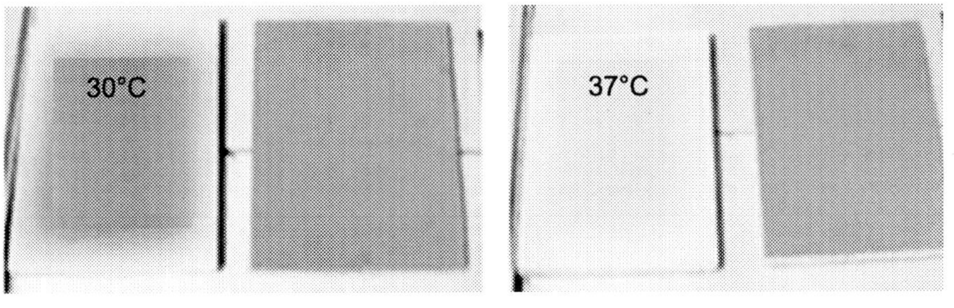

Fig. 2.2 Thermochromic coating in building [11]

However, currently this property has been accepted as a powerful tool to use in several applications like fibre, optical sciences, photo storage and optical sensors. Making polymer composite from thermochromic materials add the other advantageous features like processability, coat ability and improved stability for different applications. In an example use of poly(N-isopropylacrylamide) based hydrogel containing SiO_2 exhibits improved reversible optical behaviour at 20°C to be use in smart window with good durability and stability for modulated visibility [12]. Similarly, thermochromic effect of composite containing gold nanoparticles has been used for dual detection of protein and temperature with good sensing parameters. The basic principle of detection is based on three temperature transitions, i.e., less than 45, between 45 and 80 and more than 80°C, which are produced during denaturation of protein. The preparative mechanism of composite is shown in Fig. 2.3.

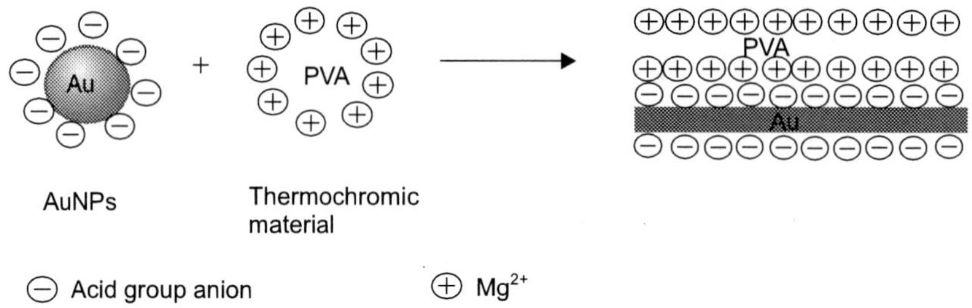

Fig. 2.3 Structural arrangement of a thermochromic composite [13]

Thermochromic materials are used in several industrial applications like:

1. Temperature indicators to determine the change in temperature during a chemical reaction.
2. Monitoring of temperature distribution in a reaction apparatus like heat exchangers and reactors.
3. Judge the feeding temperature of plastic during processing like extruder.
4. These materials are also used in temperature-sensitive paints for monitoring and indication of temperature during industrial production.
5. The automobile window made of thermochromic materials changes its colour in internal temperature beyond limits.

Thermochromism also indicates structural transition and state transitions, which can be also used to characterise the specific compounds. In example, the complex Ni(II) and 2-hydroxy-1-naphthylidene-4-aminoantipyrine (I) and salicylidene-4-aminoantipyrine (II) exhibits the colour change after heating due to structural transition. Hydration and dehydration, geometry transformation are the other reasons for thermochromism in a compound [14]. The nobility of phenomena is also promises to explore as single component emission of white light. The incorporation of chromophore is a simple strategy to develop thermochromism with specific colour, while the structural transition yield emitted colour transition like blue to white due to thermal responsiveness.

2.11 THERMOELECTRIC PROPERTIES

This class of electrical properties includes change in thermoresistance and conductance generated after heat treatment of materials due relation between heat end electricity. The basic principle controls these properties are "Seebeck and Peltier effect", which is also illustrated in Fig. 2.4, which reveals the formation of electromotive forces after heating with differential heat flow. The reason for the emf is materials characterisation due material nature like p- and n-type having ionic differences.

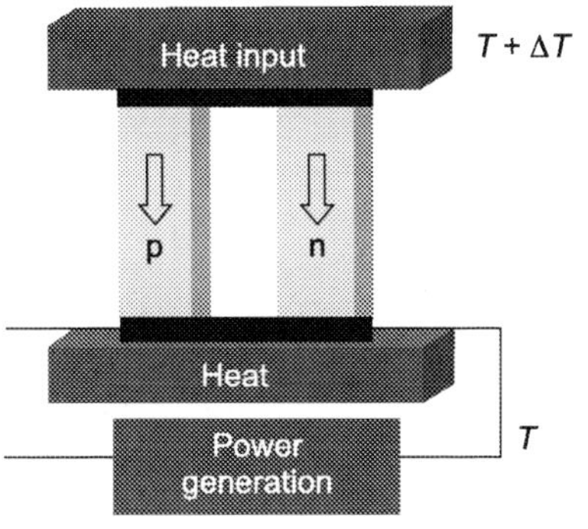

Fig. 2.4 Schematic illustration for Seebeck effect

It reveals that the differential heating of a material produces current after formation of ions with mobility. This, formation of electrons is based on thermal emission of electron as well as the reduction threshold energy level transition electron for conductance. The heat induced electrical properties has evolves in the term of electromotive force (emf), reduced resistance, capacitance and conductance. These properties have been exploited for wide range of scientific and commercial application like thermocouples, chemical sensors and semiconducting devices. The monitoring efficiency of induced emf during heating of material is responsible for the better sensitivity of techniques.

EXERCISE

1. Define thermal conductivity and give its SI units.
2. Explain the Fourier law of thermal conduction.
3. Explain the importance of thermal resistivity with examples.
4. List the factors affecting the heat capacity of materials.
5. Compare heat capacity and constant pressure, C_p, and heat capacity and constant volume, C_v.
6. Compare primary and secondary thermal changes with examples.
7. Explain the Searle's bar method for estimation of thermal conductivity.
8. Write the SI units of the following:
   (a) Specific heat
   (b) Latent heat of fusion
   (c) Heat capacity at constant pressure

9. Write a note on thermal diffusivity.
10. Explain the importance of thermal conductivity with examples.
11. Write the thermal diffusivity of the following:
    (a) Al (b) Cu
    (c) Ag (d) Nylon
    (e) Al_2O_3
12. Write notes on the following:
    (a) Thermal effusivity (b) Volumetric heat capacity
    (c) Remote sensing (d) Extensive property
    (e) Specific heat capacity

REFERENCES

1. Holman, J. P. (2002). Heat Transfer, 9th Edition., McGraw-Hill.
2. Callister, W. (2003). Materials Science and Engineering: An Introduction. John Wiley & Sons.
3. Halliday, D., Resnick, R. and Walker, J. (1997). Fundamentals of Physics. John Wiley and Sons, New York.
4. Srivastava, G. P. (1990). The Physics of Phonons. Adam Hilger. IOP Publishing Ltd., Bristol.
5. Laider, J. K. & Gerischer, H. (1995). The World of Physical Chemistry, Oxford University Press.
6. Fraundorf, P. (2003). Heat capacity in bits. *Am. J. Phys.* 71, 1142-1151.
7. Perrot, P. (1998). A to Z of Thermodynamics. Oxford University Press, NY, USA.
8. Stefan, D. and Dieter, B. (2006). Why molecules move along a temperature gradient? *Proceedings of the National Academy of Sciences (USA).* 103, 19678-19682.
9. Chiras, D. D. (2002). The Solar House: Passive Heating and Cooling. Chelsea Green Publishing Company, Australia.
10. Cui, Y., Ke, Y., Liu, C., Chen, Z., Wang, N., Zhang, L. ... and Long, Y. (2018). Thermochromic VO_2 for energy-efficient smart windows. *Joule*, 2(9), 1707-1746.
11. Francesco Asdrubali Umberto Desideri {Eds} (2019). Handbook of Energy Efficiency in Buildings, Chapter 6, Building Envelope, Pages 295-439, Elsevier. Inc.
12. Tian, J., Peng, H., Du, X., Wang, H., Cheng, X. and Du, Z. (2021). Hybrid thermochromic microgels based on UCNPs/PNIPAm hydrogel for smart window with enhanced solar modulation. *Journal of Alloys and Compounds*, 858, 157725.
13. Agyapong, D. A. Y., Jiang, H., Ni, X., Wu, J. and Zeng, H. (2021). Gold nanoparticles/thermochromic composite film on screen-printed electrodes for simultaneous detection of protein and temperature, *Journal of Biomaterials and Nanobiotechnology*, 2021, 12(2), 7-19.
14. Donia, A.M. and Ebeid, E.Z.M. (1988). Thermochromism in Ni(II) complexes with Schiff base derivatives of 4-aminoantipyrine. *Thermochimica Acta*, 131, 1-6.

3

Thermogravimetric Analysis (TGA)

3.1 INTRODUCTION

Thermogravimetric analysis (TGA) is the most widely used thermal method [1-3]. It is based on the measurement of mass change of material as a function of temperature or time. In thermogravimetry, a continuous graph of mass change against temperature is obtained when a substance is heated at a controlled rate or kept at constant temperature. A plot of weight change versus temperature (T) is referred to as the thermogravimetric curve (TG curve). For the TG curve, generally weight (W) decreasing downwards is plotted on the y-axis (ordinate), and temperature (T) or time (t) increasing to the right on the x-axis (abscissa) as shown in Fig. 3.1 for a reaction: A (Solid) $\rightarrow$ B (Solid) + C (Gas). TG curve will help in revealing the many thermochemical properties of analyzed samples and in determining the mode of their transformation within specified range of temperature.

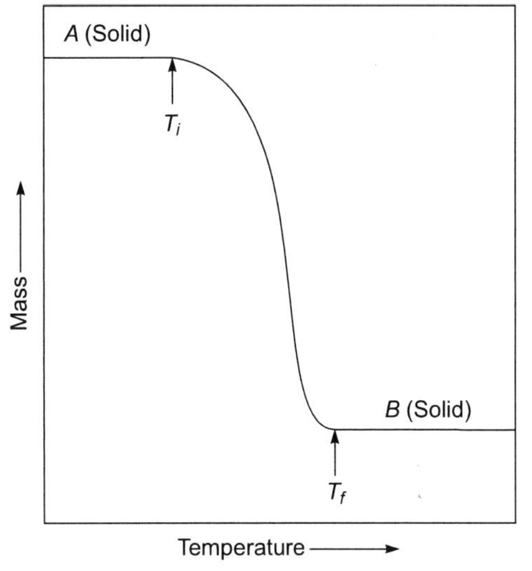

Fig. 3.1 Representative single stage TG curve

In TG curve of a single stage decomposition, there are two characteristic temperatures; the initial temperature, T_i, and the final temperature, T_f (see Fig. 3.1).

T_i is defined as the lowest temperature at which the onset of a weight change can be detected by thermo balance operating under particular conditions and T_f as the final temperature at which the particular decomposition appears to be completed. Although T_i has no fundamental significance, it can still be a useful characteristic of a TG curve and the term procedural decomposition temperature has been suggested, and T_f is the final procedural temperature. The difference $(T_f - T_i)$ is termed reaction interval. In thermogravimetry analysis, a sample is subjected to continuous increase in temperature usually linear with time whereas in isothermal or static thermogravimetry, the sample is maintained at a constant temperature for a period of time during which any change in weight is noted. The thermogravimetric analysis can be performed in three ways:

(a) **Static:** The temperature of the sample is maintained constant and change in mass is recorded as a function of time. This method is also called isothermal TG. This technique is important for the study of reaction kinetics and lifetime assessment of a sample.

(b) **Quasi Static:** The method is partially static and partially dynamic and also known as stepwise isothermal TGA. The temperature of sample changes at the desired temperature and then maintained for a certain time for observation of mass change. The advantages of the method are: Sample is held isothermal until transition is completed, thus excellent resolution of overlapping transitions; permits careful control of reaction environment. While the disadvantages are: Requires several scans to optimize experiment conditions; inappropriate parameter choices may produce artifacts; long runtime; routine analysis of similar samples.

(c) **Dynamic:** The complete change in sample condition was kept variable for desired range of temperature. The atmosphere can also be changed depending on the nature of studies. The change in mass is measured under the dynamic condition of a sample. The analysis is highly useful for assessment of thermal stability and probable thermal changes occurring in the sample.

The mass change behaviours of different reactions are different: sometimes increases and sometimes decreases. A comparison is made in Table 3.1 to understand the probable mass change in some reactions.

TABLE 3.1 Different mass change behaviours in TG

Process	Mass gain	Mass loss
Adsorption or absorption	÷	–
Desorption	–	÷
Sublimation	–	÷

Contd.

Vaporisation	–	÷
Decomposition	–	÷
Solid state reaction	÷	÷
Solid gas reaction	÷	÷

In general, the trend of weight loss is observed differently depending on the nature of thermochemical reactions. The TG analysis provides seven types of curves as given in Fig. 3.2.

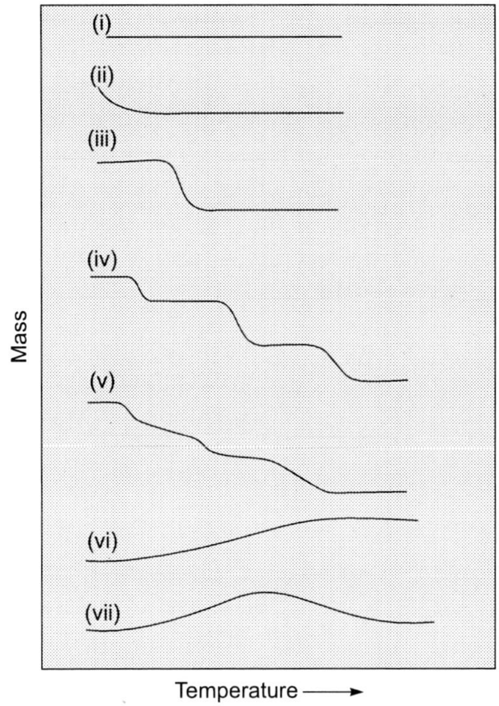

Fig. 3.2 Representative types of curve obtained in TG (Adapted from Ref. 2)

Type I: In this type of curve, the sample undergoes no decomposition with loss of volatile products over the temperature range. As such no information is obtained from this curve, however solid phase transitions, melting, polymerization or reaction involving no volatile product may occur. Use of other techniques such as DTA, DSC, etc., is necessary to eliminate these possibilities. Assuming that these possibilities are eliminated, the sample would then be known to be stable over the temperature range considered. This could be a good information if a heat-resistant material was under investigation or bad if potential explosives were being tested.

Type II: This type of curve is not often observed in TG analysis which indicates that rapid initial observed is a characteristic feature of desorption or drying. This type of change also occurs, when TG is recorded at reduced pressure and due to the effect of thermomolecular flow or convection.

Type III: Represents the decomposition in single step mass loss, which may be used to define the limit of stability of reactants, to determine the stochiometry of reaction and also investigate the kinetics of the reaction.

Type IV: Represents a multistage decomposition with the relative thermal stable intermediates. The temperature limit of stability of the sample and other intermediates can be determined from the curve, together with the more complicated stochiometry of the reaction.

Type V: Determines the multistage decomposition but no sample stable intermediates are formed. Only little information about the stochiometry of the overall reaction can be obtained. It is important to check the effect of heating rate on the resolution of such curve. At lower heating rate type V may tend to resemble type IV more closely, while at higher temperature rate both types IV and V may resemble type III curve and hence, detail complex decomposition is lost.

Type VI: Shows a gain in mass which results in a reaction of the sample with the surrounding atmosphere. The typical example would be oxidation of a metal sample or conversion of metal sulphides in metal sulphates in O_2/air atmosphere.

Type VII: This is a typical type of curve not often observed. At lower temperature, mass increases then at higher temperature mass decreases, e.g.,

$$2Ag + O_2 \rightarrow Ag_2O \rightarrow 2Ag + O_2.$$

The resolution of individual stage of TG curve is important and can be improved by examining the derivative of the curve result either with mass or time [4-5]. The resultant curve is known as differential thermal gravimetric (DTG) curves. This curve presents the rate of weight change (dW/dt) as a function of temperature, or time (t) against T on the abscissa (x-axis) as shown in Fig. 3.3. In derivative thermogravimetric analysis (DTG) the peak temperature of DTG represents the temperature at which the rate of the mass change is at a maximum, and not the temperature at which the sample begins to lose mass.

The reactions corresponding to the mass losses can be best determined or confirmed by simultaneously evolved gas analysis (discussed in Chapter 8: Complimentary Techniques). Sometimes the complimentary technique such as hot stage microscopy (HSM) may provide information and mechanism of thermal reaction or chain of the reaction/product interface inward from catalyst surface.

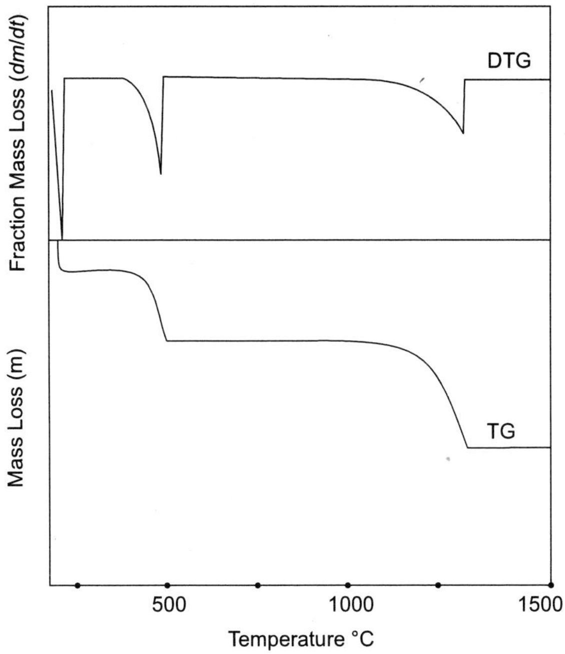

Fig. 3.3 TG and DTG curves

3.2 INSTRUMENTATION

The instrument used in thermogravimetry (TG) is called a thermobalance [6-8]. It was designed by K. Honda in 1915 as shown in Fig. 3.4.

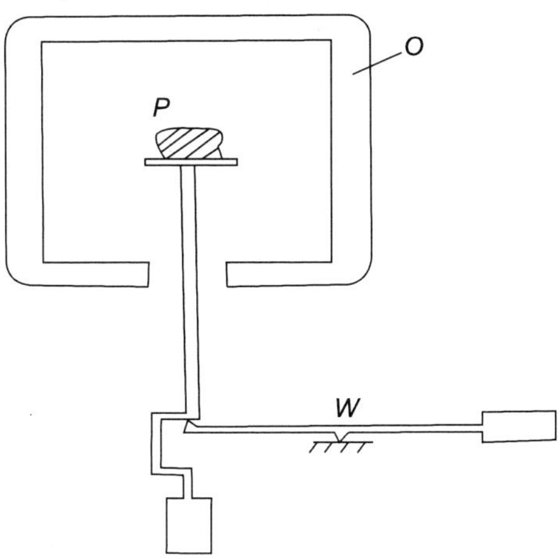

Fig. 3.4 First thermobalance developed by Honda in 1915

However, the present instruments are improvised, normally, consisting of several basic components in order to provide the flexibility necessary for the production of useful analytical data in the form of TG curve as shown in Fig. 3.5. Figure 3.6 describes minute details including attachment points in Netzsch model STA-409.

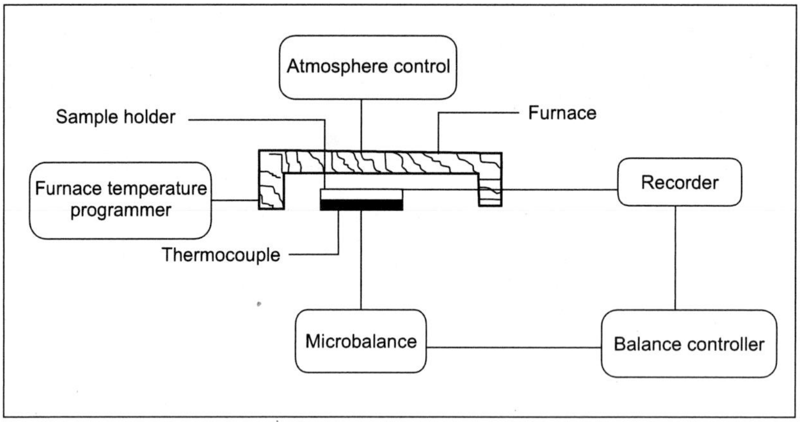

Fig. 3.5 Block diagram of a TG instrument

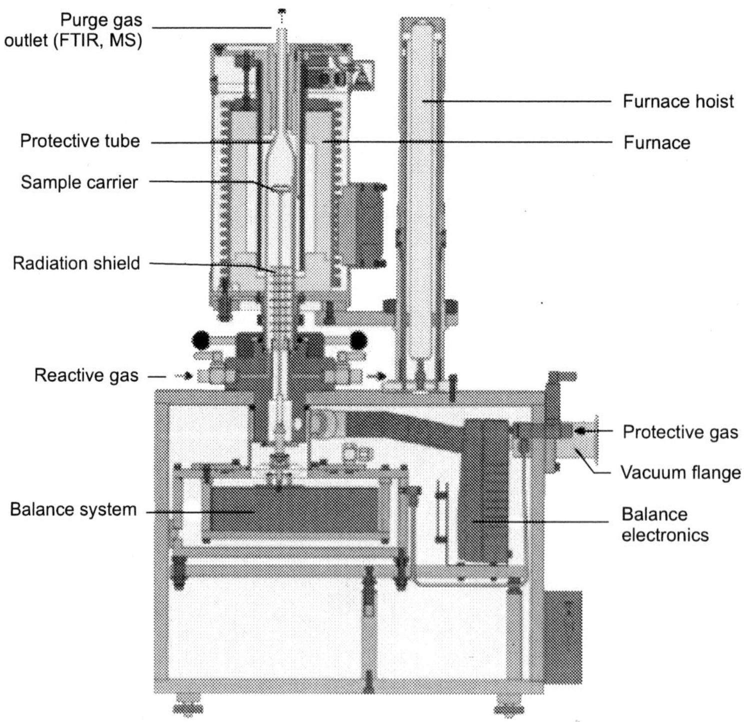

Fig. 3.6 Detailed diagram of NETSCZ STA-409 thermal analyzer (Reproduced with the permission of NETZSCH)

The components of a thermogravimetric instrument are:

(i) Balance
(ii) Furnace
(iii) Temperature measurement and control
(iv) Recorder
(v) Atmosphere control unit
(vi) Sample holder

3.2.1 Balance

The purpose of a balance is the automatic continuous monitoring of mass of a sample. The basic requirement of an automatic recording balance includes accuracy, sensitivity, reproducibility and capacity. Recording balances are generally of two types: null point and deflection type.

The null type balance, which is more widely used, incorporates a sensing element, which detects a deviation of the balance beam from its null position; horizontal for beam balances and vertical for electromagnetic-suspension types. A restoring force, either electrical of mass loading proportional to the change in weight is then applied to the beam. Though the appropriate electronic or mechanical linkage restoring it to the null position change in weight are measured by determining the magnitude of the current or voltage applied to the null point restoring device, which can be recorded as a function of mass.

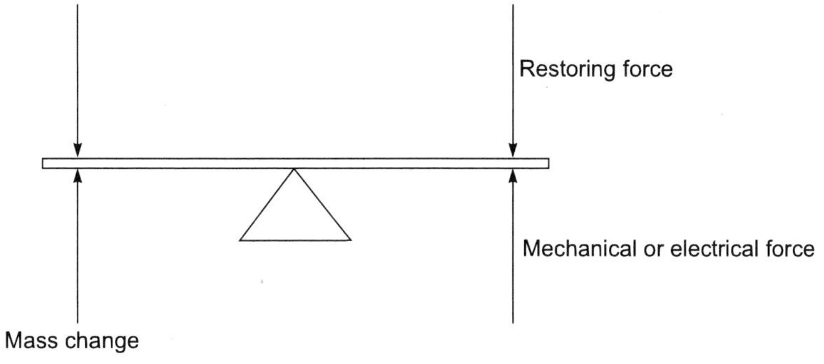

Null type balance

Deflection type balance involves the conversion of the balance beam deflection about the fulcrum into a suitable mass change by: (a) photographic recording, i.e., change in path of a reflected beam of light available of photographic recording, (b) recording electrical signals generated by an appropriate displacement measurement transducer, and (c) using an electrochemical device. The different balances used in TG instruments have measuring range from 0.0001 mg to 1 gm depending on sample and containers used during analysis.

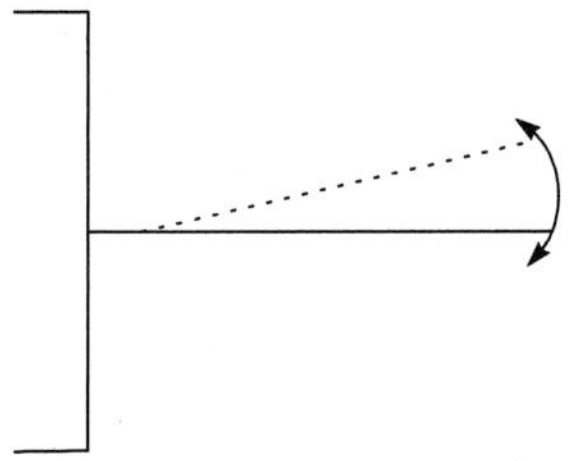

Deflection type balance

However, the present instruments provide options to select the measurable range with selected features.

3.2.2 Furnace

The furnace is base for the instrument and also decides the measuring range, as well as utility of the TG instrument. The furnace and control system must be designed to produce linear heating over the entire working temperature range. The provision must be made to maintain fixed temperature for stepwise analysis. A wide temperature range, generally –150°C to 2000°C of furnaces, is used in different instruments. The range of furnace basically depends on the types of heating elements used. Some elements with their heating are shown in Table 3.2. Further, the rate of heating of a furnace is also controlled by controlling the electric supply to the furnace from 1°C per minute to 50°C per minute. While the ultra high heating rate up to 10^3 K per minute furnace is also in use with the help of superadiabatic conditions and advance electronic modifications.

TABLE 3.2 Different types of heating elements and their heating range

Elements	Temperature (°C)	Recommended atmosphere
Kanthal	Ambient-1000	Inert, Oxid, Red, Vac
SiC	Ambient-1600	Inert, Oxid, Red, Vac
$MoSi_2$	Ambient-1750	Inert, Oxid, Red,Vac
Graphite	Ambient-2000	Inert, Oxid, Red
Graphite	Ambient-2400	N_2/V
Thermocoax	–150-500	Inert, Oxid, Red, Vac

3.2.3 Temperature Measurement and Control

Temperature measurement is commonly done using thermocouples, e.g., chromal-alumel thermocouples are often used for temperatures up to 1100°C whereas Pt

(Pt-10% Rh) is employed for temperatures up to 1750°C. Temperature may be controlled or varied using a program controller with two thermocouples arrangement.

3.2.4 Recorder

Instruments like graphic recorders, strip chart recorders were earlier used. Nowadays, *X-Y* recorders are used as they plot weight directly against temperature. Further, modern instruments facilitate microprocessor-controlled operation, digital data acquisition and processing using personal computer with different types of recorders (plotters) for better presentation of data. The temperature sensor of furnace is linked to the program to control heating rates. So that the balance output and thermocouple signal may be fed to the computer and further to the recorder to record the TG curve.

3.2.5 Atmosphere Control Unit

To monitor the atmosphere around the sample is important for critical analysis of materials. For this purpose furnace is attached with a vacuum pump and purge gas cylinder, generally N_2, Ar are used but sometimes O_2, H_2, etc., are also passed in the furnace depending on the nature of studies. The amount of gas flow decides the reaction kinetics and it is a highly regulated quantity. The gas flow rate depends on sample and nature of studies. Generally, 100 mL/min total while 40 mL/min for balance but 60 mL/min furnace separately if it is required. Currently, the use of carrier gas is receiving more importance in Pulse Gas Thermal Analysis (PGTA).

3.2.6 Sample Holder

It is a simple container for samples during thermogravimetric analysis. In general, the sample holders are made up of chemically and thermally stable materials in desired performed range of analysis. Therefore, different types of materials are used to prepare sample holder for TG analysis like aluminium up to 550°C since the melting point of aluminium is 680°C , platinum up to 1500°C due to their thermal inertness in larger temperature range. The sample holders are made also by ceramic for specific applications like chemically reactive materials. The shape and size of sample holder also play important roles, along with its covered and open types design like deep, shallow, cone shape, cylinder and loosely covered.

3.3 SOURCES OF ERRORS IN TGA

In thermal analysis, the variables affecting resolution for a specific hardware design are typically sample size, heating/cooling rate, and purge gas composition. Generally, smaller sample size, slower heating/cooling rates, and high thermal conductivity of purge gases (e.g., helium) result in improved resolution for TG curves. However, there are a number of sources of error in TG, and they can lead to inaccuracies in the recorded temperature and weight data. Some of the errors may be corrected by placing the thermobalance at proper place and handling it with care. Some common sources of errors during operation of a thermobalance are discussed as under.

3.3.1 Buoyance Effect

If a thermally inert crucible is heated when empty, there is usually an apparent weight change as temperature increases. This is due to the effect of change in buoyancy of the gas in the sample environment with the temperature, the increased convection and possible effect of heat from the furnace in the balance itself. In most modern thermobalances, this effect is negligible. However, if necessary, a blank run with empty crucible can be performed over an appropriate temperature range. The resultant record can be used as a correction curve for subsequent experiment performed in the same condition.

3.3.2 Condensation on Balance Suspension

Condensation of the volatile effluents also affect the weight of the sample and consequently, the shape of TG curve. This can be avoided by maintaining a dynamic atmosphere around the sample in the furnace so that all the condensable product may be driven by the purge gases.

3.3.3 Other Probable Errors

(i) Random fluctuation of balance mechanism
(ii) Reaction between sample and container
(iii) Convection effect from furnace
(iv) Turbulence effect from gas flow
(v) Induction effect from furnace

Error of type 3.3.1 can be avoided by proper placing of balance in the laboratory and error (3.3.2) can be avoided by sensible choice of sample container. Probable errors have to be considered in the design of the furnace, the balance and its suspension system. By avoiding excessive heating rate and proper gas flow rate, some of the above-mentioned errors may be controlled.

3.4 FACTORS AFFECTING TG CURVE

In the beginning of this chapter, we talked about the initial decomposition temperature, T_i at which the onset of a weight change can be detected by the thermobalance operating under particular conditions [9-10]. We may call this decomposition temperature, but it is not correct. Actually in TG experiments, both T_i and T_f (initial and final temperatures at which the decomposition starts and completes) do not have fundamental significance, but they can still be a useful characteristic of a TG curve and the term procedural decomposition temperature is often used for the temperature at which weight change appears to commence. This indicates that a decomposition temperature does not have a fixed value, but depends on the experimental procedure employed to get it. Similar to this, there are many factors which influence a TG curve. These factors may be due to instrumentation or nature of the sample. We have listed the main factors which affect the shape, precision and accuracy of the experimental results in thermogravimetry.

3.4.1 Instrumental Factors

(a) Furnace heating rate
(b) Recording or chart speed
(c) Furnace atmosphere
(d) Geometry of sample holder/location of sensors
(e) Balance sensitivity.

Factors affecting the curves are discussed in detail below.

(a) Furnace Heating Rate

At a given temperature, the degree of decomposition is greater at slower heating rate, and thus the shape of the TG curve can be influenced by the heating rate. For a single stage endothermic reaction, it has been found that:

(i) $(T_i)_F > (T_i)_S$
(ii) $(T_F)_F > (T_F)_S$
(iii) $(T_F - T_i)_F > (T_F - T_i)_S$

Where subscript F and S indicate fast and slow heating rates respectively. The significance of heating rate can be understood from an example; calcium carbonate would not show any weight loss below 600°C, when heated in a thermobalance at heating rate of 3°C per minute, and yet it is known that CO_2 is evolved at 250°C. Similarly, polystyrene decomposes 10% by weight when heating rate is 1°C per minute by 357°C and 10% by weight when heating rate is 5°C per minute by 394°C.

The appearance of an inflection in a TG curve at a fast heating rate may well be resolved into a plateau at a slower heating rate. Therefore, in TG there is neither no standard heating rate, but a heating rate of 3°C per minute gives a TG curve with maximum meaningful resolution. The effect of three different heating rates on the TG curves of $CaC_2O_4 \cdot 2H_2O$ is shown in Fig. 3.7. As can be seen: for each consecutive reaction, an increase in heating rate increases the procedural decomposition temperatures T_i and T_f. This effect in general, is quite small for the three heating rates studied 2.5, 5 and 10 °C/min. It should be noted that although the procedural decomposition temperatures are changed with change in heating rate, the mass loss remains unchanged. It should not be inferred that the use of high heating rate in thermogravimetry always has a deleterious effect on the TG curve. If a small sample is used, very fast heating rates may be employed and one will still be able to detect the presence of intermediate compounds formed during the decomposition reaction. The detection of intermediate compounds in the mass curve also depends on the heating rate employed. The different types of heating rate dependent hydrates are observed in the case of $NiSO_4 \cdot 7H_2O$ and $CuSO_4 \cdot 6H_2O$.

Further, the recording of more pronounced horizontal mass plateaus in the TG curve is possible by the use of a quasistatic rate mode. With this technique, provision is made for the interruption of the linear temperature rise cycle and continuation of heating at a constant fixed temperature. This method gives mass loss curves that are in general, steeper than those obtained under dynamic conditions and provide more accurate data on the final decomposition temperature.

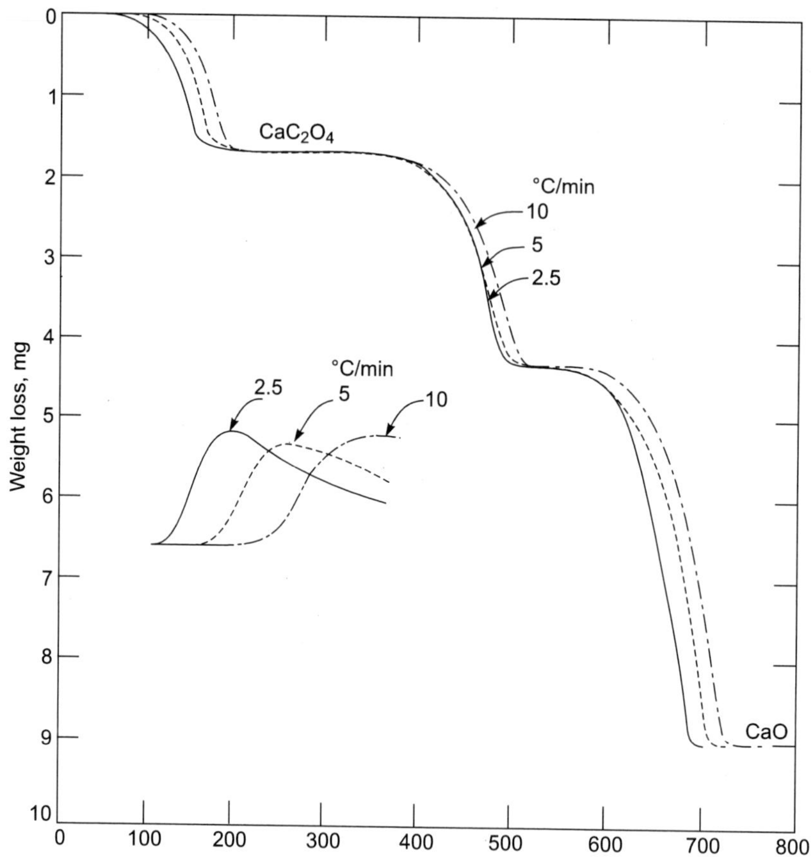

Fig. 3.7 Effect of heating rate on the TG curve

(b) Recording or Chart Speed

This effect was important in old instruments, when chart recorder was used, but, it was discussed to make the users aware. However, the same effect is related to the modern instrument by term data transfer points. The chart speed on the recording of the TG curve of rapid or slow reaction has pronounced effect on the shape of the

TG curves. For a slow decomposition reaction, low chart speed is recommended for recording the TG curve because at high chart speed the curve will be flattened and it will not show the sharp decomposition temperature. For a slow reaction followed by a rapid one at the lower chart speed, the curve will show less separation in the two steps than the higher chart speed curve. For fast reaction followed by slower one, similar observation was made in shorter curve plateaus.

(c) Furnace Atmosphere

The effect of atmosphere on the mass change curve depends on: (i) the types of the reaction, (ii) the nature of the decomposition products, and (iii) type of the atmosphere employed. The effect of the atmosphere on TG curve may be illustrated by taking the example of thermal decomposition of a sample of monohydrates of calcium oxalate in dry O_2 and dry N_2 as shown in Fig. 3.8.

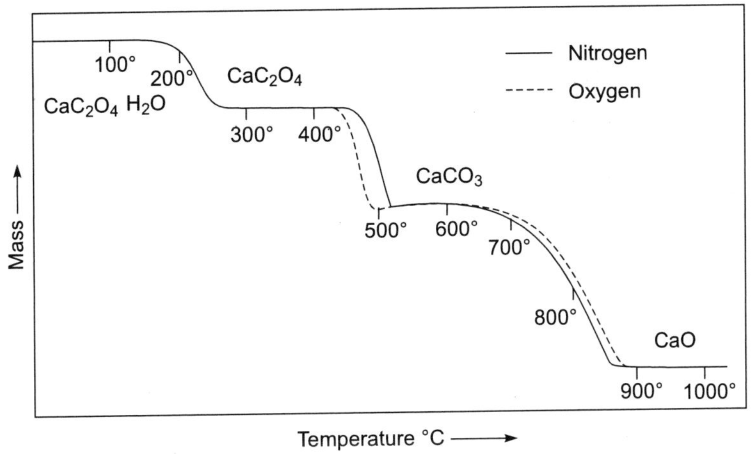

Fig. 3.8 TG curve of calcium oxalate in O_2 and N_2 atmospheres

The first step is dehydration and it is reversible.

$$CaC_2O_4 \cdot H_2O(s) \rightleftharpoons CaC_2O_4(s) + H_2O(g)$$

This is unaffected because both gases are equally effective in sweeping the evolved water vapours away from the sample surface, for the second step,

$$CaC_2O_4(s) \rightarrow CaCO_3 + CO \text{ (in } N_2)$$

The curve diverges in O_2 atmosphere because the oxygen reacts with evolved CO, giving a second oxidation reaction which is highly exothermic and so raises the temperature of the unreacted sample. The temperature accelerates the decomposition of the compound more rapidly and completely at a lower temperature as shown in the above figure in dry O_2 than in N_2 atmosphere. The third step in decomposition reaction is also reversible.

$$CaCO_3 \rightleftharpoons CaO + CO_2 \text{ (g)}$$

Hence, should not be influenced by O_2 or N_2. However, there is a slight difference for the two gases as shown in the figure. The small difference was due to the difference in the nature/composition of $CaCO_3$ formed in the two atmospheres. It was due to the particle size, surface area, lattice defects or other physical characteristics of $CaCO_3$ formed.

Further, if we want to study three types of reactions (I, II, III) either reversible or irreversible:

I. A $solid_1 \leftrightarrow$ B $solid_2$ + C gas
II. A $solid_1 \rightarrow$ B $solid_2$ + C gas
III. A $solid_1$ + B $gas_1 \rightarrow$ C $solid_2$ + D gas_2

If an inert gas is employed, its function will be to remove the gaseous decomposition product of reactions I and II and to prevent reaction III from occurring. If the atmosphere contains the same gas as is evolved in the reaction, only reversible reaction I will be affected and no effect will be observed in reaction II. In reaction III, if gas I is changed in composition, the effect on the reaction will depend on the nature of gas introduced (e.g., oxidizing or reducing will probably affect the mass change curve). The above discussion concerns only dynamic gas atmosphere, in the case of static atmosphere, the following behaviour probably takes place. If the sample evolves a gaseous product reversibly, as the temperature of the furnace and sample increases it will begin to dissociate as soon as its decomposition procedure exceeds the partial pressure of the gas or vapour in its intermediate vicinity. Since a dynamic temperature system is employed, the specific rate of decomposition reaction will increase, as well as the concentration of ambient gas surrounding the sample, due to its decomposition. If the ambient gas concentration increases, the rate of reaction will decrease. However, due to convection currents in the furnace, the gas concentration around the sample is continuously changing, which is one of the reasons why static atmosphere is not recommended. For reproducible results, dynamic atmosphere under rigidly controlled conditions is used. The effect of pressure on the TG curve has been reported. High pressure effect would be opposite to those encountered in low pressure atmosphere. Under high pressure, the initial temperature (T_i) of the reaction would be shifted to higher temperature and T_i T_f reaction interval will also increase.

(d) Sample Holder and Balance Sensitivity

The geometry of sample holder affects the heat transmission in the sample; the used sample holders range from flat plates to deep crucibles of various capacities. The shape of the TG curve will vary as the sample will not be heated in identical conditions. Generally, preconditioning temperature of the thermocouple is placed as near the sample as possible and is not dipped into the sample because it might get spoilt due to sticking of the sample to the thermocouple on heating. So actual sample temperature is not recorded; it is the temperature at some point in the furnace near the sample. Thus, it leads to source of error due to the thermal lag and partly due to the finite time taken to cause detectable weight change. If the sensitivity of recording mechanism is not enough to record the weight change of the sample, then this will also cause error in recording the weight change of the sample. If the composition of the sample container is such that it reacts either with the sample, or product formed or the evolved gases then this will cause error in recording the weight change of the sample. The shape and size of sample holder decides the temperature gradient (Y_m) of the program and detects temperature. The maximum temperature gradient between the wall of sample holder and the centre of the sample is:

(a) For disk:

$$Y_m = (\Delta HG\phi/2\lambda)^{1/2} \cdot S/2$$

(b) For cylinder:

$$Y_m = (\Delta HG\phi/2\lambda)^{1/2} \cdot \tau$$

where, S is the sample thickness, τ is the diameter, ΔH is enthalpy of the reaction, G is the heat capacity of the sample, ϕ is the heating rate and λ the thermal conductivity. For a cylindrical sample 1 mm in diameter in a silver block at heating rate of 5°C/minute, the maximum temperature gradients found were 4.8°C of Kaolin, 13.2°C for decomposition of $MgCO_3$ and 3.1°C for $CaSO_4.0.5, H_2O$. In this regard the shape of sample holder and composition are important factors.

The balance sensitivity is a critical parameter for TG analysis. It permits to improve the determination of mass plateaus and analyze smaller sample amount up to sub-micrograms. This also allows the use of faster heating rate.

Sample Characteristics

(a) Amount of the sample
(b) Particle size
(c) Heat of reaction
(d) Sample packing

3.4.2 Sample Characteristics

(a) Amount of the Sample

The mass sensitivity is a critical criteria for TG; greater sensitivity of thermal balance permits the use of smaller sample, with improved determination of mass plateau of intermediate compounds and the use of faster heating rate. The sample weight affects the TG curve in the following manner.

In general,

(i) The endothermic and exothermic reactions of the sample will cause sample temperature to deviate from a linear temperature change.
(ii) The degree of diffusion of evolved gases through the void space around the solid particles.
(iii) The existence of large thermal gradients throughout the sample particularly, if it has a low thermal conductivity.

Thus it is preferably recommended to use as small a sample as possible, depending on the sensitivity of the balance. As the amount of sample used increases, several problems may arise. The temperature of sample becomes non-uniform through slow-heat transfer and through either self-heating or self-cooling as reaction occurs. The exchange of gas within the surrounding atmosphere is also decreased in high mass of sample. These factors may lead to irreproducibility. However, the small sample mass also protects apparatus in the event of explosion.

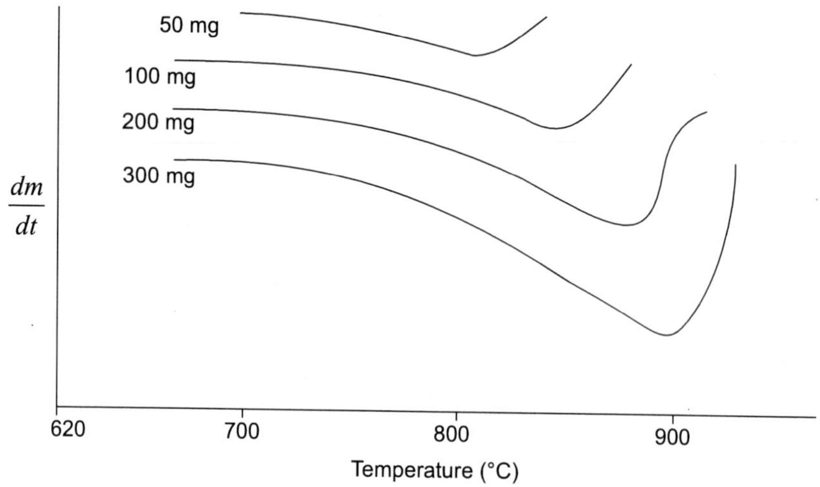

Fig. 3.9 Effect of mass on DTG curve

(b) Particle Size

The particle size of the sample causes a change in the diffusion of the evolved gases, which alters the reaction rate and hence the shape of the curve. The smaller the particle size, greater the extent of decomposition at any given temperature. The use of large crystals may result in apparent, very rapid weight loss during heating. This may be due to the mechanical loss of part of the sample by forcible ejection from the sample container, when the accumulated evolved gases within the coarse grains are suddenly released.

(c) Heat of Reaction

The heat of reaction will affect the extent to which sample temperature follows the furnace temperature. This depends on whether the reaction is exothermic or endothermic and consequently, the extent of decomposition will also be affected. The other sample characteristics such as sample packing, nature of the sample and its thermal conductivity will also affect the shape of TG curves. If the sample is packed loosely, then the evolved gases may diffuse more easily than if the sample is packed tightly. If the sample reacts with the sample container on heating, then it will not give the true weight of the product formed so the sample weight will change.

(d) Sample Packing

The packing characteristics of sample decides the diffusion of heat in sample. This is related to the decomposition of materials if material is loosely packed, slope of curve is smaller or vice versa. The difference in packing of the analysed samples leads to symmetric peak while if packing is inappropriate, diffused peaks spread in larger temperature area.

3.5 INTERPRETATION OF TG CURVES

3.5.1 Differential Thermogravimetric (DTG) Curve

If we turn our attention to quantitative aspects of TG as discussed earlier, TG curve represents the variation in the weight (W) of the sample with the temperature (T) or time (t). Normally, we plot weight loss downward on the ordinate (y) axis and weight gain upwards as shown in Fig. 3.1. Sometimes we also record derivative thermogravimetric (DTG) curves. A DTG curve presents the rate of weight change (dW/dt) as a function of temperature, or time (t) against T on the abscissa (x-axis) as shown in Fig. 3.10. Actually dotted curve of Fig. 3.10 is the derivative of the curve. The curve can be interpreted to have quantitative mass change with respect to temperature. While DTG gives information about the rate of change of mass with temperature and time. The curve is also useful for improving the resolution of the thermal gravimetric changes.

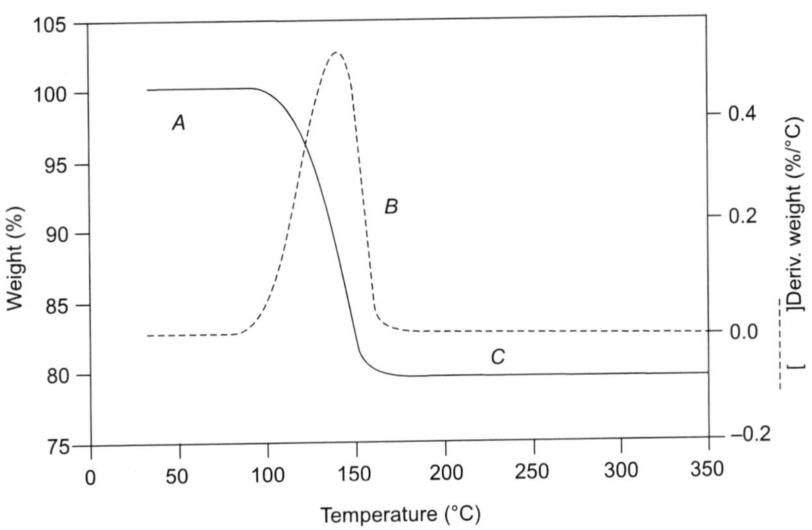

Fig. 3.10 A representative TG and DTG curve. Note the plateau of constant weight (region A), the weight loss portion (region B), and another plateau of constant weight (region C)

In thermogravimetry, the mass of a sample m is continuously recorded as a function of temperature, T or time t.

$$m = f(T \text{ or } t)$$

The quantitative measurement of mass changes is possible by determining the distance on the mass curve axis between the two points of interest or between the two horizontal mass levels. In derivative thermogravimetry, the derivative of the mass change with respect to time dm/dt is recorded as a function of temperature or time. The curve obtained in the first derivative of mass change curve:

$$dm/dt = f(T \text{ or } t)$$

A series of peaks are now obtained, instead of stepwise curve in which the area under the curve is proportional to the mass change of the sample. A comparison between a conventional TG and a derivative curve of mass loss is given in Fig. 3.10. The derivative curve may be obtained manually or by electronic differentiation of the TG signal. The advantageous features of DTG are:

(a) On the TG curve, changes following each other very closely cannot be distinguished, as the corresponding stages coincide. The DTG curve gives sharp maxima of some change of TG and hence can be distinguished clearly.

(b) The DTG curve is exactly proportional to the derivative of the TG curve, therefore, the area under the curve gives the change in mass precisely and DTG can give exact quantitative analysis.

(c) DTG shows overlapping changes better and permits the reaction temperature to be defined more exactly.

(d) Comparison of DTG and DTA enables distinction between peaks due to mass loss and those due to other thermal changes. However, DTG technique has been also criticized for not being subjective of extensive critical analysis.

The peak temperature of DTG represents the temperature at which the rate of the mass change is at a maximum, and not the temperature at which the sample begins to lose mass.

3.5.2 Stoichiometry Calculation

TG curves of a pure compound are characteristic of that compound. Thus, through the TG curve, we can relate the weight changes to the stoichiometry involved. This can often lead us directly to the quantitative analysis of samples whose composition is known. To further illustrate, let's consider the example of $CaCO_3$, Fig. 3.11. This curve indicates that $CaCO_3$ decomposes in a single step between 800°C and 950°C to form stable oxide CaO and the gas, carbon dioxide. This can be explained as the chemistry of $CaCO_3$ when it is heated

$$CaCO_3(s) \rightarrow CaO(s) + CO_2(g)$$

Now again consider Fig. 3.11, it indicates that the % weight lost by the sample is 44 at 950°C. This exactly corresponds to the weight changes calculations based on stoichiometry of the decomposition of $CaCO_3$ expressed by the chemical equation. As in this case, percentage weight loss of $CaCO_3$ will be

$$W\% = \frac{M_r\,(CO_2)}{M_r\,(CaCO_3)} \times 100 = \frac{44}{100} \times 100 = 44$$

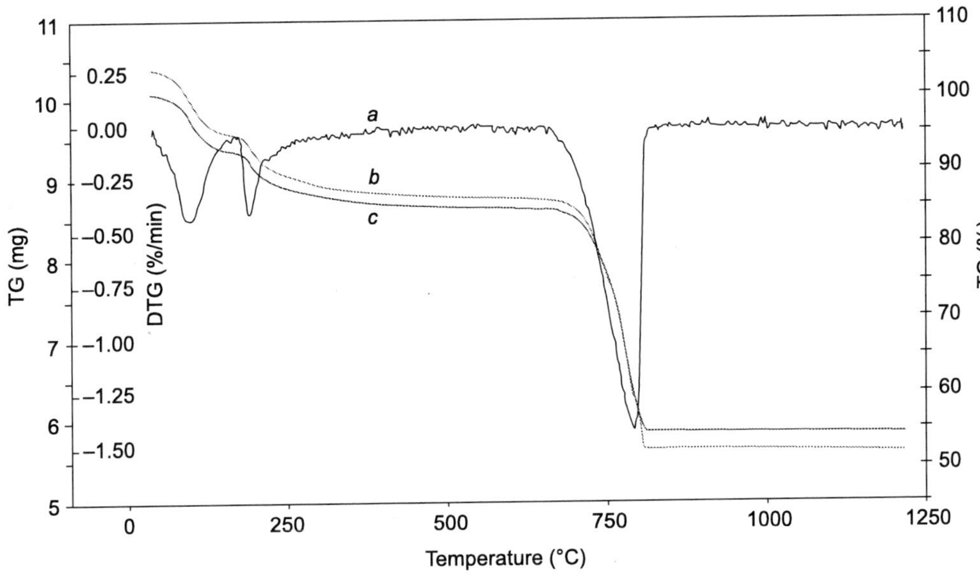

Fig. 3.11 TG and DTG curves of $CaCO_3$ (DTG = Rate of change of mass, dm/dt)

3.5.3 Qualitative Interpretations

Further, TG curves can be used to compare thermal stability of materials (qualitative interpretation). Such information can be used to select material for certain end-use application, predict product performance and improve product quality. Figure 3.12 gives TG curves of some polymers. TG curves clearly indicate that polymer (PVC) is the least thermally stable and polymer (PI) is most thermally stable. Polymer (PI) loses no weight at all below about 500°C and then decomposes abruptly by about 600°C. The other three polymers have all decomposed by about 450°C. Polymer (PMMA) decomposes more slowly overall than the others as indicated by slopes of TG curves. TG curve of polymer (PMMA) has less slope than the others.

Another important use of TG curve is to calculate the kinetic parameters of different thermal decomposition reactions taking place in sample.

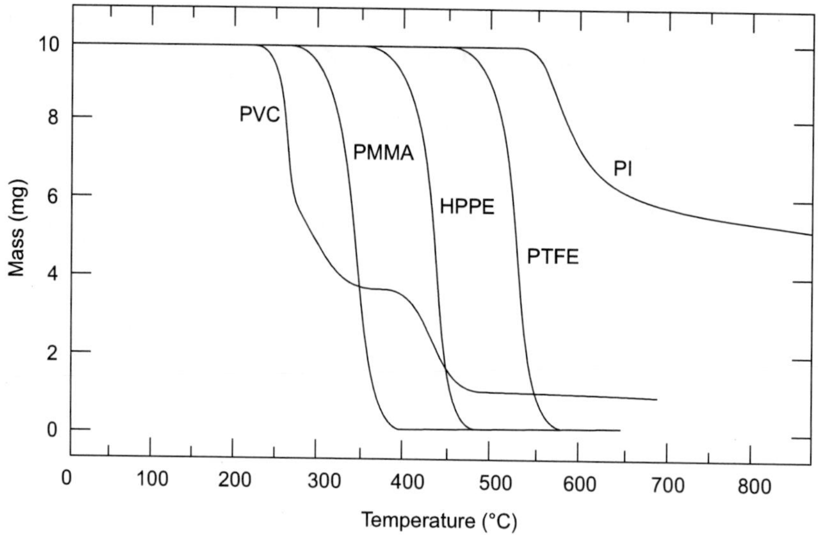

Fig. 3.12 TG curves of different polymers (Polyvinyl Chloride (PVC), Polymethyl methacrylate (PMMA), High Density Polyethylene (HDPE), Polytetrafloroethylene (PTFE) and Poly pyromellitimide (PI))

3.6 APPLICATIONS OF THERMOGRAVIMETRIC ANALYSIS

In the previous sections we have discussed how TG results can be used to understand the thermal chemistry of a particular compound [9,11-12]. The qualitative and quantitative interpretation of TG curves explain the use of the technique. Some specific applications of the technique are listed below:

(i) Purity and thermal stability
(ii) Solid state reactions.
(iii) Decomposition of inorganic and organic compounds.
(iv) Determining composition of the complex mixture
(v) Corrosion of metals in various atmospheres
(vi) Pyrolysis of coal, petroleum and wood
(vii) Roasting and calcination of minerals.
(viii) Reaction kinetics studies.
(ix) Evaluation of gravimetric precipitates.
(x) Oxidative and reductive stability.
(xi) Determining moisture, volatile and ash content.
(xii) Desolvation, sublimation, vaporization, sorption, desorption, and chemisorptions.

Now it is worth describing some of the above applications which are commonly used. While it is not possible to discuss all these applications but efforts have been made to incorporate maximum types and examples.

3.6.1 Estimating Water of Hydration

Water is one of the important properties in deciding chemical and biochemical components including edible items. Estimation of water content determines the physical and chemical properties of food products, their texture, preservation, storage and assimilation abilities by an organism. The various methods used for estimating the water content are distillation, chemical, and refractometric methods. The methods are suitable for individual food items, but TG is useful for all types of chemicals and food products. Here two examples are discussed, one is of interest to analytical chemists and other to food chemists. Water is present in most of the natural foods, at levels up to 90-95% by weight in some fruits (oranges) and vegetables (tomatoes). Most beverages contain high proportions of water. Water exists in foods in various forms: free water, water droplets, water adsorbed on a surface, chemically bound water, crystal water and composition water. Often, water is removed from processed foods as it affects keeping quality, or to reduce weight and volume of products. In most cases, however, part of this water is extremely difficult to remove therefore even dehydrated foods may contain 2-3% residual water.

The nature bonding is related to the binding energy or energy required to break the bonds. This also relates to the thermal stability or decomposition temperature. The higher the decomposition temperature, stronger will be the bonding behaviour of the molecule. The thermogram of hydrated cupric sulphate ($CuSO_4 \cdot 5H_2O$) is shown in Fig. 3.13.

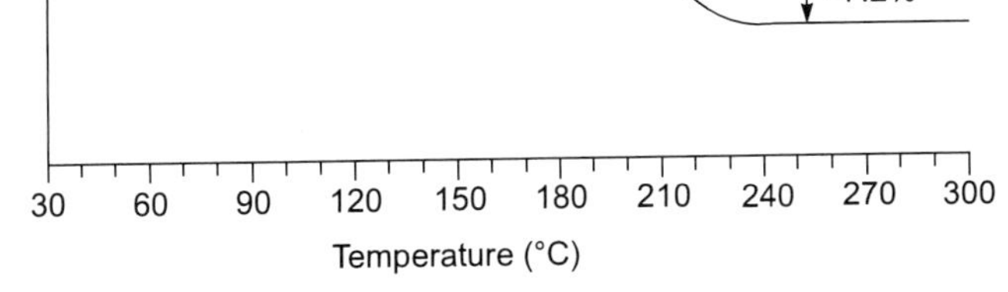

Fig. 3.13 TG curve of penta hydrated cupric sulphate

The hydrated copper sulphate exists as several forms which differ in regard to their degree of hydration. When hydrated forms are left at room temperature, they begin to lose weight. The TG results of the copper sulphate pentahydrate salt shows the per cent mass as a function of sample temperature. The TG results show that the sample loses its water of hydration in the following ratio of 2:2:1 at three temperatures 62.7°C, 108.7°C and 275°C because of three kinds of attachment of H_2O. This is what is expected for the copper sulphate penta hydrate salt, the first two lose water at 62.7°C and 108.7°C whereas the third at 275°C.

Maslowska et al. [15] analysed the wheaten flour, cake flour, corn flour and potato flour by using TG technique. The TG diagram of wheaten flour, cake flour and potato flour is shown in Figs. 3.14-3.16. In inference, all the products show dehydration in two stages from 35-200°C with two DTG peaks. However the rate of dehydration varies from (wheaten flour), (cake flour) to (potato flour). Further, the data was also used to calculate activation energy for dehydration and it was found to be 29.851 kJ mol^{-1} for wheaten flour, 29.085 for cake flour and 27.825 for potato flour. The information can be used to fix the processing and storage conditions. Similarly, methods can be employed for other food products.

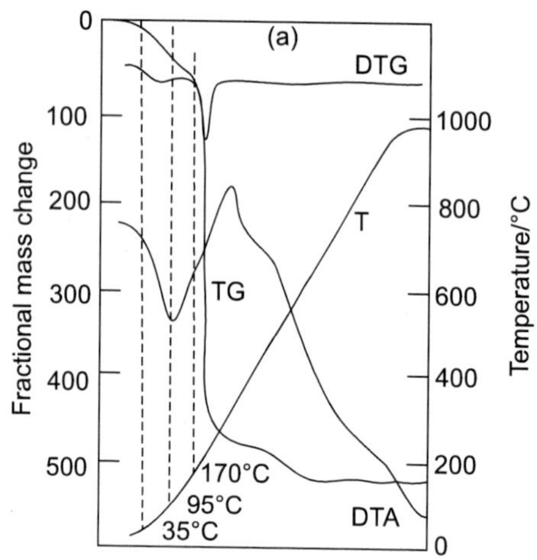

Fig. 3.14 Thermal curves of wheaten flour

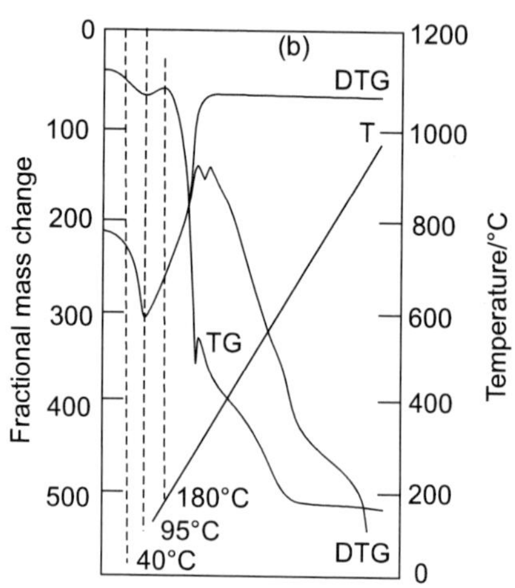

Fig. 3.15 Thermal curves of cake flour

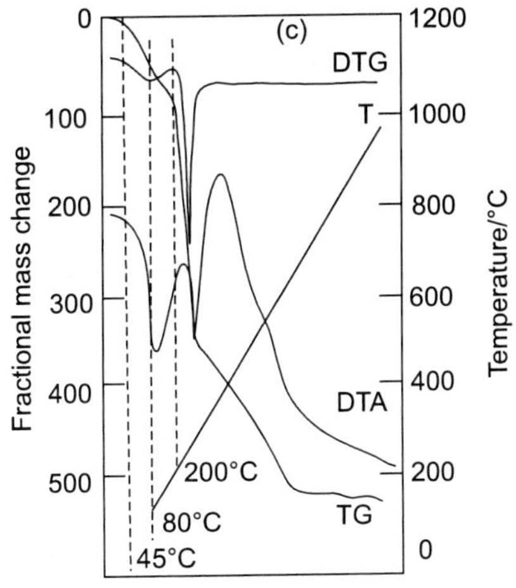

Fig. 3.16 Thermal curves of potato flour

3.6.2 In Analytical Chemistry

We have already seen that a single and pure compound gives characteristic TG curves. Now, we will see how TG curves can be used in predicting relative quantities of the components of a complex mixture.

(a) Binary Mixtures

Consider a mixture of two compounds *MX* and *NY* with their characteristic TG curves, which are different from each other as shown in Fig. 3.17

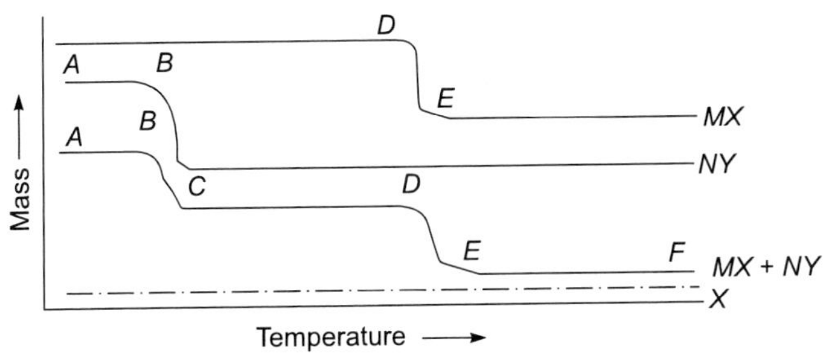

Fig. 3.17 Thermogravimetric curves of two compounds *AB* and *CD* and their mixtures

The decomposition of *MX* occurs at T_1 and is labelled as *DE* in Fig. 3.17. The decomposition of *NY* occurs at about temperature T_2, and is labelled as *BC* in this figure. Curve *MX* + *NY* represents the TG curve of the mixture. In this figure, the plateaus commence at about the same temperature as they do in the TG curves for the pure compounds *MX* and *NY*. Thus, from the mixture curve we can determine the relative quantities of *MX* and *NY* in the original binary mixture. Such a mixture of calcium and magnesium carbonates is discussed below.

A typical TG curve of a mixture of calcium and magnesium carbonates is shown in Fig. 3.18. One can notice that a significant weight loss occurs before 210°C. This is due to the moisture present in the mixture. Another weight loss at about 480°C is due to the following reaction:

$$MgCO_3(s) \rightarrow MgO(s) + CO_2(g)$$

Earlier we mentioned that $CaCO_3$ decomposes at about 800°C. In Fig. 3.18, at the weight loss between 600°C and 900°C can be interpreted due to decomposition of $CaCO_3$.

$$CaCO_3(s) \rightarrow CaO(s) + CO_2$$

Portion of the curve *EF* represents a mixture of MgO and $CaCO_3$ and GH represents a residue of the mixture MgO and CaO. Both these plateaus EF and GH represent weight w_1 and w_2 respectively. In fact, weight w_1 is due to $CaCO_3$ + MgO.

Thus, $w_1 - w_2$ is the loss of CO_2 between 600°C and 900°C due to the decomposition of $CaCO_3$. Using TG curve we can relate the weight of different components formed during TG experiment.

The weight of CaO (w_3) formed:

$$w_3 = 1.28\ (w_1 - w_2)$$

This can be obtained as follows:

The CaO is formed by the evolution of CO_2 on the decomposition of $CaCO_3$,

$$\underset{Mw\ 100}{CaCO_3(s)} \rightarrow \underset{56}{CaO(s)} + \underset{44}{CO_2(g)}$$

From the above equation, 1 mole of $CaCO_3$ gives 1 mole of CO_2 and 1 mole of CaO. Thus moles of CO_2 formed in the given examples $= \dfrac{(w_1 - w_2)}{44}$

Thus the amount of CaO will be

$$w_3 = \frac{(w_1 - w_2)}{44} \times Mw(\text{CaO})$$

$$w_3 = \frac{(w_1 - w_2)}{44} \times 56\text{ g}$$

$$w_3 = 1.27(w_1 - w_2)\text{ g}$$

We know the weight of residue left, w_2, the weight of MgO (w_4) can be calculated.

$$w_4 = w_2 - w_3$$

Here w_3 is the weight of CaO formed, which is equal to 1.27 ($w_1 - w_2$). Thus,

$$w_4 = w_2 - 1.27\ (w_1 - w_2)$$

Weight of the Ca (w_{ca}) in the original sample can also be related to w_1 and w_2 by the following formula

$$w_{ca} = 0.91\ (w_1 - w_2)$$

This can be obtained as follows:

You know amount of Ca in $CaCO_3$ and CaO will be equal in moles, therefore amount of Ca in

$$w_{Ca} = w_3 \frac{A_w(\text{Ca})}{M_w(\text{CaO})} = 1.27(w_1 - w_2) \times \frac{A_w(\text{Ca})}{M_w(\text{CaO})}$$

$$= 1.28\ (w_1 - w_2) \times 40.1/56.1 = 0.91\ (w_1 - w_2)$$

Similarly, weight of magnesium in the original sample can be related to w_1 and w_2:

$$w_{Mg} = (w_2 - w_3) \times \frac{A_w(\text{Mg})}{M_w(\text{MgO})} = (w_2 - w_3) \times \frac{24.3}{40.3}$$

$$w_{Mg} = 0.60\ (w_2 - w_3)$$

Thus, the weight of Mg in original sample

$$= (\text{weight of residue} - \text{weight of CaO}) \times \frac{A_w(\text{Mg})}{M_w(\text{MgO})}$$

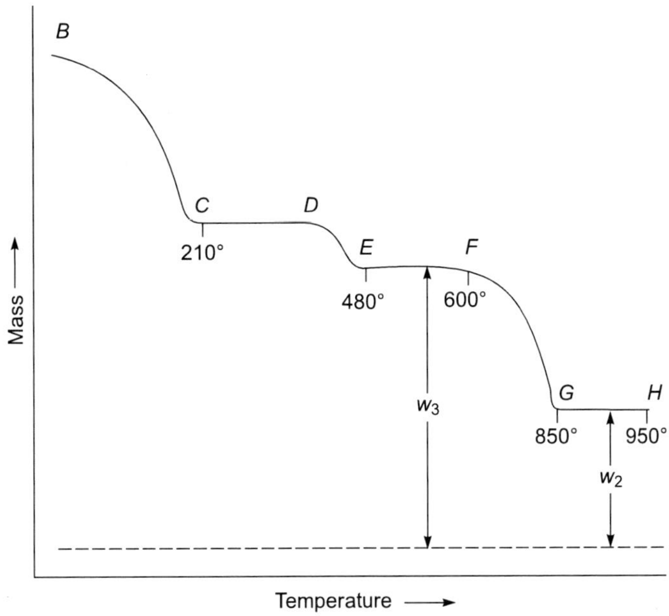

Fig. 3.18 TG curve of a mixture of calcium and magnesium carbonate

Similarly, in another example calcium oxalate monohydrate is in unsatisfactory weighing form for determining calcium as oxalate because its tendency to retain excess moisture even at 110°C, co-precipitated ammonium oxalate that remained un-decomposed. The anhydrous calcium oxalate is hygroscopic but excellent stable form, if heated to 500 ± 25°C. Above 635°C, the decomposition of $CaCO_3$ commences to become significant under usual laboratory conditions and completely converts to CaO at 900°C. Thus, calcium oxalate monohydrate decomposes in three steps, first dehydration, second removal of CO and formation of $CaCO_3$, then $CaCO_3$ decomposes to CaO.

$$CaC_2O_4.H_2O \rightarrow CaC_2O_4 + H_2O \quad \text{at about 135-220°C}$$

$$CaC_2O_4 \rightarrow CaCO_3 + CO \quad \text{at about 400-480°C}$$

$$CaCO_3 \rightarrow CaO + CO_2 \quad \text{at about 635-900°C}$$

But magnesium oxalate dihydrate decomposes in two steps instead of three steps. First, dehydration then removal of CO and CO_2 simultaneously and forming MgO, there is no horizontal corresponding to $MgCO_3$ as it is thermally unstable at this temperature.

$$MgC_2O_4.2H_2O \rightarrow MgC_2O_4 + 2H_2O \quad \text{at about 135-220°C}$$

$$MgC_2O_4 \rightarrow MgCO_3 + CO \rightarrow MgO + CO_2 \quad \text{at about 500°C}$$

The TG curve for the mixture of $CaC_2O_4{\cdot}H_2O$ and $MgC_2O_4{\cdot}2H_2O$ shows two weight losses up to 500°C, first at about 200°C due to loss of water and the second mass loss occurs between the 390 and 500°C range which is due to decomposition of both calcium and magnesium oxalates. Thus, at 500°C, the composition of the mixture will be $CaCO_3$ and MgO. Third weight loss after 600°C is due to the decomposition of $CaCO_3$ to CaO. The w_1 and w_2 are the weights of the mixture at 500°C and 900°C. If x and y are the weights of calcium and magnesium in the mixture, we will get two equations at 500°C and 900°C:

$$\frac{100}{40}x + \frac{40.32}{24.32}y = w_1 \quad \text{at 500°C}$$

$$\frac{56}{40}x + \frac{40.32}{24.32}y = w_2 \quad \text{at 900°C}$$

Solve the above two simultaneous equations for x and y as w_1 and w_2 are known from the TG curve.

One more example of a mixture of $AgNO_3$ and $Cu(NO_3)_2$ is discussed here. The TG curves of pure $AgNO_3$, $Cu(NO_3)_2$ and their mixture are given in the Fig. 3.19. Curve A is the curve of dry crystalline $AgNO_3$. A horizontal extends to 340°C and is followed by descent as far as 473°C. At this temperature, decomposition sets in abruptly and nitrous fumes are expelled up to 610°C. After that there is much slower weight loss from 610°C to 810°C (decomposition of $AgNO_2$), which is not observed when CuO is present, no doubt the latter catalyses the decomposition. Above 810°C, the weight is again constant due to the formation of pure Ag metal.

$$2AgNO_3 \rightarrow 2AgNO_2 + O_2$$

$$2AgNO_2 \rightarrow 2Ag + 2NO_2 \text{ (above 600°C)}$$

While the copper nitrate hexahydrate gives a quite different curve *B*. Water and nitrogen oxide are driven off up to 150°C to 200°C there is a horizontal, the existence of zone of a new compound which has been analysed and which corresponds to a basic nitrate $Cu(NO_3)_2\ 2Cu(OH)_2$. Then between 200°C and 610°C this compound decomposes vigorously from 250°C, subsequently more slowly. The residue is CuO which becomes constant.

$$3Cu(NO_3)\cdot 6H_2O \rightarrow Cu(NO_3)_2\cdot 2Cu(OH)_2 + H_2O + NO \rightarrow CuO$$

If a mixture of $AgNO_3$ and $Cu(NO_3)_2$ is placed in thermobalance, the curve (iii) is recorded as a resultant of the two preceding ones. The horizontal related to the basic copper nitrate is, nevertheless, well marked. From 240°C to 400°C there is a residue keeping constant weight ($AgNO_3$ + CuO), while above 900°C a mixture of Ag + CuO is present.

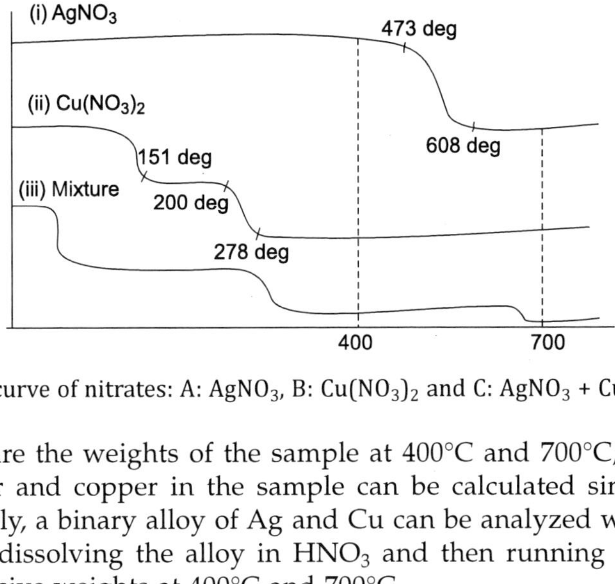

Fig. 3.19 TG curve of nitrates: A: $AgNO_3$, B: $Cu(NO_3)_2$ and C: $AgNO_3$ + $Cu(NO_3)_2$ mixture

If w_1 and w_2 are the weights of the sample at 400°C and 700°C, respectively, the amount of silver and copper in the sample can be calculated similar to previous example. Similarly, a binary alloy of Ag and Cu can be analyzed with +/.3 error by this method by dissolving the alloy in HNO_3 and then running thermogram and recording successive weights at 400°C and 700°C.

(b) Ignition of Precipitates for Gravimetric Analysis

Another important application of TG is the determination of correct drying or ignition temperature for precipitates used in gravimetric analysis. It also provides information whether a gravimetric precipitate should be dried or ignited. For familiar precipitates ignition temperature is generally based upon practical experience of the chemical properties of the substance, such as AgCl is dried at 130-150°C and $BaSO_4$ is ignited at 800-900°C. A better procedure is to make use of TG curves. It is assumed that the horizontal portion of the TG curve gives the range of temperature over which the composition of the precipitate is constant and therefore, the weight is also constant and the precipitates can be dried in this temperature range for gravimetric analysis. It was earlier reported that the precipitate of $Al(OH)_3$ should be ignited at 1200-1300°C to get Al_2O_3 but with the help of TG curve, we come to know that this pricipitatate may be ignited at 1030°C in order to get Al_2O_3. Aluminium oxinate [$Al(C_9H_6NO)_3$] may be dried at 105-110°C as indicated by TG curve. The precipitate of $TlAg_2AsO_4$ furnishes a horizontal in the thermogram between 20 and 846°C. It can therefore be weighted after drying or ignition in this range of temperature. On the other hand the precipitate of calcium picolinate, $Ca(C_{10}H_7 O_5N_4)_2{\cdot}7H_2O$ begins to decompose at 300°C, so it must be dried in a current of dust-free air. Beryllium is precipitated as ammonium beryllium phosphate with ammonium hydrogen phosphate in the presence of ammonia.

$$Be^{2+} + (NH_4)_2HPO_4 \rightarrow Be(NH_4)PO_4 \cdot 6H_2O$$

Its TG curve indicates the distinction in the crystal, i.e., the elimination of water and ammonia at 30°C so that there is no possibility of weighing anhydrous beryllium phosphate. The thermogram shows a continuous loss in weight up to 600°C, above this temperature the weight remains constant, this weight corresponded to the composition $Be_2P_2O_7$. Thus, ammonium beryllium phosphate decomposes easily as compared to beryllium hydroxide. Thus, this is a better method for the determination of beryllium. Thus, TG curve helps in finding the temperature at which the precipitate should be dried or ignited. If beryllium hydroxide is precipitated from $BeSO_4$ with ammonium hydroxide. The precipitate is washed and ignited to give BeO.

$$BeSO_4 + NH_4OH \rightarrow Be(OH)_2 + (NH_4)_2SO_4$$

$$Be(OH)_2 \rightarrow BeO + H_2O \qquad \text{at about } 900°C$$

$Be(OH)_2$ is converted to BeO only at about 900°C. Thus, ammonium beryllium phosphate decomposes easily to $Be(OH)_2$. Thus, it is better to precipitate beryllium with ammonium hydrogen phosphate and then ignite it at 640°C to form $Be_2P_2O_7$. Hence, TG curves help in finding the temperature at which the precipitate should be dried or ignited.

3.6.3 Polymeric Materials

The thermogravimetric analysis has a versatile application in polymer analysis, the acquired information by TG is confined to the detection of change in mass of the sample as its temperature or time duration alters. The technique is used to study the decomposition, oxidation reactions and also some physical processes such as vaporization, sublimation and desorption. The main conclusion drawn by TG are: thermal stability, compositional analysis lifetime assessment and degradation kinetics. The TG curve provides information about the decomposition of various polymers, which may be used for identification purposes and also for assessment of thermal stability. The basic route by which a polymer degrades can be categorized according to the following mechanisms:

1. Main chain scission
2. Side chain scission

3. Elimination
4. Depolymerisation
5. Cyclization
6. Cross-linking

Out of the above categories (1-6) of reactions, only 1-4 lead to change in mass and can be detected in TG. Thus, it provides information about polymerization types, polymers and copolymers.

TG continuously monitors the weight of a polymer sample during isothermal or dynamic temperature scans over the range from 30 to 1000°C in an air, nitrogen, oxygen, or any specific atmosphere. The TG results can be used to estimate thermal stability, compositional analysis, filler content and comparative analysis. The comparative TG diagram of some polymers: polyvinyl chloride (PVC), High Density Polyethylene (HDPE), Poly Methyl Methacrylate (PMMA), Polytetrafluoroethylene (PTFE) and Polypyromellitimide (PI) are shown in Fig. 3.12 and by comparing these diagrams, one can conclude about their thermal stability and the path of their thermal decomposition. TG can be correlated to the cross-linking; nature in polymeric chain can also be concluded. Figure 3.20 shows TG of polymers with different structures and their characteristic thermograms.

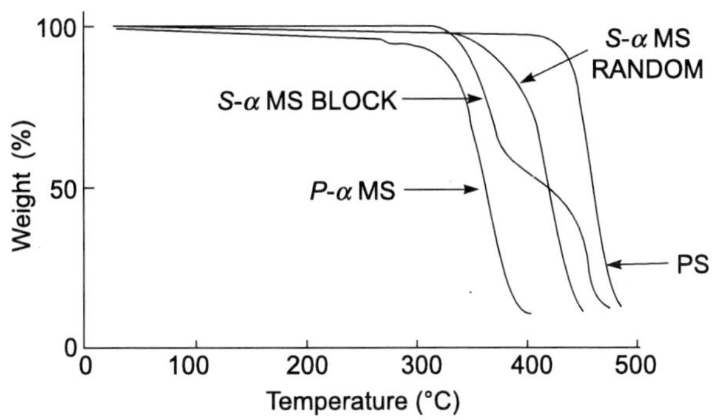

Fig. 3.20 Thermogram of different types of copolymers (Graph adapted from the website of TA instrument) [13]

Further, TG can be used in the determination of stoichiometric ratio of constituent monomers in copolymers, e.g., the amount of vinyl acetate in copolymers of vinyl acetate and polyethylene. When vinyl acetate is heated, it loses acetic acid at about 340°C in nitrogen atmosphere with heating rate of 5°C per minute. TG curve of polyethylene and vinyl acetate (a commercially important polymer used in shoe industries) shown in Fig. 3.21 for vinyl acetate polyethylene copolymers, clearly indicates the loss of acetic acid at about 340-390°C due to the decomposition of vinyl acetate.

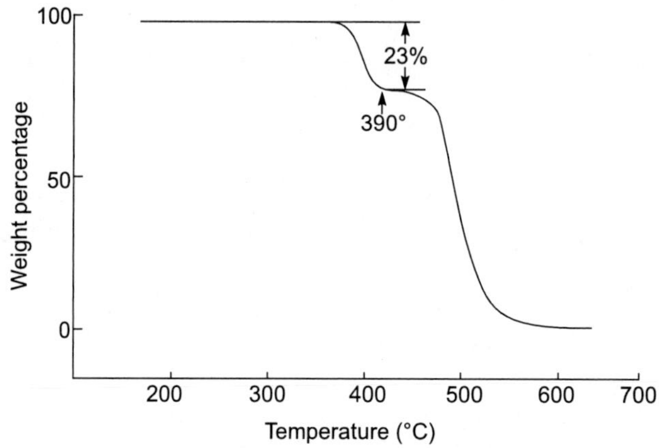

Fig. 3.21 TG curve of Ethylene-Vinyl acetate copolymers

Since, each mole of vinyl acetate loses one mole of acetic acid. Therefore, amount of vinyl acetate in the polymers can be calculated from evolved mass of acetic acid (23%) as indicated in Fig. 3.21. If the observed weight loss occurs due to evolution of acetic acid, multiplied with molecular weight of vinyl acetate and divided by molecular weight of acetic acid, it gives the amount of vinyl acetate. The simple calculation is shown below:

Vinyl acetate (VA)% = Weight loss of acetic acid × *Mw* of *VA*/*Mw* of acetic acid

$$\text{VA}(\%) = 23\% \times \frac{86.1}{60.1} = 33\%$$

3.6.4 Composites and Blends

The amount of additives present in a compound is most accurately assessed by TG and is very important for polymer and elastomers characterization. These additives, e.g., carbon fillers are generally added in rubber for attaining requisite mechanical strength and other properties. TG results of the Acrylonitrile - Butadiene - Styrene (ABS) copolymer under standard conditions (heating rate of 20°C/min) are shown in Fig. 3.22. The TG results show that the ABS resin undergoes thermal degradation at ~400°C with a total mass loss of 95.6%. Although the mass loss curve really shows no indication that two separate phases exist in ABS. However, the derivative curve (DTG) shows two peaks at 468°C and 487°C. The first peak corresponds to the degradation of the ABS copolymer, while the smaller peak at 487°C is reflective of the decomposition of the butadiene rubber phase. Further, weight loss (4.1%) observed at 700°C represents the oxidation of the fixed carbon additives after the purge gas is switched from nitrogen to oxygen.

The resolution of TG experiments can be improved by adopting slow heating rate, however, this increases the overall experimental time. Further, the auto stepwise approach provides outstanding resolution between the ABS copolymer and butadiene rubber components. The levels of the two phases in the ABS resins can be quantitatively assessed using the auto stepwise approach.

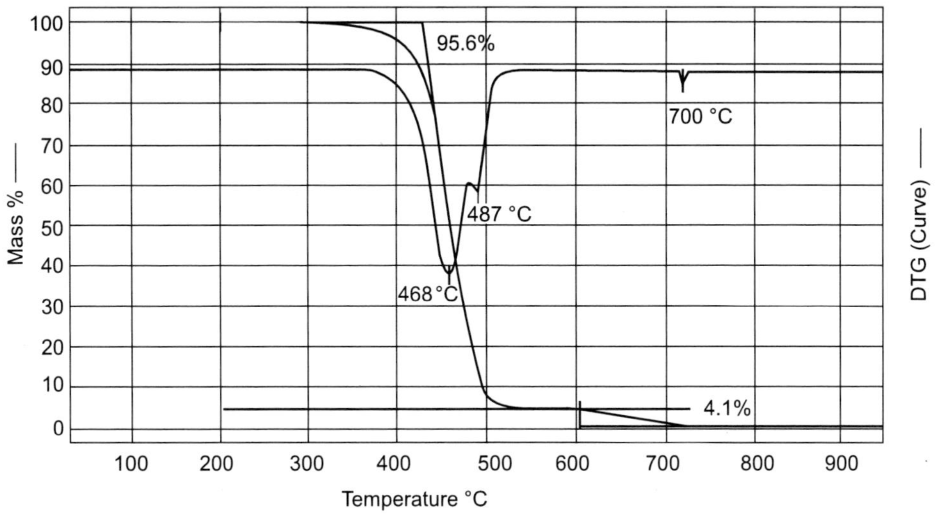

Fig. 3.22 The thermogram of Acrylonitrile-Butadine-Styrene Resin (ABS) (Adapted from website of PerkinElmer) [14]

3.6.5 Analysis of Elastomers

Elastomeric materials are class of materials used in tyre and other rubber industries. Their properties are modified with additives in various compositions which includes oil or plasticizer, carbon black, colorant, antioxidants, blowing agents, etc. The particular end-use application of a given elastomeric material is determined by the nature of its specific composition. The levels of the various components will have a major effect on the resulting end-use characteristics of the elastomer. The multi-component nature of elastomers requires accurate analytical procedures which can be used to provide both quality assurance as well as R&D information to manufacturers and users of the products. TG, provides valuable information regarding the composition of elastomers due to measured weight loss (or gain) of a material as it is heated or held under isothermal conditions. Various purge gases (inert, oxidizing or reducing) can also be used with TG, for more complete characterization purposes. Use of reduced pressure (or vacuum) aids in the volatilization of the oil extraction from the elastomer before the TG measurement.

TG curve of the representative elastomer used in tyre manufacturing is shown in Fig. 3.23. Curve shows several weight losses between 200°C and 500°C. The first weight loss at about 300°C is due to the evaporation or volatilization of the oil and not decomposition. The weight loss occurring at about 450°C is the thermal degradation of the polymer, after that the carbon filler decomposes in the end. Further, it is difficult to clearly define the levels of oil and polymer in the elastomer due to some overlap of these two transitions obtained by standard heating TG, but much better alternative is auto stepwise TG. The results shown in Fig. 3.23 yield the following compositional information on the tyre elastomer as: Oil = 21.8%, Polymer = 43.9%, Carbon black = 32.2%, Inert filler = 2.11%.

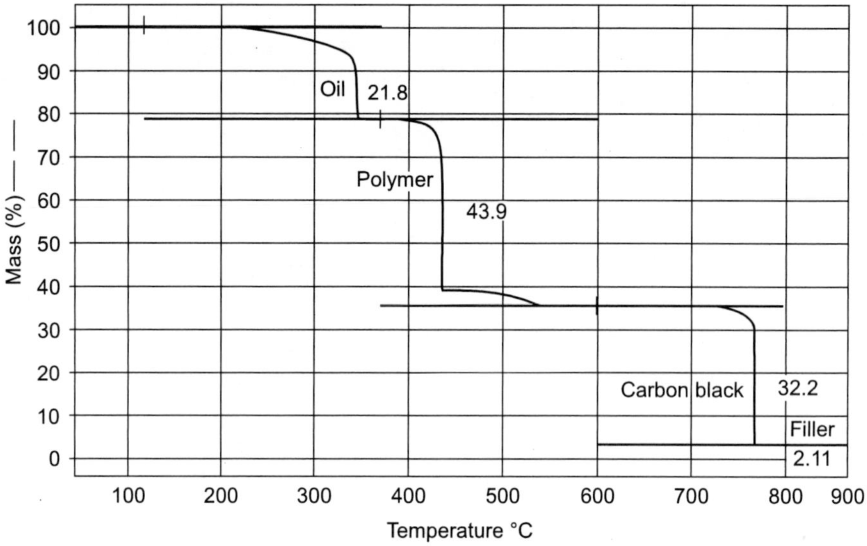

Fig. 3.23 TG results of tyre elastomers (Adapted from website of PerkinElmer)

3.6.6 Energetic Materials

In general, this class of materials provides energy providing materials along with gaseous product after thermal decomposition at a particular temperature and conditions for use as fuel, propellant and explosives. The information provided by TG in terms of decomposition temperature and release of gaseous products provide significant information regarding its use and impact. The energy releasing process of energetic materials are important for release of energy as well as constituent gaseous products. The density for formation of gaseous product is also called smoke density, which has its own impact in explosives. However, the release of gaseous and non-combustible gases affect the fuel effectiveness. Another significant information obtained from TG is thermal decomposition temperature (TD), the TD is important to select the working temperature for fuel in automobile and rocket propellants. The addition of metallic additives and composition significantly changes the decomposition temperature along with releasing mechanism. In example the decomposition rate and pattern of ammonium perchlorate-based rocket propellant significantly changes in the presence of iron by 117% and 8.27% under close and open conditions. The catalytic effect of transition metal has also significantly influenced the decomposition behaviour ammonium perchlorate like 2% by weight presence of copper oxide nanocomposite reduces the decomposition temperature from 425°C to 324.9°C.

Decomposition temperature of petroleum fuel has significant impact on the use of fuel in commercial applications. For example, petroleum coke is residual product of crude oil refinery with potential to be used as substitute of coal in different applications. Further, the decomposition pattern of pristine coke is multiple stage process and unstable. Therefore, it is used in combination gangue in different composition for controlled decompositions. The TG curve of petroleum coke with different percentage of coal gangues are exhibited in Fig. 3.24.

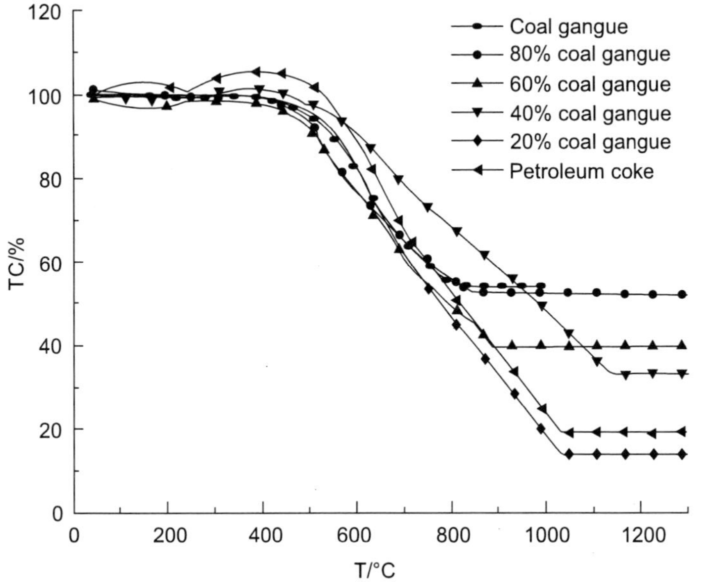

Fig. 3.24 TG curve of coke with different gangue [16]

Waxes are another class of widely used organic compounds comprised of multiple components like long chain hydrocarbon and lipid with wide range thermal stability. TG analysis of waxes were used for compositional analysis of different components on the basis relative mass changes. For example, these waxes were also treated for cross-linking using potassium persulphate (PPS) in order to make it suitable for different applications. The degree of cross-linking changes the trend in change in mass as shown in Fig. 3.25.

3.6.7 Food Science

Thermogravimetric results, i.e., change in mass loss under static as well as dynamic conditions has been used by food scientists for characterisation and preservation of food products. The mass loss in food occurs due to drying, baking, roasting and baking in the form of moisture and pyrolyzed gases under controlled conditions. The food items consist of water, lipid, protein, vitamins and minerals. In general, large amount of water is present in fruits and vegetables in different form, i.e., free water, water droplets, adsorbed water, chemically bound water, crystal water and

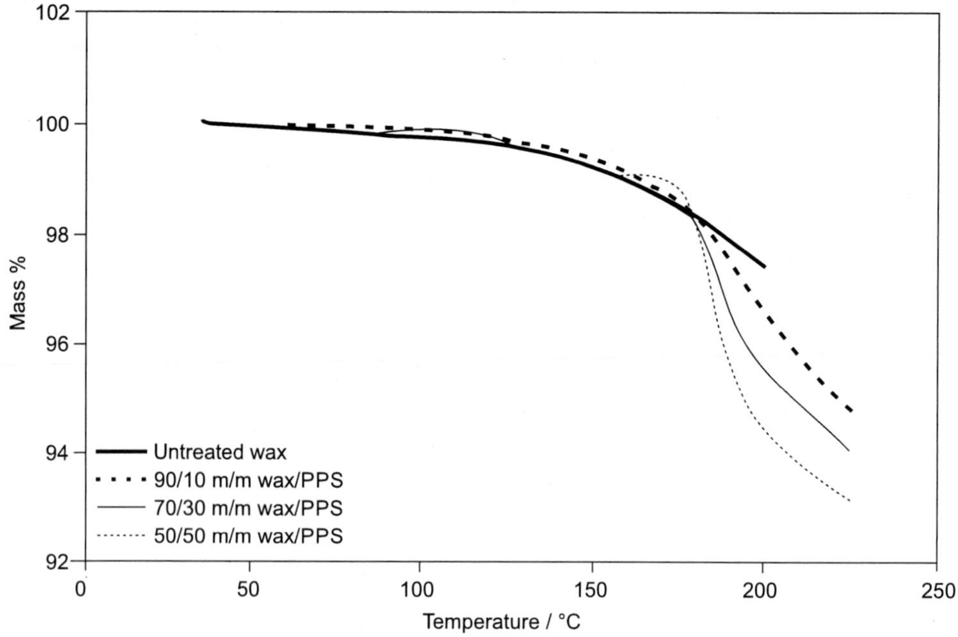

Fig. 3.25 TG curve of untreated and treated wax [17]

composition water. Here, the role of different of water for diverse types of activity as transporting medium to maintain of osmotic pressure in order to maintain the freshness of fruits and vegetables. The presence of water activity is also responsible for chemical stability of constituents along with reduced non-enzymatic browning and spontaneous autocatalyzed lipid oxidation process. Therefore, the dehydration from an edible fruits and vegetable deteriorates their quality as well as indicator of its shelf life and it needs to be optimised. For example, the edible coating comprised several biopolymers like chitosan and gum on fruits and vegetable reduces the rate of weight loss, which used to preserve the quality of different fruits like apple and cucumber. The monitoring of mass loss of water from fresh apples coated with rabinoxylan and β-glucan stearic acid ester significantly reduces the rate of water loss than uncoated applies from 7.9% to 5.6-5.8% after 30 days. Thus, the reduction in mass loss supports the better shelf life of apples after coating with esters [18].

The presence of volatile components and triglyceride in fruits, vegetables and herbal products are indicators for the quality of items. The trend and quantity removal of volatile components are also used as biomarkers for quality of the different products. The high temperature heating of vegetables removes different products at different temperatures due to multicomponent composition of vegetable and the process also explores for pyrolytic extraction of different products. The TG curve of Chinese cabbage scrap is shown in Fig. 3.26 along with DTG. The graph shows five stages for thermochemical transition, i.e., first is dehydration, second is I, II and III are indicating the major thermal degradation, while last stage reveals the char formations.

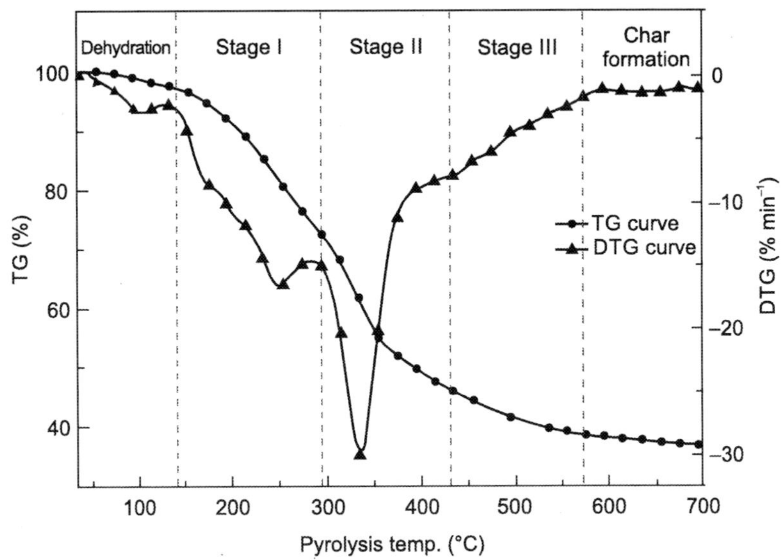

Fig. 3.26 TG/DTG curve Chinese cabbage scrap [19]

The decomposition of components also depends on characteristic of component, for example, hemicellulose degrades in first stage, while the cellulose and protein decompose in stage I and II temperature. Thus, the weight loss in these stages is also used to estimate quantity of constituents.

3.6.8 Solid State Synthesis

In general, solid state synthesis involves the formation stable oxide after thermal decomposition associated with structural rearrangement. The chemical transition of mass during the different chemical reactions are the basis for preparation of different significant industrial materials like ceramics with simplicity and large-scale production. The heating of precursor materials is an advantageous feature to prepare the highly stable ceramic materials as represented in Eq. 1.

$$R(s \text{ or } l) \leftrightarrow A(s) + B(g) \quad (1)$$

The amount of gaseous products $B(g)$ indicates proportional conversion of the reactants into products along with yield of reactions. The transition of temperatures during mass loss provides important information for heat treatment to perform a thermochemical reactions. For example, the thermal degradation of copper oxalate nanorods shown in Fig. 3.27, is revealing the transition between 200 and 305°C leads to the formation of copper oxide after mass loss at 62% along with evolution H_2O and carbon dioxide. [20]

Currently, this technique is used in the study of several solid state reactions like metathesis, which is an important process for efficient and environment-friendly synthetic reaction. The effect of impurity and rate are also important criteria for designing of different solid state synthetic methods. TG study performed during

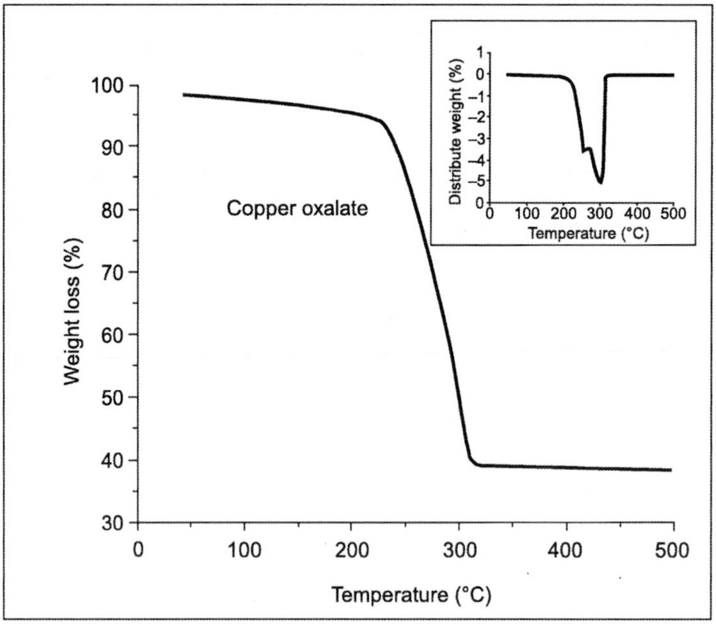

Fig. 3.27 TG curve of copper oxalate nanorod and DTG curve in inset [20]

solid state metathesis gives information of quantitative evolution and formation of different products along with rate and kinetic parameters. The atmosphere of thermal decomposition also influences the final products and its chemical structure. For example, heating of metal sulphide like ZnS in the presence of oxygen, yields metal sulphate as well as oxide, i.e., ZnO along with evolution of sulphur oxides as illustrated in Eqs. 1 and 2.

$$ZnS + 2O_2 \rightarrow ZnSO_4 \tag{1}$$

$$2ZnS + 3O_2 \rightarrow 2ZnO + 2SO_2 \tag{2}$$

Further, the presence of a particular metal also influences the properties like kinetics due catalytic effect of metals. The heating of two oxides also yields mixed oxide after heat induced structure evolutions, as shown in Eqs. 3 and 4.

$$ZnO + Fe_2O_3 \rightarrow ZnFe_2O_4 \tag{3}$$

$$2ZnO + SiO_2 \rightarrow Zn_2SiO_4 \tag{4}$$

These heat induced solid state reactions are used for preparation of several metal sulphates, metal oxide, oxalates, acetate and carbonates.

3.6.9 Reaction Kinetics

The most obvious use of TG is to predict lifetime degradation kinetics under actual service conditions and other kinetic parameters like order of reaction, activation energy, etc. Both isothermal and dynamic methods are in use, but the isothermal measurement at elevated temperature and measurement of time taken for a certain extent of mass loss to occur is simpler. The purpose of measuring reaction kinetics is to obtain information about reaction mechanism for which measurements were made. The aims are rarely unrelated, because reactions are based on chemical steps involved and can be inferred from the overall picture constructed from results of kinetic analysis (kinetic model) and complimentary evidences.

In general, temperature dependence of chemical processes may be expressed in terms of Arrhenius equation,

$$K = Ae\,(-E_a/RT) \tag{1}$$

where K is the rate constant, R the gas constant and T is temperature (Kelvin). The values of the Arrhenius parameter (E_a and A) provide measures of the magnitude of the energy barrier to reaction (the activation energy, E_a) and the frequency of the occurrence of a condition that may lead to reaction (the frequency factor, A), the rate constant (K) is defined by the relationship between the rate of reaction ($d\alpha/dt$) and the extent of conversion or fraction reacted (α). The general relation to describe the relationship between ($d\alpha/dt$) and (α),

$$(d\alpha/dt) = k\alpha^m(1 - \alpha)^n \tag{2}$$

Therefore, from Eq. (2) it is possible to derive various sub-classes of rate equations such as first order decay, nucleation, growth, etc., by changing the values of m and n. Isothermal experiments that provide the means of determining the form of kinetic equation through discrimination between different models is not straightforward. A simpler approach is to substitute the reciprocal of the isothermal lifetime for the rate constant in Eq. (1) and extrapolate the data to the region of interest.

Isothermal measurement suffers from the drawback of being rather time consuming. There are also difficulties in bringing the sample and apparatus to the required temperature without some decomposition of the sample having already taken place. One way of avoiding this problem is to change the rate controlling variable during the course of measurement. A useful method is temperature jump TG. The rate of decomposition can then be measured on either side of the change in temperature and the activation energy determined from Eq. (3):

$$E_a = R\ln\left[(dm/dt)T_1/dm/dt\;T_2\right]\left[1/T_1 - 1/T_2\right] \tag{3}$$

where, $(dm/dt)T_1$ and $(dm/dt)T_2$ are the rates of mass loss at T_1 and T_2 on either side of temperature jump. The study of kinetics using data from conventional rising temperature TG have been proposed by many scientists but the most popular method is described by Ozawa, Flynn and Wall, which has been incorporated into the American Society of Testing Materials standard methods. In this method, essentially separate measurements are carried out at different linear heating rates and temperatures at which a set of percentage mass loss occurs is noted. These are then plotted as functions of heating rate (dT/dt) and the activation energy determined by an iterative process. We often assume, without appropriate justification, that decomposition reactions are first order processes as in Eq. (4):

$$(d\alpha/dt) = k(1 - \alpha) \tag{4}$$

Thus, it is possible to predict the lifetime for say 10% mass loss as a function of temperature. The advanced method for studying degradation kinetics is known as modulated temperature TG. It uses a temperature profile in which a sinusoidal temperature fluctuation upon conventional linear rising temperature. The curves showing the heating rate (dT/dt) and rate of mass loss ($-m/dt$) make the effect of the temperature modulation apparent. E_a for the chemical decomposition is calculated according to Eq. (5):

$$E_a = R[\,T_{av}^2 - (0.5T_{amp})^2]\,L/T_{amp} \tag{5}$$

where, T_{av} is the average thermodynamic temperature, T_{amp} is the amplitude of the temperature modulation, and L is the logarithm of the amplitude of the rate of mass loss over one modulation plot of mass and E_a as a function of temperature. The mt-TG is a relatively new technique which promises to simplify the acquisition of kinetic data. Some general kinetic models used are tabulated in Table 3.3. [9]

TABLE 3.3 Important rate equations used in kinetic analysis

	$g(\alpha) = k(t^t - t_0) = kt$	$f(\alpha) = (1/k)\,(d\alpha/dt)$	$\alpha = h(t)$
1. **Accelerating α-time curves**			
P1 power law	$\alpha^{1/n}$	$N(\alpha)^{(n-1)/n}$	
E1 exponential law	$\ln\alpha$	α	
2. **Sigmoidal α-time curves**			
A2 Avrami-Erofeev	$[-\ln(1-\alpha)]^{1/2}$	$2(1-\alpha)(-\ln(1-\alpha))^{1/2}$	
A3 Avrami-Erofeev	$[-\ln(1-\alpha)]^{1/3}$	$3(1-\alpha)(-\ln(1-\alpha))^{2/3}$	
A4 Avrami-Erofeev	$[-\ln(1-\alpha)]^{1/4}$	$4(1-\alpha)(-\ln(1-\alpha))^{3/4}$	
A*n* Avrami-Erofeev	$[-\ln(1-\alpha)]^{1/n}$	$n(1-\alpha)(-\ln(1-\alpha))^{n-1/n}$	$1-\exp[-(kt)^n]$
B1 Prout-Tompkins	$\ln[\alpha/(1-\alpha)]$ $g(\alpha) = k(t^t - t_0) = kt$	$[\alpha/(1-\alpha)]$ $f(\alpha) = (1/k)\,(d\alpha/dt)$	$1 + 1/\exp[-(kt)^n]^{-1}$ $\alpha = h(t)$
3. **Deaccelerating α-time curves**			
3.1 Geometrical Model			
R2 Contracting Area	$1-(1-\alpha)^{1/2}$	$2(1-\alpha)^{1/2}$	$1-(1-kt)^2$
R3 Contracting Volume	$1-(1-\alpha)^{1/3}$	$3(1-\alpha)^{2/3}$	$1-(1-kt)^3$

3.2 Diffusion model			
D1 one-dimensional	α^2	$1/2\alpha$	$(kt)^{1/2}$
D2 two-dimensional	$(1 - \alpha)\ln(1 - \alpha) + \alpha$	$[-\ln(1 - \alpha)]^{-1}$	
D3 three-dimensional	$[1 - (1 - \alpha)^{1/3}]^2$	$3/2\,(1 - \alpha)^{2/3}$ $(1 - (1 - \alpha)^{1/3})^{-1}$	$1 - (1 - (kt)^{1/2}))^3$
D4 Ginsting Brounshtein	$1 - (2\alpha/3) - (1 - \alpha)^{2/3}$ $g(\alpha) = k(t^t - t_0) = kt$	$3/2[(1 - \alpha)^{1/3} - 1]^{-1}$ $f(\alpha) = (1/k)\,(d\alpha/dt)$	$\alpha = h(t)$
3.3 Order of reaction model			
F0 Zero order	α	1	kt
F1 First order	$-\ln(1 - \alpha)$	$1 - \alpha$	$1 - \exp(-kt)$
F2 Second order	$[1/(1 - \alpha)]$	$(1 - \alpha)^2$	$1 - (kt + 1)^{-1}$
F3 Third order	$[1/(1 - \alpha)^2]$	$0.5\,(1 - \alpha)^3$	$1 - (kt + 1)^{-1/2}$

However, the TG data are used by different scientists differently to calculate the various kinetic parameters. The important methods are discussed below.

(a) Newkirk Method [21]

Newkirk obtained rate constant for the decomposition reaction from a single TG curve, as illustrated in Fig. 3.28. For a series of temperatures, T_1 and T_2, the sample mass remaining, $a - x_1$ and $a - x_2$, and the reaction rates, $(dx/dt)_1$ and $(dx/dt)_2$, were obtained by the tangent to the curve at points 1 and 2.

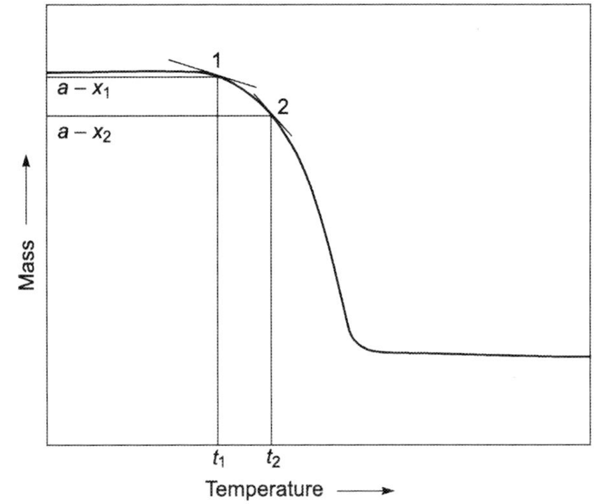

Fig. 3.28 Determination of extent of reaction from TG curve

If the reaction is of the first order, then the logarithm of the reaction rate constant k, $dx/dt = k(a - x)$, when plotted against $1/T$ should yield a straight line. The results of measurements under isothermal and nonisothermal conditions yielded straight lines that were approximately parallel.

(b) Freeman and Carroll Method [22]

This is a widely used kinetic method, and advantage of this method is that considerably fewer experimental data are required than in the isothermal method. The kinetics can be obtained over an entire range in a continuous manner without any missing regions. In addition, where a sample undergoes considerable reaction in being raised to the temperature of the interest, the results obtained by an isothermal method are often questionable. The order of reaction is calculated from the equation:

$$-(E_a/2.3R)\Delta\ (1/T)/\Delta \log W_f = -n + \Delta \log (dw/dt)/\Delta \log W_f \tag{6}$$

where $W_f = W_e - W_t$, in which W_c is the maximum mass, and W_e in the total mass lost upto time t.

(c) Coats and Redfern Method [23]

This method is used for a reaction in which the order of reaction is unknown. The derived expression is:

$$\log \{1 - (1 - \alpha)^{1-n}/T^2\,(1 - n)\} = \log AR/aE\ [1 - 2RT/E] - E/2.3RT \tag{7}$$

where, α is the fraction of the sample decomposed at time t, and a is the heating rate. A plot of either $\log \{1 - (1 - \alpha)^{1-n}/T^2(1 - n)\}$ against $1/T$, or when $n = 1$, $\log \{-\ln (1 - \alpha)\}/T^2$ against $1/T$ should result in a straight line of slope $E/2.3R$ for the correct value of n. The quantity $\log (AR/aE)\ 1 - (2RT/E)$ appears to be reasonably constant for most of the values of E and in the temperature range over which most reactions occur.

(d) Horowitz and Metzer Method [24]

The reaction in which the rate constant involves concentrations expressible as mole fractions and the total number of moles are constant, the expression:

$$\ln (W_0 - W_t/W - W_t) = E\theta/RT^2 \tag{8}$$

may be used, in which W is the mass remaining at a given temperature, W_0 and W_t are initial and final masses respectively, and θ is the reference temperature.

(e) Miscellaneous Methods

Some other simplified methods for the calculation of rate constants from TG curves have been described by various workers:

(i) Dave and Chopra [25] calculated the specific rate constant by use of equation:

$$k = (A/m_0)^{n-1}\,(dx/dt)/(A - a)^n \tag{9}$$

where, A is the total area under the derivative TG curve (DTG); a is the area for the reaction up to time t; dx/dt is the height of the curve a at time t; m_0 is the initial mole fraction of the reactant; and n is the order of reaction.

(ii) A similar approach was used by Papazian et. al. [26] in which a specific rate constant was calculated from DTG curve employing following expression in Eq. (6)

$$K = (dx/dt)/(a - x)^n \tag{10}$$

where $(a - x)$ is the mass of reactant not decomposed. This method yields kinetic results similar to isothermal measurements.

(iii) The importance of individual steps needed to calculate the kinetics of the thermal decomposition of high molecular weight dimethyl silioxane polymer. Calculation was made simply Eq. (11): [27]

$$-dW/dt = kpW \tag{11}$$

where, $-dW/dt$ is the rate at which mass is decreasing, kp is the procedural rate constant, and W is the sample mass.

A method [28] was developed for the pyrolysis:

$$\ln 1/y = -(E/R)(1/T) + \text{Constant} \tag{12}$$

where y is the number of the initial molecules not yet decomposed. Thus, a plot of $\ln(1/y)$ versus $(1/T)$ should yield a straight line from which E can be calculated from the slope.

3.7 CALIBRATION

The TG instrument is necessarily calibrated in routine for obtaining proper results. The precision of a measurement of a quantity depends upon factors such as instrumentation, the homogeneity of the sample, and experimental techniques. Good precision is required for determination of the true value of the accuracy, but the systematic errors may produce constant offsets in the experimental measurement, rather than random fluctuations that determine the precision. To detect and eliminate these systematic errors, calibration is required. The calibration is done for the following.

3.7.1 Baseline

It is specially important for measuring small weight losses associated with volatilization or small amounts of residue calibrated for reproducibility of TGA baseline and effect of heating rate on baseline. For baseline calibration: **Run the instrument in clean empty pan, over temperature range of interest at the desired heating rate. Plot weight vs. temperature, which is further used to correct the analysis of samples.** The basic factors that influence baseline are stability of table, hang down wire condition, hang down tube condition, leveling of TG, cleanliness of furnace, and purge gas flow rate. Most TG instruments have baseline drift of +/−0.025 mg which is 0.25% of a 10 mg sample.

3.7.2 Mass

Calibration standards (mass) in a wide range of sizes and several classes of accuracy may be obtained from the national standards organizations or commercial suppliers. Two-point mass adjustment is important for TG apparatus; 100 mg range (use 100 mg weight), 1000 mg range (use 1000 mg weight). Run TG weight calibration routine. Follow screen instructions to tare and mass calibrate using two calibration weights (if known, enter exact mass of calibration weights). Change in buoyancy resulting from changing the temperature usually has to be determined from blank runs under essentially identical conditions. The very high temperatures (above 1300°C) vaporization of metal from the sample suspension system may give an unexpected mass loss error.

3.7.3 Temperature Calibration

For a meaningful TG measurement, careful calibration of temperature of the sample position is essential. Temperature calibration of TG is carried out with standard samples with a known reproducible transitions temperature. Curie points of a wide range of metal or alloy on heating a ferromagnetic material loses its ferromagnetism. At the Curie point, the magnetic force is reduced to zero and an apparent mass loss is observed. This calibration is also an example of the use of thermomagnetometry. A few materials are certified and marketed for this purpose. Such as Nickel showing transition at 361.14°C, theoretical values for Curie Point Transition. Standard practice for calibration of temperature scale for thermogravimetry is paramagnetic: a material that is susceptible to attraction by a magnet. Curie Point Temperature is a temperature where the material loses its magnetic susceptibility (defined as offset point), requires a magnet and well-characterized transition materials.

An alternative method, known as fusible link method, uses inert mass pieces that are suspended from the balance in the place of the sample by link of thin fusible wire. When the temperature is raised through the melting point, the link melts and the drop from its support causes momentary detectable disturbance in mass signal.

Whichever of the above methods is used, the final stage of calibration is to correlate with the known transition temperatures with the 'apparent' temperature measured using the instrument's temperature sensor in its normal operating position.

Even with careful temperature calibration, T, it may still not be separately known, because slow heat transfer may cause self-heating or self cooling from strongly endothermic or exothermic processes in relatively large samples.

The early attempts to use the mass losses associated with some thermal decomposition reactions for calibration were unsuccessful because of the slow rate of most of such reactions. Some peroxo coordination compounds, which undergo explosive decomposition over a very narrow transition temperature range are recommended as possible standards.

Simultaneous TG-DTA instruments can be calibrated by measuring temperature directly in the DTA system. A suitable pair of pure metals, whose melting points and magnetic transition are known, are placed in the sample pan alongside the magnetic material. Some standards for temperature calibration are melting points of In, Sn, Pb with their theoretical melting points as given in Table 3.4.

TABLE 3.4 Recommended standards for calibration

Metals	M.P.	Heat
Indium (In)	156.6°C	28.5 J/g
Tin (Sn)	231.9°C	60.1 J/g
Lead (Pb)	327.5°C	23.0 J/g
Zinc (Zn)	419.6°C	107.5 J/g
Aluminium (Al)	660.3°C	397.0 J/g

The measure of melting points is used to construct a correction curve for the observed Curie temperatures, T_C. Some magnetic materials used as temperature standards are listed in Table 3.5.

TABLE 3.5 Recommended T_c of standards for calibration

Materials	Recommended T_c (°C)
Alumel	163
Nickel	354
Iron	780
Nicroseal	438
Hisat-50	1000
Numetal	393
Monel	65
Perkalloy	596
Trafoperm	753

In a series of TG-DTA experiments nickel gave the best results, but its standard deviation of 5.4°C was not good enough for reference standards. The cobalt and nickel base alloy provides a series of standards in the temperature range of 160 to 1130°C. A typical example is shown in Fig. 3.29 for nickel. However, the reported literature value is 354°C.

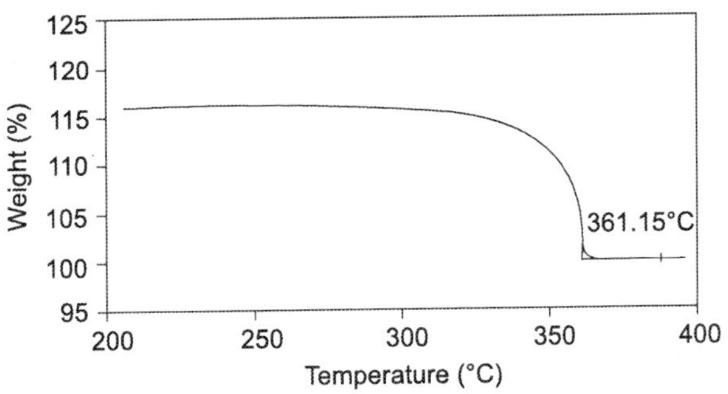

Fig. 3.29 Standard transition temperature of nickel

3.7.4 Purge Gas

Nitrogen and helium are most commonly used as purge gases and often provide the best baseline. Dry air can also sometimes improve resolution because of differences in the oxidative stability (versus thermal stability) of components. Purge gas flow rates are generally: 100 mL/min; total: 10 mL/min in balance and 90 mL/min in furnace. During TG cool-down (between scans), normally select air to cool the furnace. When furnace cools to 500°C, then the time furnace opens, it allows the air to cool. However, the purging of a gas in an experimental condition is required to be selected after careful observations.

EXERCISE

1. What is TG? Explain its applications.
2. What is differential thermal gravimetric (DTG) analysis? List its advantageous features.
3. Describe the importance of atmosphere control units in TGA.
4. List the factors affecting the TG curve.
5. Select the properties which cannot be tested by TG analysis:
   (a) Melting point
   (b) Decomposition temperature
   (c) Thermal stability
   (d) Glass transition temperature
   (e) Order of a reaction
   (f) Activation energy
   (g) Curing temperature
6. Figure below was obtained by subjecting $CuSO_4 \cdot nH_2O$ to thermal gravimetric analysis (TGA). The transition from *A* to *D* represents the loss of all water of crystallization. The sample masses at points *A* and *D* were 12.484 and 7.980 mg, respectively. Estimate the value of *n* at each of the points *A*, *B*, *C* and *D*.

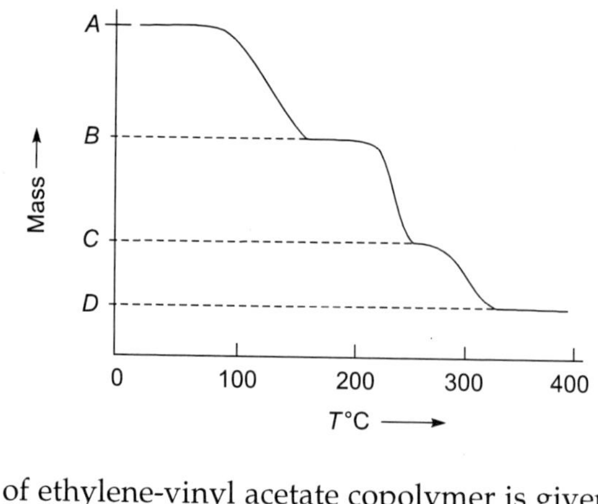

7. The TG curve of ethylene-vinyl acetate copolymer is given below. Evaluate the relative percentage of ethylene-vinyl acetate in copolymer.

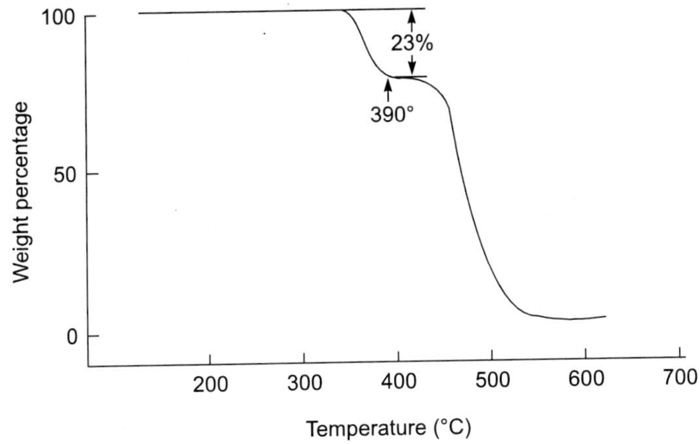

8. Write notes on the following methods used in TG analysis.
   (a) Horowitz and Metzer'
   (b) Coats and Redfern
   (c) Freeman and Carroll
9. Explain why calcium oxalate is the most suitable material for calibration of TGA.
10. Discuss the applications of TGA in rubber and cement industries.
11. What happens on TG curve if we
    (a) Change the heating rate from 5 to 10°C per minute?
    (b) Change the nitrogen flow from 50 mL per minute to 100 mL per minute?
12. Write short notes on the following:
    (a) Furnace atmosphere
    (b) Thermobalance
    (c) Evolved gas analysis
    (d) Thermochemical reactions

REFERENCES

1. Wandlandt, W. W. (1986). Thermal Analysis. Wiley, New York.
2. Brown, M. E. (2001). Introduction to Thermal Analysis. Kluwer Academic Publisher, London.
3. Willard, H. H., Merrit Jr., L. L., Dean, J. A. and Settle Jr., F. A. (1986). Instrumental Method of Analysis. Wadsworth Publishing Company, USA.
4. Duval, C. (1962). Inorganic Thermogravimetric Analysis. Elsevier, Amsterdam.
5. Garn, P. D. (1965). Thermoanalytical Methods of Investigations. Academic Press, New York.
6. Blazek, A. (1972). Thermal Analysis. Van Nostrand Reinhold Comp., London.
7. Gunzzler, H. and Williams, A. (2001). Handbook of Analytical Techniques. Wiley–VCH, Weinheim, Vol-2.

8. Ewing, G. W. (1990). Analytical Instrumentation Handbook. Marcel Dekker Inc, New York.
9. Meyer, R. A. (2000). Encyclopedia of Analytical Chemistry. John Wiley & Sons Ltd., Vol 15.
10. Skoog, A. D., Holler, F. J. and Nieman, T. (1998). Principles of Instrumental Analysis. 5th Edition, New York.
11. Dean, J. A. (1995). The Analytical Chemistry Handbook. McGraw Hill, Inc., New York.
12. Brydson, J. A. (1999). Plastics Materials. Butterworth-Heinemann. 7th Edition.
13. https://www.tainstruments.com/
14. https://www.perkinelmer.com/
15. Mastowska, J. and Wicdtocha, M. (1995). Thermal dehydration of some food products and kinetic parameters of this process. *J. Thermal Anal.* 43, 123.
16. Song, C., Liu, K., Gong, Z. and Liu, Y. (2019). Thermogravimetric analysis of combustion characteristics of coal gangue and petroleum coke mixture. In: *Journal of Physics: Conference Series* (Vol. 1324, No. 1, p. 012077). IOP Publishing.
17. Luyta, A.S., Ishripersadh, K. and Timm, D.N. (1999), Comparative thermoanalytical investigation of the cross-linking behaviour of three different paraffin waxes in the presence of potassium persulphate, *Thermochimica Acta* 339, 55-67.
18. Ali, U., Kanwar, S., Yadav, K., Basu, S. and Mazumder, K. (2019). Effect of arabinoxylan and β-glucan stearic acid ester coatings on post-harvest quality of apple (Royal Delicious). *Carbohydrate Polymers*, 209, 338-349.
19. Yin, H.S., Liu, H.M. and Liu, Y.L. (2017). Structural characterization of lignin in fruits and stalks of Chinese Quince. *Molecules*, 22 (6), 890.
20. Kumar, B., Saha, S., Basu, M. and Ganguli, A. K. (2013). Enhanced hydrogen/oxygen evolution and stability of nanocrystalline (4–6 nm) copper particles. *Journal of Materials Chemistry A*, 1(15), 4728-4735.
21. Newkirk, A. E. (1960). Thermogravimetric measurements. *Anal. Chem.* 32, 1558-1563.
22. Freeman, E. S. and Carroll, B. (1958). The application of thermoanalytical techniques to reaction kinetics: The thermogravimetric evaluation of the kinetics of the decomposition of calcium oxalate monohydrate. *J. Phy. Chem.* 62, 394-397.
23. Coats, A. W. and Redfern, J. P. (1964). Kinetic parameters from thermogravimetric data. *Nature*. 201, 68-69.
24. Horowitz, H. H. and Metzer, G. (1963). A new analysis of thermogravimetric traces. *Anal. Chem.* 35, 1464-1468.
25. Dave, N. G. and Chopra, S. K. (1966). A method for the evaluation of kinetics of solid state reactions from thermogravimetric curves. *Z. Physsik. Chem.* 48, 257-266.
26. Papazian, H. A., Pizzolato, P. J. and Orrell, R. R. (1972). The thermal decomposition of aluminium sulfate and hafnium sulfate. *Thermochim. Acta.* 4, 97-103.
27. Magnuson, J. A. (1964). Tabulation for deriving procedural rate constants from dynamic thermograms. *Anal. Chem.* 36, 1807-1810.
28. Broido, A. (1969). A simple, sensitive graphical method of treating thermogravimetric analysis data. *J. Polym. Sci. Part A-2.* 7, 1761-1773.

4

Differential Thermal Analysis (DTA), Differential Scanning Calorimetry (DSC) and Thermometric Titrations

SECTION A: DIFFERENTIAL THERMAL ANALYSIS (DTA)

4.1 INTRODUCTION

Differential thermal analysis is the most widely used analytical technique [1-3], suitable for the identification of all types of chemical or physical changes which are even not accompanied by the change in weight on heating and thus are not indicated in thermogravimetric analysis. This is because all the changes must be associated with changes in energy and temperature so observed in DTA.

In this technique, the temperature of the substance under investigation is compared with the temperature of a thermally inert material, such as β-alumina. This is recorded as a function of furnace temperature as the substance is heated or cooled at a predetermined uniform rate. In DTA a sample of material under investigation (specimen) is placed side by side with the sample, thermally inert material (the reference sample) usually calacite or β-alumina in suitable sample holder or block, the temperature difference between the two is continuously recorded as they are heated. The block is heated in an electric furnace, i.e., both are heated under identical conditions as represented in Fig. 4.1.

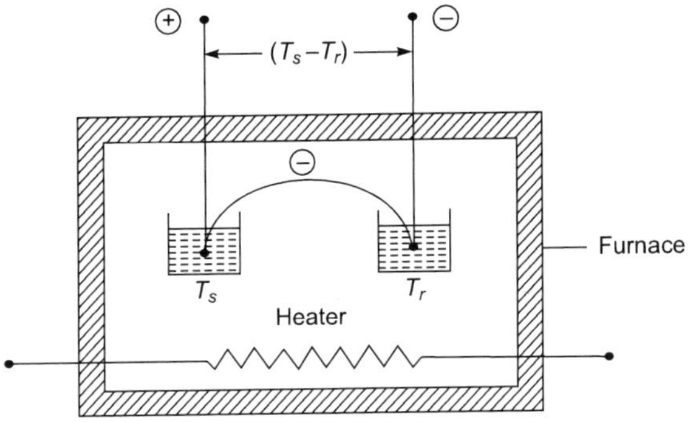

Fig. 4.1 Measuring principle of DTA

ΔT is the difference in temperature between the two cavities (i.e., $T_s - T_r$), one for reference and other for specimen sample. If there is no reaction then $\Delta T = 0$, when reaction takes place it is either exothermic or endothermic. In endothermic, the temperature of the cavity containing the sample will fall because heat is absorbed and ΔT will be negative. However, in the case of exothermic reactions, vice versa happens. A typical graph is shown in Fig. 4.2 representing peak upper (+) side exothermic and below (–) side endothermic.

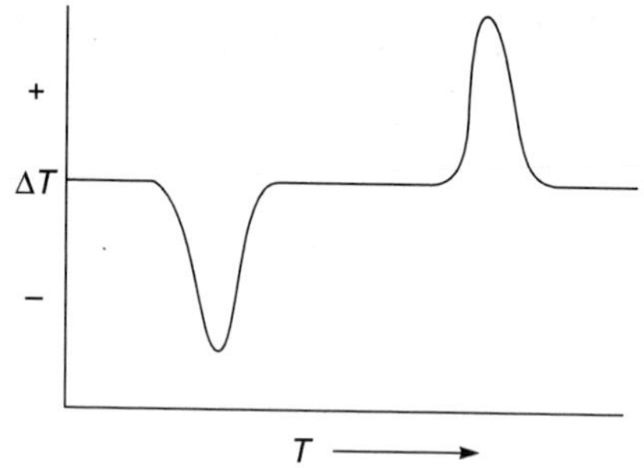

Fig. 4.2 A typical DTA curve

The range of phenomenon measurable in the course of DTA is much larger than TG determination [4-6]. Because, during TG, pure fusion reactions, crystalline transition, glass transition and crystallization and solid state reactions with no volatile product would not be indicated because they provide no change in weight of the specimen. However, these changes are indicated during DTA by endothermal or exothermal shift from the baseline. Thermal changes due to fusion, crystalline structure inversions, boiling, dissociation or decomposition reactions, oxidation and reduction reactions, destruction of crystalline lattice structure and other chemical reactions are generally accompanied by an appreciable rise or fall in temperature and hence are accounted in DTA. Generally speaking, phase transitions, dehydration, reduction and some decomposition reactions produce endothermic effects whereas crystallization, oxidation and some decomposition reactions produce exothermic effects.

4.2 HISTORICAL DEVELOPMENT

4.2.1 Introduction

The first DTA instrument (shown in Fig. 4.3) was introduced by Le Chatelier in 1887 and perfected by Roberts-Austen in 1899.

Instrument is devised from two thermocouple designs suggested by Robert and Austin in 1989. Since then many modifications have been made to design different types of DTA instruments, each slightly different in the design of the furnace, temperature programmer, recording equipment, sample holder, etc. Some specific designs are discussed here, which possess some novelty in design or have made important contributions to the development of this technique.

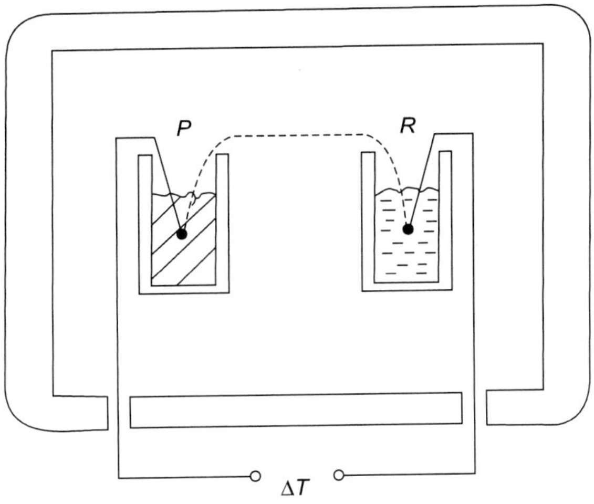

Fig. 4.3 First DTA introduced by Le Chatelier in 1887, perfected by Roberts-Austen in 1899

4.2.2 High Temperature DTA

Problem of DTA apparatus works above 1200°C are quite different from those operated below this temperature. Because the electrical leakage increased electrical conductivity of refractories component and of the air in the furnace become important. Thus, adequate shielding of the low level thermocouple circuit is more critical. The high temperature also often melts the sample, which destroys the thermocouple assembly as well as the sample holder. Even with these problems, DTA instruments which operate up to 3000°C or higher have been invented by incorporating better heating element, insulating materials and thermocouple sustaining at higher temperatures. The NETZSCH is marketing DTA model STA 429 CD-Simultaneous TG-DTA serves up to 2400°C. Even DSC 404 F3 Pegasus temperature range: –150°C- 2400°C using Tungsten furnace (RT-2400°C), High-speed furnace (RT-1250°C), Heating rates: 0.001 K/min – 50 K/min (dependent on furnace), measuring sensors for DSC and DTA, variable types of thermocouple such as W/Re, and atmospheres: inert, oxidizing, static, dynamic.

4.2.3 Sealed Tube DTA

In this technique, the sample is enclosed in a sealed and sometimes evacuated chamber or tube prior to analysis. The technique is useful for the phase diagram investigations, especially those involving corrosive materials, e.g., chalcogenides, metal salt hydrates, metallurgical problems, molten salt equilibrium.

4.2.4 High Pressure DTA

The technique is designed to study the effect of pressure on solid → liquid and solid$_1$ → solid$_2$ equilibria. The equipment used in high pressure studies employs pressure generating systems in the range of 60-80 K bar, while low pressure equipment involves external pressurization by gases. This can be used at a pressure up to 3000 psi/g at maximum temperature of 500°C.

4.2.5 Miscellaneous DTA Instrument

Other modifications made for running a sample are vacuum or inert atmosphere in the instrument, a low temperature DTA apparatus with the Dewar filled with liquid nitrogen, the heating rate controlled by increasing the voltage in the heater coils, while cooling was accomplished by varying the pressure of gas present in the outside of the chamber.

A controlled pressure system for a DTA apparatus has been developed, which permits the determination of sample vapour pressure by DTA. The DTA is used to evaluate the catalytic properties of a material.

4.2.6 Micro DTA Instrument

The recording of DTA curves from microgram quantities of a sample and up to micro calorimeter range was developed by Millers and Sommr [7-8]. In this instrument the thermocouple junction acts as a heater and temperature detector, as well as the sample holder. The thermocouple heated by high frequency current, has been amplitude modulated by low frequency signal to achieve the optimum sensitivity. A highly sensitive DTA apparatus is able to determine the phase transition temperature of ±0.5°C over a wide range of heating rate.

4.3 INSTRUMENTATION

The DTA instrument heats a compound under controlled temperature and environment. For an accurate measurement, it is essential that the rate of heating should be also absolutely reproducible and uniform over the whole range analysis. This reproducibility is difficult to attain by supply of fixed voltage or a fixed current to apparatus and hence, automatic control is more or less essential for adequate reproducibility and this is only possible with correct design of the apparatus.

The block diagram of commercially available DTA apparatus is shown in Fig. 4.4. It consists of the following basic components:

1. Furnace assembly
2. Sample holder
3. Temperature programmer and detecting system
4. Recorder
5. Atmosphere control equipment for furnace and sample holder

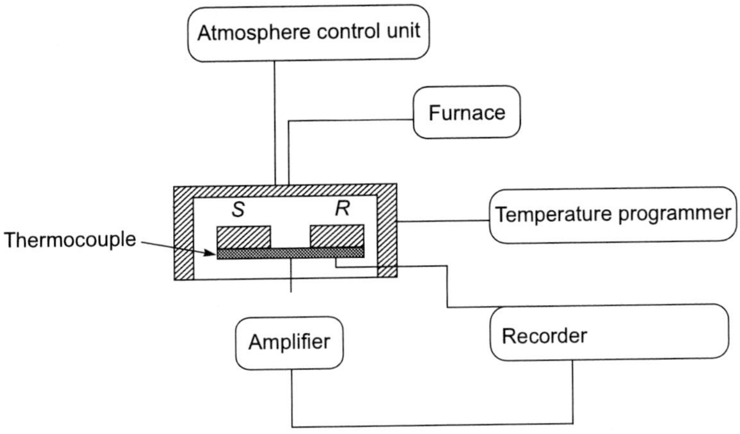

Fig. 4.4 Block diagram of differential thermal analysis

The instrument measures the differential temperature of the sample as a function of temperature or time since the temperature rises at a constant linear rate.

4.3.1 Furnace Assembly

The details of furnace assembly is quite similar to the TG, as discussed in TG chapter.

The different types of heating element are required to design furnace up to different ranges such as: nichrome furnace can be used up to 1300°C, platinum and its alloys up to 1750°C and molybdenum (Mo) for higher range up to 2000°C. However, special types of higher frequency induction heating are also used for higher temperature furnaces.

4.3.2 Specimen Holder

It is designed to accommodate even a small quantity of the material and to give maximum thermal effect. It can be made of Al, Pt, Ni, Ag, alloy like Pt-Rh, ceramic materials such as sintered alumina, silica, fire clay, heat-resistant glass and even graphite have been recommended as materials specimen holders used for the sample under investigation and reference material like (α-alumina). The materials of sample holders are selected on the basis of analysis range and probable thermal change, e.g., Al is used up to 500°C since Al melts after 500°C and Pt is recommended for higher temperature.

4.3.3 Temperature Regulating System and Detecting System

Uniform rate of heating of the furnace is ensured through electronic temperature regulators. When a sample is heated at a constant rate, the temperature function, T, is linear up to the moment when the sample undergoes change, its slope representing the rate of heating which remains constant. The temperature measurement was performed by employing thermocouples. Rare metal complex like Pt-Rh (Pt-10-13% Rh) is commonly used as thermocouple for measuring temperature. For temperatures up to 2000°C W-Mo thermocouple may also be used. Very thin thermocouple is inserted in the sample and reference holders. However, visual galvanometric observations, though inconveniently can be used when a few samples are to be investigated.

4.3.4 Recorder

Automatic pen and ink electronic recorder has been found to be more convenient and was generally used in earlier days. At the moment, when an exothermic or endothermic change can takes place, shape of the curve changes. An exothermic reaction causes an increase in the heating rate while an endothermic reaction causes a decrease in the heating rate. The main disadvantage of this method is its sensitivity, because small changes in the temperature cause small deviations in linearity of the curve which are sometimes not observable. So, small temperature changes occurring in the sample are generally not detected by this method. The present instrument includes some advanced microprocessor based computer, appropriate for recording and processing the data.

4.3.5 Atmosphere Control Unit

This unit is important and similar to the one discussed in previous chapter.

4.4 FACTORS AFFECTING DTA CURVES

DTA is a dynamic temperature technique, so there are large number of factors which can affect the resulting experimental curves. These factors can be divided into the following general categories of variables:

- **Instrumental Factors** like furnace atmosphere, size and its slope, sample holder material and geometry, heating rate of the sample, sensitivity of recording system and location of thermocouple in the sample.
- **Sample Characteristics** like particle size, packing density, amount of the sample, heat capacity and thermal conductivity, degree of crystallinity, effect of diluents, swelling and shrinkage of the sample.
- **Furnace Atmosphere:** The nature of gases present in the furnace also decides the DTA curve. The oxygen atmosphere supports oxidation processes but inert and reducing gases such as nitrogen, carbon monoxides reduce the oxidizing change. Similarly, the vacuum atmosphere also affects the result.
- **Heating Rate** is generally higher heating rate shifts the thermal transition towards higher temperature side and lower one towards lower temperature. Therefore, deciding the heating rate is an important parameter.

The rather empirical nature of the method gives rise to many difficulties originating from small differences in technique, e.g., the peak temperature used for reporting differential thermal results as well as identification is rather variable depending upon the rate of heating, amount of active material, packing of specimen, type of specimen holder, etc. Thus, standardization of technique is essential. In general, it is essential that each apparatus should be calibrated from the mineral expected to be the sample under investigation.

4.5 SOURCES OF ERRORS AND THEIR MINIMIZATION

There are a number of sources of error in DTA, and they can lead to inaccuracies in the recorded temperature and weight data. Some of the errors may be corrected by placing the thermocouple at proper place and handling it with care. We are discussing some common source of errors during operation of a DTA.

4.6 EVALUATED PARAMETERS

4.6.1 Estimation of Transition Temperature

The precise determination of a transition temperature (e.g., M.P. and B.P.) is up to a precision of ± 0.3°C over a wide range of heating rate. The temperature estimated for M.P. T_m or B.P. T_B, were often selected from the portion peak as shown in Fig. 4.5. Point *A* is the intersection of extrapolated straight line portion of the low temperature side of the peak with the baseline, and point *B* is the inflection point of low temperature side of the peak. Point C is the extrapolated peak, while *D* is the extrapolated return to the baseline. Generally, melting point (T_m) measured by use of a sample is closest to the temperature of point *C* for sample and temperature at *B* for reference, while exception is closer to point *D*.

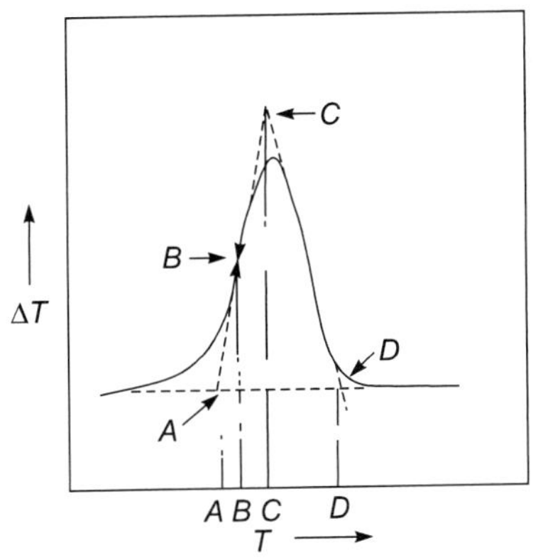

Fig. 4.5 Representation for evaluation of transition temperature

Further, transition temperatures from DTA curves of pure compounds are characteristic feature for their physical-chemical changes [9-11]. These transition temperatures are characteristics of an individual compound, element, alloy or other mixed matrix that can be used as fingerprint to identify the presence of a material in sample in qualitative manner. Such as the melting point of organic compounds or polymers. Figure 4.6 is differential thermal analysis (DTA) curve of seven components of a polymeric mixture showing their melting points: Polytetrafloroethylene (PTFE), High Density Polypropylene (HPPE), Low Density Polypropylene (LPPE), Polypropylene (PE), POM, Nylon 6, Nylon 66. The graph is showing the characteristics peaks of all the polymers and hence confirms the presence of individual polymers in the analyzed sample by comparing the reported values with recorded one. Some literature reported data is given in appendix for reference.

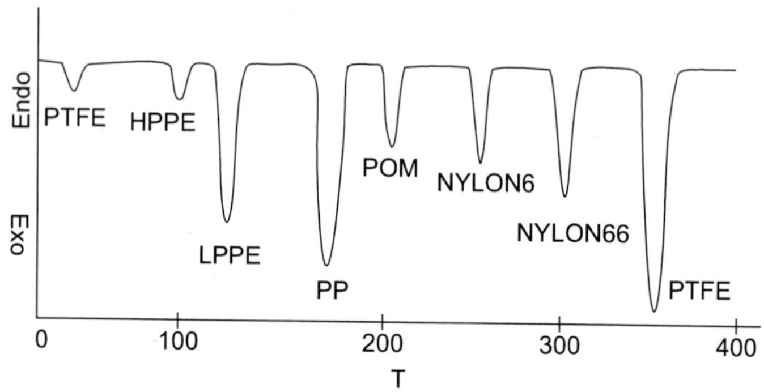

Fig. 4.6 DTA curve of a typical polymeric mixture

4.6.2 Heat of Reaction

The area under DTA curves is related to the amount of energy released or absorbed in a physico-chemical change [12-13]. It has been found that under certain conditions, the area under peak is proportional to the amount of heat involved in reaction. So this area under the curve is used for stochiometric ratio of the analyzed compounds (quantitative interpretation). Now we see in the next example how it can be used to compare thermal stability of materials, chemical identification of a material (qualitative interpretation). Such information can be used to select material for certain end-use application, predict product performance and improve product quality. Thus, DTA curves of a polymeric mixture and an copolymer sample clearly indicate individual peaks for the melting points of individual polymers. Thus, the ratio of area under the curve is simply dividing the enthalpy of heat of decomposition and reported values provide the ratio of individual monomers in the analysed copolymer sample.

The DTA results are very good supplement for TGA. To further illustrate, let's consider the example of $CaC_2O_4 \cdot H_2O$. This curve indicates three DTA peaks; first is endothermic in nature, second is exothermic and third one is again endothermic in nature. The correlates of the TGA result confirm the nature of reaction occurring in the endothermic because of desolvation and decarboxylation while exothermic is due to decomposition followed by oxidation and finally formation of stable oxide CaO and the gas, carbon dioxide. This can be explained in the chemistry of decomposition of $CaC_2O_4 \cdot H_2O$ when it is heated more satisfactorily. Even by comparing TGA and DTA results, one can conclude the nature of change (physical or chemical), for example, if melting process occurred. The DTA will show an endothermic peak while TG will show no weight loss. Similarly, the change between surface desorption and decomposition can be made by diffuse nature of DTA and TGA curve.

4.6.3 Phase Diagram

The coexistence behaviour of two or more than two phases of miscible or immiscible materials, also known as phase diagram, can be well studied by this technique. The information is useful for material scientists in order to establish the synthesis condition of a mixed phase, intermetallic salts, etc. The diagram can be made by measuring the transition temperatures of individual and mixed phases as a function of concentration. Further, the phase diagram can be constructed between both the variables.

Differential thermal analysis (DTA) measurements are a standard method of determining transformation temperatures of materials. Accurate interpretation of these measurements is essential when they are used to infer the location of equilibrium phase diagram boundaries. The accuracy of the measurements is affected by the details of heat transfer within the measurement cell and by transformation kinetics. DTA results from samples with unknown characteristics. A heat flow of a DTA and thermal characteristics obtained by fitting experimental results for a pure metal, the response of the DTA is modeled for melting and solidification of alloys phase diagram. In a report, K. Endoh et al. [14], explored the use of TG-DTA techniques to determine the solid-solution equilibrium curve between certain inorganic salts and water. In the experiment the rate of evaporation of water through a pin hole of an aluminium pan, containing a small amount of aqueous solution of the salt, was measured by TG and DTA. A slight change in the rate was found to occur when the solution crossed the solubility curve. The composition and temperature at which the vaporization rate changed were determined by differentiating the TG curve with respect to time and then comparing with the literature value. Thus, the solubilities of NaCl, KCl, and Na_2SO_4 crystals into water were determined rapidly with more accuracy for a small amount of sample solution. The observed diagram is shown in Fig. 4.7.

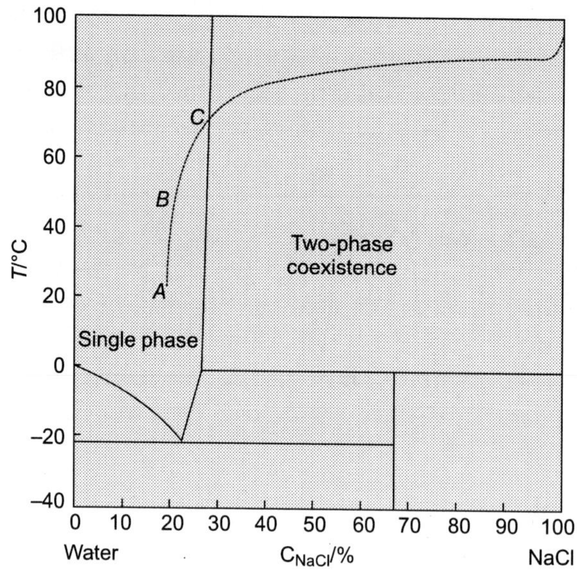

Fig. 4.7 Phase diagram of the NaCl ± H_2O system by determining the solubility equilibrium

4.6.4 Kinetic Studies (Discussed in Section 4.8)

4.7 APPLICATIONS

4.7.1 Identification of Minerals and Matter

For the detection of any minerals, it is essential that it undergoes a major energy change in the temperature range used. Since most of the minerals undergo such changes at such temperature choice of an appropriate temperature range. The DTA provides a rapid method for estimating the presence or absence of a particular phase in a mixture.

Figure 4.8 show peaks at 120°C, 670°C and 930°C, which are due to dehydration of gypsum, decomposition of Kaolinite (magnesium basic carbonate) and calcite respectively. Similarly, the DTA transition temperature also confirms the existence of polymorphism. A case of glaserite may be taken, which shows the transitions as listed in Table 4.1.

TABLE 4.1 The transition temperature of glaserite

Component	Transition temperature (°C)
Picromerite	435
Picromerite to biodite	437
Biodite to mirabilite	440

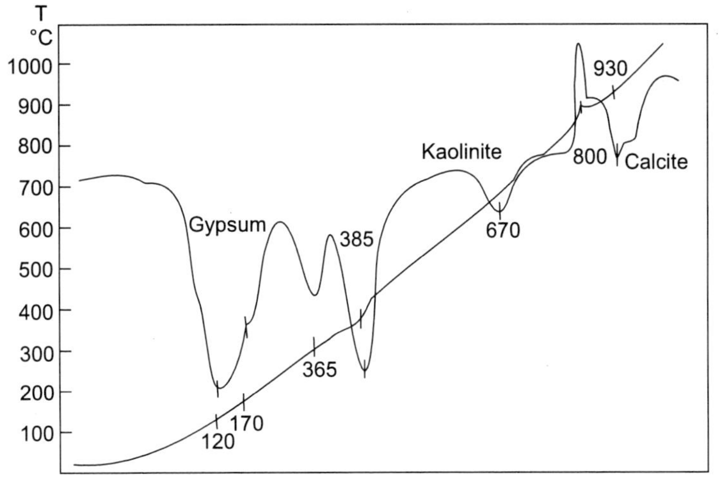

Fig. 4.8 DTA curve of gypsum, kaolinite and calcite

In another example the DTA curve of a mixture of sodium sulphate (thenardite) and gypsum, clearly shows the endothermic effect due to dehydration of gypsum at 120°C and 190°C and polymorphism transition of sodium sulphate at 240°C.

The complication arises from the fact that certain minerals giving reasonably characteristic thermal effect, when pure sometimes do not show similar results in clays. Another difficulty may arise by the presence of organic matters. Oxidation of organic materials if present masks the thermal effect of substance, consequently the analysis of such cases should be preferentially carried out in inert atmosphere or in vacuum. Organic material sometimes should be removed by a suitable solvent or may be oxidized by H_2O_2 treatment. When clays or ore contain organic material then a broad exothermic peak appears which is between 200-600°C and may mask the thermal effect of the substance under investigation. The appearance of broad exothermic peak between 200-600°C indicates the presence of organic material.

In practice however, difficulties do arise since the peak areas for two distinguishing samples of same mineral giving peak at two different temperatures may not necessarily be identical. It is therefore necessary for quantitative work to know the peak area given by pure mineral identical to that in the sample under investigation which is sometimes not possible. Another limitation in the quantitative work is the occurrence of overlapping of peaks, e.g., when Kaolinite and Illite are present in the same sample almost complete overlapping takes place.

4.7.2 Analysis of Polymeric Materials

DTA is a useful technique for characterization of polymeric materials in the light of identification of thermophysical, thermochemical, thermomechanical and thermoelastic changes or transitions. It provides important parameters for polymer processing and its end use. DTA also provides useful information about quantitative ratio, degree of fusion, crystallinity, phase equilibrium, heat of polymerization, degree of curing, etc. The information can be calculated by comparing the observed heat of reaction (ΔH) with reported heat of reaction in the pure form.

A typical DTA curve of a polymer is shown in Fig. 4.9 with four labelled transitions: glass transition, crystallisation, melting and oxidation, which are also abbreviated as T_g, T_c, T_m and T_d respectively.

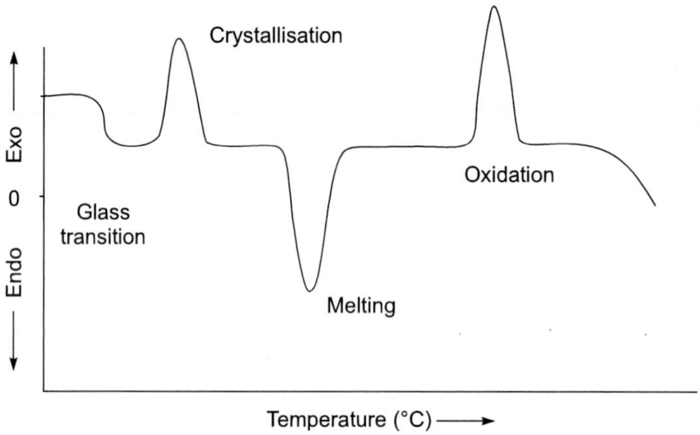

Fig. 4.9 DTA curve of a typical polymeric sample

4.7.3 Determination of Transition Temperatures

The technique is appropriate to evaluate the transition temperatures such as glass transition (T_g), melting point (T_m), etc. Polymers are unique in the dominance of the glass transition as the decisive factor in their mechanical properties. Polymers are the only materials for which the equilibrium ground state is often glassy rather than crystalline. This is because topological and stereochemical constraints prevent the formation of crystals in many cases. The glass transition is often called a pseudo-second order transition because of the dominance of kinetics. Slower cooling rates, for instance, lead to lower measured values of the glass transition. The glass transition is an important parameter for polymer scientists and engineers to recommend the processing condition and end-use of polymers. The glass transition temperature, T_g, is the temperature at which an amorphous solid, such as glass or a polymer, becomes brittle on cooling, or soft on heating. More specifically, it defines a pseudo-second order phase transition in which a super cooled melt yields, on cooling, a glassy structure and properties similar to those of crystalline materials, e.g., of an isotropic solid material. T_g is usually applicable to wholly or partially amorphous solids such as common glasses and plastics (organic polymers), although there is an analogous phenomenon in crystalline metals called the ductile-brittle transition temperature. Below the glass transition temperature, T_g, amorphous solids are in a glassy state and most of their joining bonds are intact. In inorganic glasses, with increased temperature more and more joining bonds are broken by thermal fluctuations so that broken bonds begin to form clusters. Above T_g, these clusters become large facilitating the flow of material. In organic polymers, secondary, non-covalent bonds between the polymer chains become weak above T_g. Above T_g, glasses and organic polymers become soft and capable of plastic deformation without fracture. This behaviour is one of the things which make most plastics useful. It is important to note that the glass transition temperature is a kinetic parameter, and thus parametrically depends on the melt cooling rate. Thus, the slower the melt cooling rate, lower is the T_g. In addition, T_g depends on the measurement conditions, which are not universally defined. T_g of some polymers are listed in Table 4.2.

TABLE 4.2 List of glass transition of some materials

Polymer	T_g (°C)
Polyethylene (LDPE)	−105 or −30 also cited
Rubber	−72
Polypropylene (atactic)	−20
Poly(vinyl acetate) (PVAc)	28
Polyethylene terephthalate (PET)	69
Poly(vinyl alcohol) (PVA)	85
Poly(vinyl chloride) (PVC)	81
Polystyrene	95
Polypropylene (isotactic)	0
Poly-3-hydroxybutyrate (PHB)	15
Poly(methylmethacrylate) (atactic)	105
Poly(carbonate)	145
Chalcogenide AsGeSeTe	245
ZBLAN	235
Tellurite	279
Avatrel; Polynorbornene	215
Fluoroaluminate	400
Soda lime glass	520-600
Fused quartz	1175

The T_g is evaluated by DTA employing different methodologies by scientists such as: To determine T_g two lines are drawn parallel to the baseline above and below the inflection at T_g. The midpoint of the departure from the left and the intersection with the baseline nearest to T_g on the right is determined by the instrument. This midpoint is taken as the glass transition temperature. The residual enthalpy associated with T_g is sometimes also recorded from the high temperature baseline. This residual enthalpy will decrease with heating rate and has a strong dependence on the processing conditions of the sample and the length of time the sample is allowed to anneal near T_g. Glass transition is a second order transition since it displays a discontinuity in the derivative of the enthalpy (heat capacity). It is often also termed a pseudo-second order transition because it displays a finite range of temperatures over which it occurs at finite heating rates.

The first evidence that a glass transition could be detected by DTA was provided around 1957, where a transition at 28°C was shown but not interpreted. Currently H, a large data bank is available about this transition. As such this transition is not associated with latent heat, but rather a sudden change in specific heat to bring a liquid into the glassy state. The condition is necessary to cool it rapidly to approximately two-thirds of its melting point.

$$T_g/T_m = 0.66$$

The glass transition of a polymer can be obtained using number average molecular weight, M, ranging from 8×10^3 to 3×10^6 from the expression:

$$T_g = [(96.5 +/-1.0) - (2.8 +/-0.1)] \times 10^3/M$$

The deduction about the slope of the glass transition as depicted by DTA curve:

The slope of the curve ΔT surface temperature should be sigmoidal.

The maximum value of ΔT at the glass transition temperature should be linearly dependent on the heating rate. The inflection point, T_g, should rise in temperature with the heating rate. Although many scientists have shown the glass transition to be associated with a sudden shift in baseline.

The T_g depends upon the heating rate, volume fractions, and molecular weight. The DTA results provide important information about polymerization reaction, specifically heat of polymerization reaction, degree of curing, effect of catalysis, decomposition reaction and radiation effects.

The melting point is a first order transition, since it involves a discontinuity in the first derivative of the Gibbs free energy with respect to temperature (entropy) or pressure (volume). There is a discontinuity in volume, for instance, at the melting point.

The melting point (T_m) is determined, for broad melting polymers, by the temperature at the maximum in the (dH/dT) plot near this transition. The enthalpy of melting is determined by constructing a baseline above the melting and extending it to below any cold crystallization phenomena (exothermic peak below T_m and above T_g). Generally, the enthalpy of crystallization is taken as the area above this baseline. Sometimes the enthalpy of crystallization, which has occurred in the measurement, is subtracted from this value to adjust the enthalpy as an estimate of the enthalpy associated with the original sample. The onset of melting is taken as the initial rise of the curve above the baseline. The maximum melting temperature at this heating rate is taken as the final point which deviates from the baseline. Various principles in practice report the melting point for different types of materials. However, majority of people recommend the peak temperature as melting point.

4.7.4 Measurement of Crystallinity

Another common application of DTA is the measurement of the weight fraction of crystalline material in semi-crystalline polymers. The method is based upon the measurement of the polymer's heat of fusion, ΔHf, and the plausible assumption that this quantity is proportional to the crystalline content. If by some process of extrapolation the heat of fusion, ΔHf^*, of a hypothetical crystalline sample is known then the weight fraction of a crystallinity is $\Delta Hf/\Delta Hf^*$.

Thus, crystallinity of a polymer sample can be determined by measuring the total energy absorbed by the sample per gram and subtracting the amount of energy which would be absorbed by one gram totally amorphous in the temperature interval, and then dividing by the heat of fusion of one gram of a perfectly crystalline sample. The following equation is used to measure crystallinity:

$$X = \Delta H - \Delta Hfa/\Delta Hfc.$$

The properties of plastics are critically affected by their crystallinity. The more crystalline the moulded part, the more rigid and stronger it is but the more brittle it also is. The crystallinity of a polymer is influenced by its chemical structure, cooling conditions during processing, and any thermal post-treatment. Crystallinity is influenced by the chemical structure and thermal history. Figure 4.10 shows the melting of a moulded polyoxymethylene (POM) part in the first heating curve. The crystallinity simply is determined by dividing the experimental heat of fusion, H_m = 173 J/g by the literature value for POM H_m = 326 J/g). For this specimen, the degree of crystallisation computes to wc = 53%, melting point peak at 167°C.

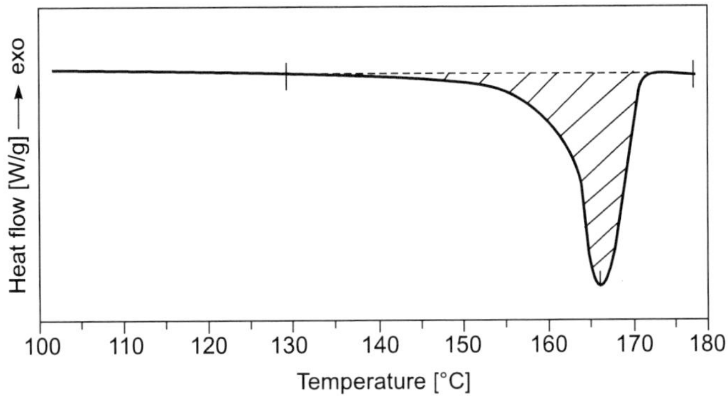

Fig. 4.10 DTA curve of POM

Another method for determination of polymer crystallinity is based upon the ability of the instrument to cool a molten sample rapidly and reproduce it to a selected temperature where isothermal crystallization is allowed to occur. A number of crystallization curves may be obtained at different temperatures. The difference in crystallinity may be caused by branching, nucleation and molecular effect.

4.7.5 Heat of Polymerization

The area observed in DTA curve is directly related to the heat of polymerization reaction and can be modified in terms of per mole or per gm. The DTA graphs of polymerization process of trialkylcyanurate and triallylisocyanurate samples were prepared by mixing two parts of Al_2O_3, one part of monomer, 0.1 part catalyst as a 50% paste in tricresyl phosphate, heating the mixture at a rate of 8°C per minute. From the peak area, the heat of polymerization was estimated by some researchers as shown in Table 4.3. This polymerization reaction occurs in two distinct steps, thus two ΔH values are reported in the table below.

TABLE 4.3 Estimation of heat of polymerization

Materials	Heat of polymerization (ΔH_1) K Cal/mole	Heat of polymerization (ΔH_2) K Cal/mole
Trialkylcyanurate	39	107
Triallylisocyanurate	56	–

The wide data about heat of polymerization from DTA results are available in the literature. In another example the area (size) of peak appearing in DTA curve had great value in assessing the degree of curing; this is done by the residual cure remaining in a polymer system after various treatments. This approach has been applied for estimation of degree of curing in an unsaturated polyester-styrene copolymer cured at ambient temperature. The variation in the size of the curve represents the percentage of curing in 2 hours (63%), 3 hours (68.6%), 4 hours (74.3%), 5 hours (77%) and 6 hours (78.2%). Similarly, the relative change in the heat of reaction measured by DTA also gives information regarding the role of catalyst, degree of crystallinity and decomposition of polymer samples.

4.7.6 Biological Materials

The technique has immense use in the determination of thermal characteristics of bioorganic molecules, the main constituent of body of the living being. The biomaterials are heterogeneous which is a significant feature. It is described as multicomponent, molecular non-homogenous material, in which the component exists as continuous, separate and intermixed structures. The complexity of biological system even in static state and then facing difficulty in obtaining meaningful results and of establishing correlation between thermal characteristics and other physical-chemical properties presumably accounts for the paucity of DTA studies in these systems. However, the investigations carried out under variable experimental conditions indicate that DTA curve should be of value as 'fingerprint' of biological materials. The thermal characteristics of a number of biological materials such as fresh biological materials and decomposed materials determined under different atmospheric conditions has been reported. Fresh biological material means constituent of materials of active body of plants and animals. It is always difficult to elucidate the chemical constitution of the plant materials and information obtained from the study of the simple molecules cannot be necessarily applied directly to heterogeneous system. In the study of such system indirect method are applied, e.g., complete combustion, dilution effect, environmental effect, etc. Figure 4.11 shows DTA of four-leave samples of different plants in oxygen exhibit two pronounced exothermic effects in the range of 240-470°C region.

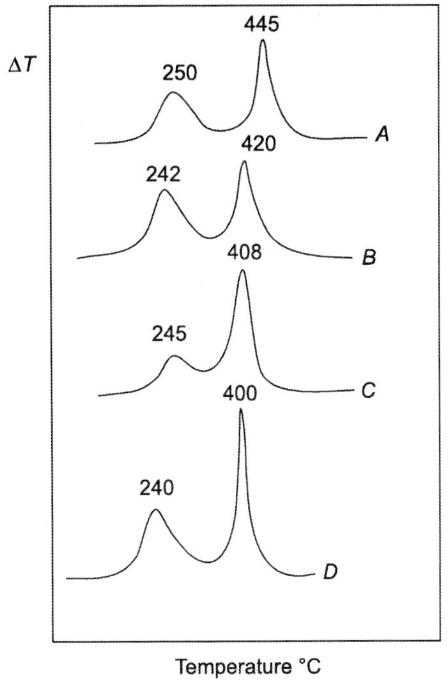

Fig. 4.11 DTA peaks of four leave samples

The first peak is invariably smaller than the second one at least in height always, not in area and apart from variation peak temperature. The general similarity in these curves may be inferred that leaves of the plants have at least same micro-chemical composition.

The DTA curves for some plant materials have also been obtained in a static or air or dynamic oxygen atmosphere, since it is difficult to distinguish between decomposition and oxidation effects. Which may overlap, endothermic reactions such as dehydration or volatilization with exothermic reaction accompany combustion under inert atmosphere condition can be detected. Similarly, the analytical method is useful for analysis of high energy materials, explosives, ceramic, cement and pharmaceuticals.

Sometimes, the first derivative of change in temperature between the sample and reference with reference to furnace temperature or time is calculated, which yields a graph used for better inference about thermochemical change called Derivative Differential Thermal Analysis (DDTA). This improves the resolution of thermal transition and can be used for calculation of kinetic parameters. An example of DTA and DDTA is shown in Fig. 4.12. The record of DDTA should be plotted derivative on the ordinate, and time or temperature on the abscissa.

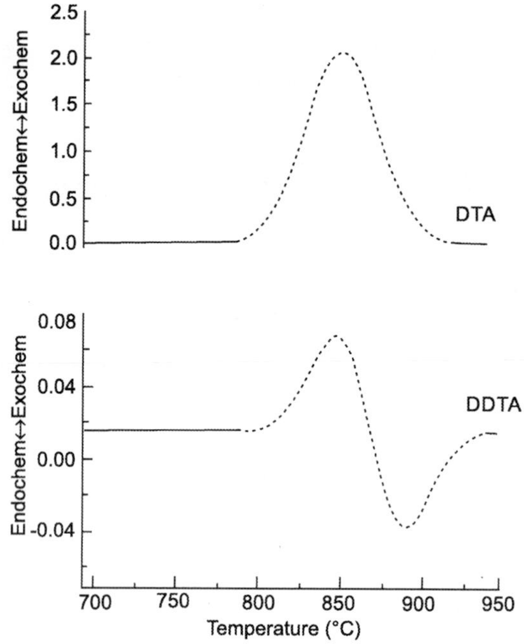

Fig. 4.12 DTA and DDTA curves of representative materials

SECTION B: DIFFERENTIAL SCANNING CALORIMETRY

4.1 INTRODUCTION

The term "differential scanning calorimetry" (DSC) refers to the measurement of energy involved in a physico-chemical reaction with time or temperature [15-17]. The instrument is based on the measurement of heat flow recording $(dQ/dt)_{\Delta=0}$, or $(dT/dt)_{\Delta=0}$. In DSC, the temperature difference between the reference and sample is maintained isothermal to each other by supplying extra electrical energy through a secondary heater attached below the sample pan as they are heated or cooled at a linear heating rate (Fig. 4.13). The provision is also attached to record the energy with time called isothermal differential scanning calorimetry.

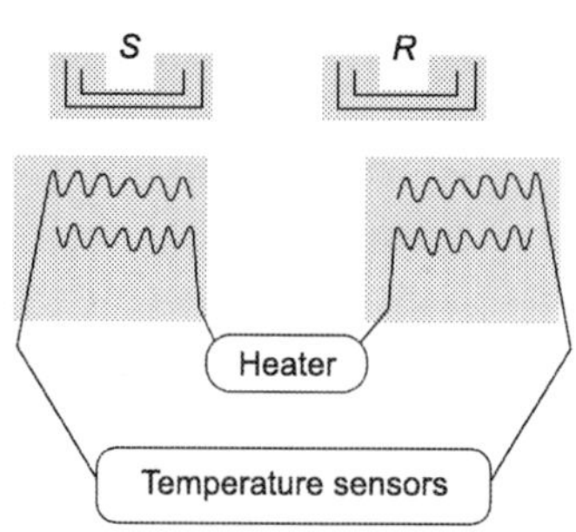

Fig. 4.13 Simple DSC cell

The science of calorimetry obeys only two laws: (a) the law of conservation of energy, and (b) the law of heat transfer. Therefore, the art of calorimetry is still growing, on the basis of knowledge and the craft required to make accurate measurements of heat and rate of heat exchange.

There are three methods for measuring heat:

1. Measurement of a temperature change which is then multiplied by a thermal equivalent (the apparent heat capacity) determined in a calibration experiment.
2. Measurement of the power required to maintain isothermal conditions, with the power being supplied either through an electronic temperature controller or by an isothermal phase change in a substance in contact with the calorimeter.
3. Measurement of a temperature difference across a path of fixed thermal conductivity, the thermal conductivity being determined in a separate calibration experiment.

The above methods developed the three techniques: Conventional DTA, power compensated DSC, and heat flux DSC.

4.1.1 Comparison of DTA and DSC

The information obtained from DTA and DSC is identical but DSC is more sensitive because of the measuring principles. Since it measures the ΔH directly while DTA measures ΔT, which is affected by sample quantity, heat conduction, etc. An explanatory illustration (Figs. 4.14 and 4.15) compares DTA and DSC together with the definitions of the terms needed for the comparison of classical DTA and both the types of DSC (heat flux and power compensated) (Fig. 4.14).

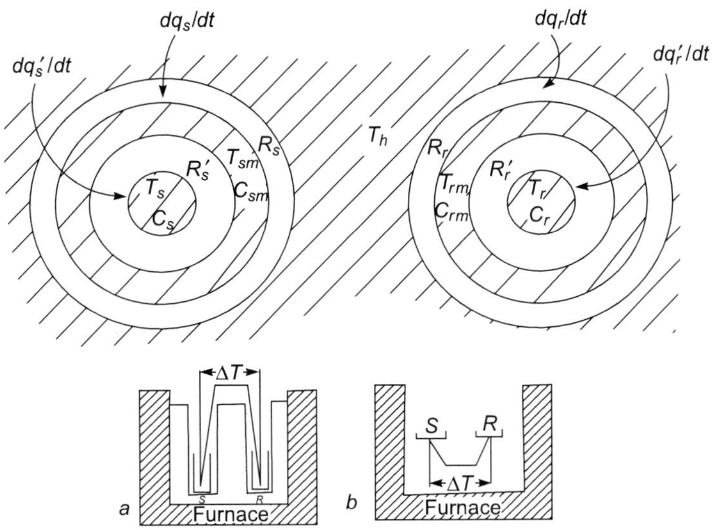

Fig. 4.14 Comparison among DTA, Hf-DSC and Pc-DSC (Figure adapted from Ref. 2)

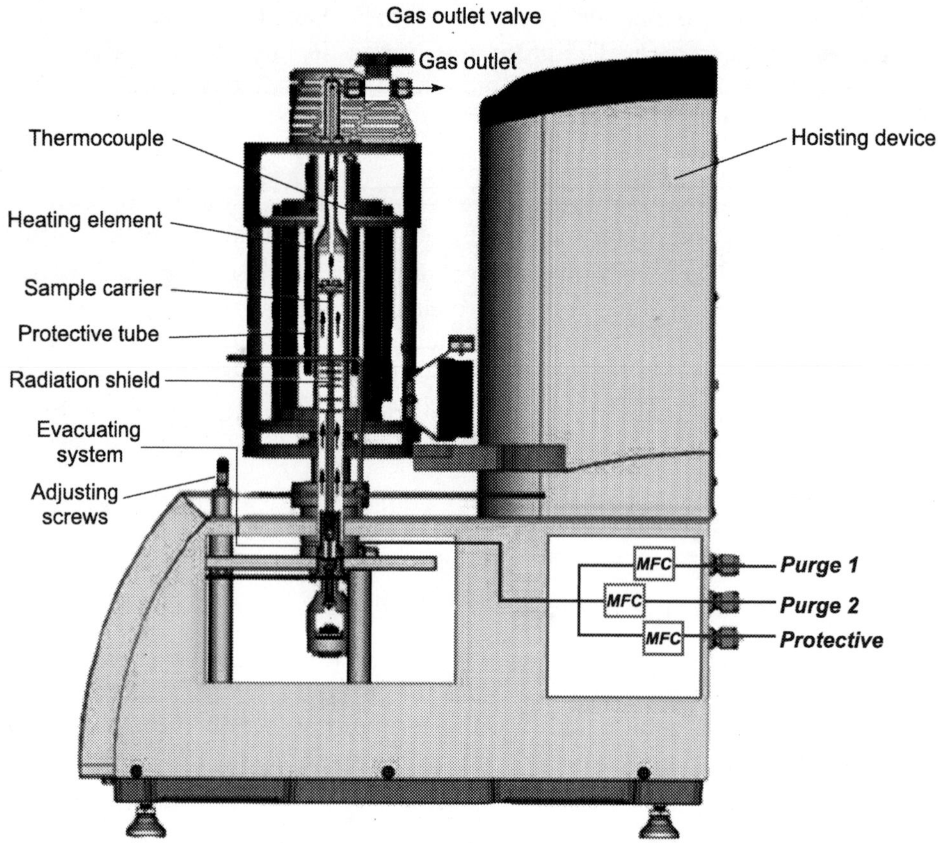

Fig. 4.15 A detailed diagram of Netssch DSC 404 F1 Pegasus (Reproduced with the permission from Netzsch)

For an ideal instrument, the heat capacities and thermal resistances would be matched, $C_{rm} = C_{sm}$; $R_r = R_s = R$ and $R_r{}' R_s{}' = R'$, note that $R \neq R'$ and $C_s \neq C_r$. It is further assumed that $C_s > C$, i.e., the heat flow is governed by Newton's law:

$$dQ/dt = (1/R)\Delta T$$

Heat flow to the sample side, heats both (i) the sample monitoring station, and (ii) the sample:

$$dq_s/dt = C_{sm}(dT_{sm}/_{dt}) + C_s(dT_s/dt)$$

Also,

$$dq_s'/dt = C_s(dT_s/dt)$$

So,

$$dq_s/dt = C_{sm}(dT_{sm}/dt) + dq_s'/dt$$

Applying Newton's law:

$$dq_s/dt = (1/R)(T_h - T_{sm}) \text{ and}$$

$$dq_s'/dt = (1/R')\,(T_{sm} - T_s)$$

Similar expression holds for the reference side.

The above conclusion helped scientists develop the appropriate model for DSC. The mains are discussed here.

4.1.2 Classical DSC

Suppose, thermocouples are in the sample and in the reference material. Then $T_{sm} = T_s$ and $T_m = T_r$, i.e., $R' = 0$ and the signal is:

$$\Delta T = R\,(dT/dt)\,(C_s - C_r)$$

Thus, temperature difference depends upon the difference in heat capacities, the heating rate and thermal resistance (R). Further R is difficult to determine, depends on the instrument and the properties of the sample and reference.

4.1.3 Power Compensated DSC

The power is varied to make $T_{sm} = T_{rm}$ thus $T_h = T_{sm} = T_m$ and $R = 0$, indicates that there is no thermal resistance. The signal is then:

$$\Delta T\,(dT/dt) = (dT/dt)\,(C_s - C_r)$$

4.1.4 Heat Flux DSC

The conditions are that:

$$T_h \neq T_{sm} \neq T_s \quad \text{and} \quad R \neq R' \neq 0$$

Then the signal is:

$$\Delta T = T_{rm} - T_{sm} = R\,(dT/dt)\,(C_s - C_r)$$

This is similar to classical DTA, except that R depends only on the instrument and not the characteristic sample.

4.1.5 Modulated Temperature Differential Scanning Calorimeter (MTDSC or mt-DSC)

In MTDSC, the conventional linear heating programme is modulated by superimposing a sine wave (or other periodic waveform) of small amplitude on the linear rise. Portions of each cycle then involved heating while other portions involved cooling. The overall trend, however remains a linear change in the average temperature with time. The resultant heat flow signal is analyzed to separate the response to the perturbation from the response to the underlying heating programme.

The modulated heating programme may be considered Eq. (1):

$$T = T_0 + \beta t + B\sin(\omega t) \tag{1}$$

where, T_0 is the starting temperature, β is the heating rate, B the amplitude of the modulation and ω its angular frequency (= $2\pi f$, where f is the frequency).

The contributions to the resulting heat flow can be written in the form Eq. (2):

$$dq/dt = C_{p,t}\,(dT/dt) + f(t, T) \tag{2}$$

where, dq/dt is the heat flow into the sample, $C_{p,t}$ is the heat capacity of the sample, and $f(t \cdot t)$ is the heat flow arising as a consequence of a "kinetically hindered" event. The form of $f(t, T)$ will be different for different types of processes.

Combination of Eqs. (1) and (2) gives:

$$dq/dt = \{\beta C_{p,t} + F(t, T) + \{\omega BC\, C_{p,t}\cos(\omega t) + D\sin(\omega t)\} \tag{3}$$

Underlying component Cyclic component

or

$$dq/dt = \{\beta C_{p,t} + F(t, T) + \{\omega BC\, C_{p,t}\cos(\omega t) + D\sin(\omega t)\} \tag{4}$$

The total or underlying signal is $\{\beta C_{p,t} + F(t, T)\}$, where, $F(t, T)$ is the average of $f(t, T)$ over the interval of at least one modulation, and the cycle signal is $\{\omega\ BC\ C_{p,t} \cos(\omega t) + D\sin(\omega t)\}$, where, D is the amplitude of the kinetically hindered response to the temperature modulation. Both $C_{p,t}$ and D will be slowly varying function of time and temperature but can be considered effectively constant over the duration of a single modulation. $F(t, T)$ can also give rise to a cosine contribution.

The kinetically hindered responses are usually assumed to follow Arrhenius type behaviour. The cosine response of $f(t, T)$ can be made negligible by adjusting the frequency of modulation and the underlying heating rate, to ensure that there are many cycles over the course of the transition. The amplitude of the temperature modulation is usually restricted to a degree or less so that the kinetic response can be considered to be linear.

Thus, signal derived from MTDSC experiment can be used to derive varied information as shown in Fig. 4.16:

(i) the average or underlying signal, equivalent to conventional DSC;
(ii) the in-phase cyclic component from which $C_{p,t}$ can be calculated; and
(iii) the out-of-phase signal, D, $C_{p,t}$ can be multiplied by the heating rate, β, to give (iv) the reversing contribution to the underlying heat flow. Subtraction of (iv) from the underlying signal (i) gives (v) the nonreversible heat flow.

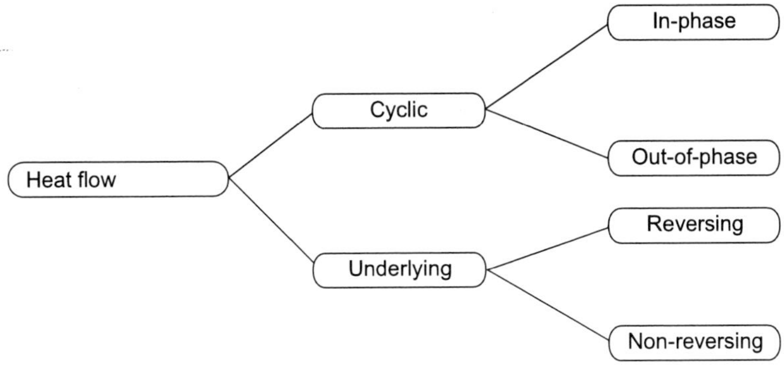

Fig. 4.16 Representative diagram of MTDSC

Some scientists and manufacturers use the square sine wave and Fourier transform analysis to get identical type results.

Advantage of MTDSC

The use of MTDSC with appropriate frequencies and amplitude of modulation allows: separation of reversing processes, such as glass transitions, from non-reversing process such as cure reactions (chemical process). The baseline curvature on the cyclic signal is generally very low, thus making it easier to distinguish between baseline effect and real transitions. The signal-to-noise ratio of the cyclic measurement of heat capacity is generally greater, because all drift or noise at frequencies other than that of modulation is ignored by use of Fourier transform analysis. Thus, resolution of process can be improved because very low underlying heating rates can be used.

4.1.6 Pressure DSC

The DSC measurement can also be performed under the optimized required gas pressure which is necessary to check the relative reactivity of a mixed sample in different chemical atmospheres. The technique is found quite appropriate to check the effectiveness of additives, e.g., antioxidants in processed food and polymer composites. The technique is sometimes called Oxidation Induction Temperature/ Time (OIT) technique.

4.2 INSTRUMENTATION

A DSC instrument essentially works on the temperature control of two similar specimen holders assembly. The left half of the block diagram (Fig. 4.17) represents the circuit for differential temperature control, while right-hand side indicates that for average temperature control. In the average temperature control circuit, an electrical signal proportional to the dialed temperature of the sample and reference holder can be generated through the programmer. The dialled temperature was initially fed to the recorder on the chart but nowadays it is entered into the computer. This programmed signal is compressed with the average signal from the temperature measuring probe in the two holders and the resultant signal difference is amplified by the average temperature amplifier. As discussed above two principles work: one based on power compensation and the other on heat flow method. In power compensation method smaller secondary heater is attached to equalize the generated energy difference between sample and reference materials. While in heat flow technique, heat flux passing through the sample and reference are evaluated and their difference is related to energy consumed or released in the thermal reactions. A comparison is shown in Fig. 4.18.

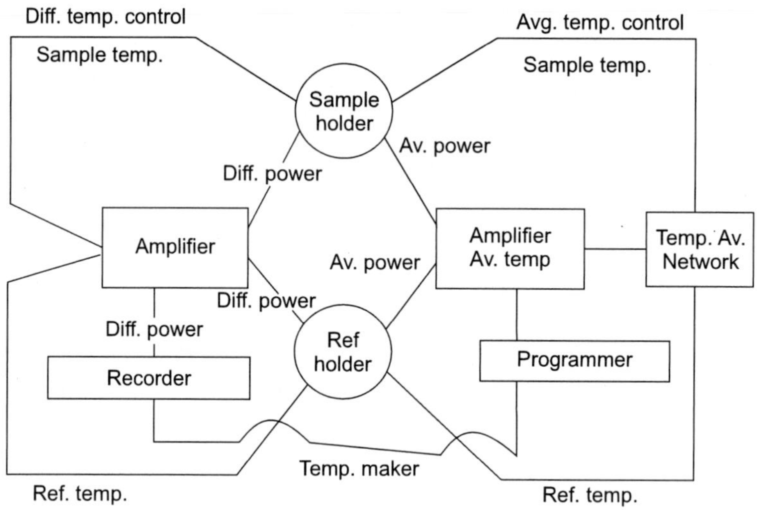

Fig. 4.17 Block diagram of a DSC instrument

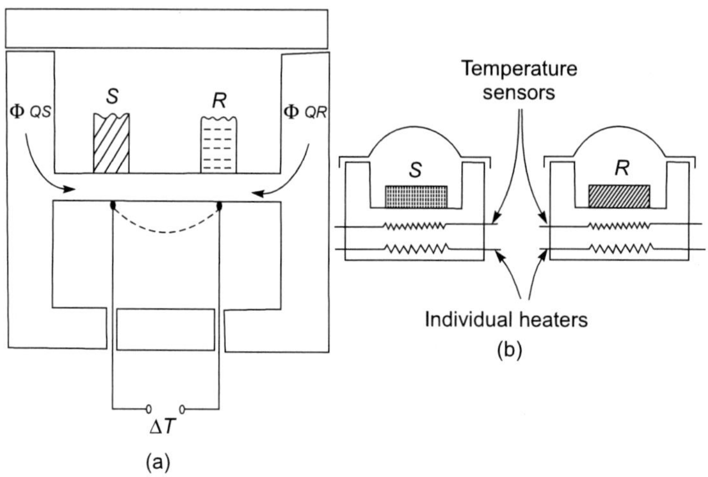

Fig. 4.18 Block diagram of: (a) Heat flux and (b) Power compensating sample holder

The area under the DSC melting curve gives the heat of the reaction for the sample in millicalories and the heat of reaction per mole reading is computed. T_0 is directly from the curve obtained by plotting $1/F$ (F = fraction of melt) at each temperature and this is plotted against temperature [if necessary to linearise (or straight line) curve, make necessary correction]. From this plot T_0 is read directly at zero intercept and by knowing the slope of this line, we can get ΔT.

Volume of its sample must follow the given temperature programmed dT/dt, determined by the carrier, after the termination of the effect. The duration of the temperature effect depends on the amount of the sample and on its thermal conductivity. The rate of heat flow in the sample is expressed by the equation,

$$\frac{dH}{dt} = \frac{mC_p dT}{dt} \tag{1}$$

where dH/dt is rate of heat flow (Cal Sec^{-1}), C_p is specific heat (Cal $gm^{-1}K^{-1}$), dT/dt is rate of change in temperature (K Sec^{-1}) and m is the mass of the sample (gm). The area under the dH/dt curve between two temperatures corresponding to the change in the heat content, i.e.,

$$H_{T_2} - H_{T_1} = \int_{T_1}^{T_2} \frac{dH}{dt} \times dt \tag{2}$$

In order to use Eq. (2) in a specific heat calculation, the ordinate calibration and the temperature program rate must be known with at least the precision requirement of the final result. However, the two parameters may be eliminated from the calculation, if a material with a known specific heat is used to calibrate the instrument. Such a material is α-aluminium oxide, or synthesized sapphire, for which the specific heat has been determined to five significant figures in the temperature range 0°K to 1200°K. After the baseline and sample program, a third program is run with a weighed sapphire structure. At any temperature (T), the equation applied is:

$$Ky = mC_p\, dT/dt \tag{3}$$

$$Ky' = m'C_p'\, dT/dt \tag{4}$$

where, y and Y' are ordinate deflection in the inch due to the sample and the standard, respectively $m'C_p'$ are mass and specific heat of the standard and K is the ordinate calibration factor (cal $inch^{-1}$ sec^{-1}); dividing Eq. (3) by (4), we get

$$C_p/C_p' = m'y/my' \tag{5}$$

Thus, calibration requires only comparison of the two ordinate deflections at the same temperature. So one can easily calculate value of C_p on putting rest of the values in Eq. (5). The basic components of DSC are quite similar to DTA except the differential energy measuring system. However, the most advantageous feature of DSC is its sensitivity and capability to resolve two peaks, because of the measuring principle as seen in Table 4.4.

TABLE 4.4 Measuring parameters of DTA and DSC

Techniques	Measuring parameters
DTA	ΔT $(T_s - T_r)$
Heat flux DSC	dQ/dT
Power compensated DSC	$dQs/dT - dQr/dT\ d(\Delta Q/dT)$

Thus, measuring efficiency of DSC can be optimized by factors listed in Table 4.5.

TABLE 4.5 Factors optimizing measuring efficiency of DSC

Parameters	Increase Sensitivity	Increase Resolution
Sample weight	Increase	Decrease
Heating rate	Increase	Decrease
Purge gas	N_2	He

The heat flow also relates as condition:

$dQ/dT = C_p \times dT/dt + F(T, t)$ or $dQ/dT = C_p$ (heat capacity) × heating rate + Kinetic component is function of temperature and time.

4.3 FACTORS AFFECTING CURVE

In Chapter 3, the lowest temperature, T_i at which the onset can be detected by the instrument operating under particular conditions was discussed. We may like to call this a transition temperature, which is not correct. Actually, in DSC experiments, T_i, T_f and T_c (initial, final and complete temperatures at which the decomposition is completed) do not have fundamental significance, but they can still be a useful characteristic of a DSC curve and the term procedural thermogram, is often used for the temperature at which temperature change appears to commence. This indicates that start of a thermal reaction, temperature does not have a fixed value, but depends on the experimental procedure employed to get it. Similar to this, there are many factors which influence a DSC curve. These factors may be due to instrumentation or nature of sample. The main factors which affect the shape, precision and accuracy of the experimental results are listed below:

1. **Instrumental factors**
   Furnace heating rate
   Recording or chart speed
   Furnace atmosphere
   Geometry of sample holder/location of sensors
   Sensitivity of recording mechanism
   Composition of sample container

2. **Sample Characteristics**
   Amount of sample
   Solubility of evolved gases in sample
   Particle size
   Heat of reaction
   Sample packing
   Nature of sample
   Thermal conductivity

4.4 SOURCES OF ERROR

There are a number of sources of error in DSC, and they can lead to inaccuracies in the recorded data of heat. Some of the errors may be corrected by placing the thermobalance at proper place and handling it with care. The common source of errors during operation or common are discussed in DTA except the one in accuracy caused by secondary heaters and thermostats. These errors can be avoided by proper placing of instrument in the laboratory, maintaining operating temperature, and constant power supply. By avoiding excessive heating rate and proper gas flow rates, many errors can be also avoided.

4.5 EVALUATED PARAMETERS

DSC curves of pure compounds are fingerprints of that compound in the context of transition temperature as well as heat required for that transition. Therefore, using DSC curve one can infer about the presence of a particular compound and its thermal behaviour. A typical DSC curve is shown in Fig. 4.19. The peak above the baseline is exothermic while the peak below is endothermic.

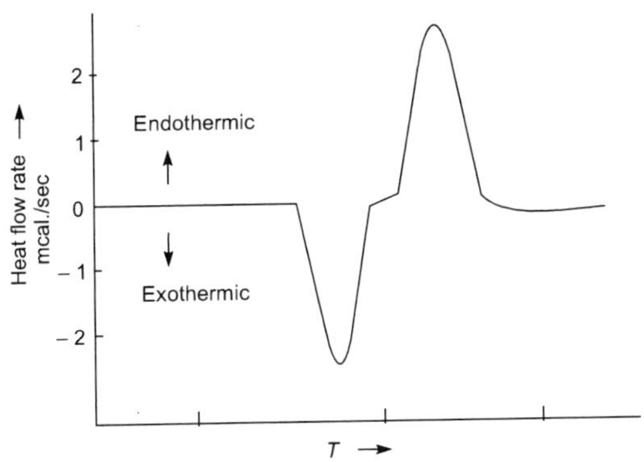

Fig. 4.19 A typical DSC curve

We have discussed earlier how area under DSC curves is related to the amount of energy released or absorbed in a physico-chemical change. It has been shown that under certain conditions, the area under peak is proportional to the amount of heat involved in reaction. The information obtained in DSC is quite identical to DTA, only with difference of sensitivity. This is because the measuring principle makes the technique more informative and useful.

4.5.1 Measurement of Enthalpy Change

Enthalpy change of a reaction (ΔH) can be measured by area under the curve:

$$\Delta H \approx \frac{A \times K}{m}$$

where, A = area, K is calibration constant, m = mass. The value of K can be determined by relating known enthalpy change of measured peak area. Usually, the melting point of Indium (In) metal is used for determination of K. The value of K also depends on geometry and thermal conductivity of sample. As per basic principle, in power compensated DSC, K is proportionality constant and virtually independent of temperature, while in heat flux DSC, K markedly depends on temperature. Thus, it is advisable to calibrate the instrument in the entire temperature range with different standard materials. Further, heat of reaction observed in DSC can be further used to calculate molar enthalpy of reactions by using following formula:

$$\Delta H_m = \frac{\Delta H_x \times M_w}{M}$$

(ΔH_m = molar enthalpy of reaction, ΔH_x = observed enthalpy of reaction, M_w = molecular weight of analysed compound, and M is mass of substance used for analysis.)

Thus, area under the curve is used for stoichiometric ratio of analyzed compounds (quantitative interpretation). Now we see in next example how it can be used to compare thermal stability of a material for physical states and chemical states. This can be used for chemical identification of a material (qualitative interpretation). Such information can be used to select material for certain end-use applications, predict product performance and improve product quality. DSC curves of a polymeric mixture and probable transitions are shown in Fig. 4.20 for illustration about probable change in behaviour of a polymer sample.

The DSC technique is more sensitive than DTA and it provides clear presence of thermal events occurring during the course of heating of time and ageing of materials. Thus, the information acquired by DSC is more realistic. The technique is used for presence of polymorphism, degree of crystallinity, curing fraction, etc. Curves clearly indicate that Fig. 4.20 which shows the peaks for the glass transition, ordering, melting and decomposition of individual polymers, is not possible by DTA. The ratio of area under the curve is obtained simply by dividing the enthalpy of heat of decomposition and reported values provide the ratio of individual monomers in analysed copolymer sample.

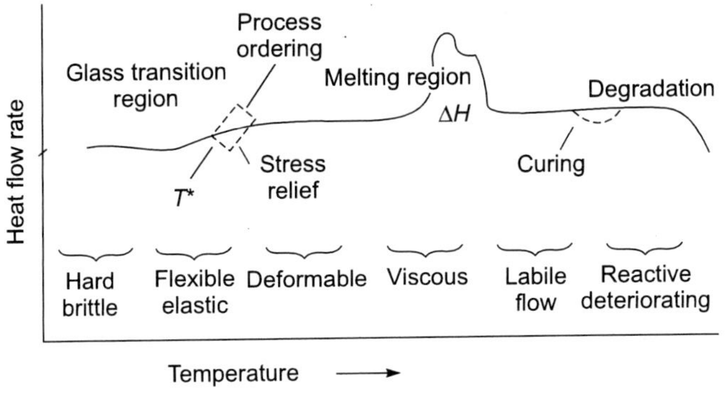

Fig. 4.20 Change in behaviour of polymeric materials in DSC

4.5.2 Measurement of Heat Capacity

This technique is used for evaluation of heat capacity (C_P), entropy (S) with the following formulas:

$$\Delta H = \int_{T1}^{T2} C_p dT$$

$$\Delta S = \int_{T1}^{T2} \frac{C_p}{T} dT$$

$$\Delta C = \Delta H - T\Delta S$$

In the experiment, the baseline is displaced in the endothermic direction upon which the sample is heated. The displacement (h) is proportional to heat capacity. Hence, $h \alpha\ C_p$ and $h = B\beta C_p$, where β is heating rate and B is calibration factor. The value of B can be determined by standard substance such as sapphire. So, for heat capacity measurement the area of displacement between baseline with empty pan and filled with sample is measured as shown in Fig. 4.21.

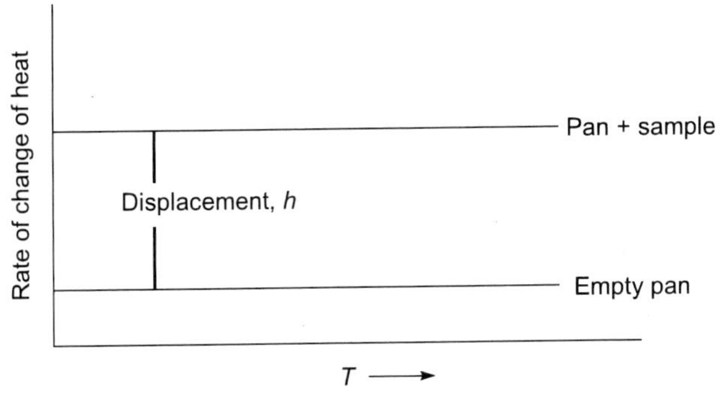

Fig. 4.21 Diagram for calculation of heat capacity

4.5.3 Determination of Specific Heat

The DSC curve is conveniently used to measure the specific heat. For this purpose, first experimental procedure is same as in heat capacity. The absolute change in temperature (ΔT_x) has been measured and put in the equation for calculations:

$$(C_p)_T\,(\text{mcal/mg°C}) = (\Delta T_x + \Delta T_{\text{blank}})\,Et/M\beta$$

where $(C_p)_T$ is the specific heat at temperature T, ΔT_x is the absolute temperature difference for sample in °C, ΔT blank is absolute temperature difference of empty container, Et is the calibration coefficient at temperature T in mcal °C min, M is the sample mass in mg and β is the heating rate in °C per minute.

4.5.4 Measurement of Thermal Conductivity

Thermal conductivity (λ) is effectively measured by DSC, apart from other methods discussed earlier. For this purpose a special DSC cell is used, in which both top and bottom temperatures of sample can be monitored. The λ is measured by comparing result with a standard calibrants material, as per the formula below.

$$\lambda_s = \lambda_c\,(h_s \times l_s \times d_c^2 \times \Delta T_c) \times (h_c \times l_c \times d_s^2 \times \Delta T_s)$$

where l = length, d = diameter of sample, c and s represent calibrant and sample respectively. Further, if density of the sample is known, thermal diffusivity can also be determined by the formula, $D = \dfrac{\lambda}{\rho C_p}$.

4.6 APPLICATIONS

Differential scanning calorimeter can be used to measure energy change as a function of temperature or time. Using this technique it is possible to observe a number of characteristic properties of a sample like fusion, crystallization, glass transition temperatures (T_g) as well as the progress of a thermochemical reaction. The glass transitions may occur as the temperature of an amorphous solid is increased. These transitions appear as a step in the baseline of the recorded DSC signal. This is due to the sample undergoing a change in heat capacity; no formal phase change occurs. As the temperature increases, an amorphous solid will become less viscous. DTA can also used to measure T_g but sensitivity is better in DSC. At some point, the molecules may obtain enough freedom of motion to spontaneously arrange themselves into a crystalline form. This is known as the crystallization temperature (T_c). This transition from amorphous solid to crystalline solid is an exothermic process, and results in a peak in the DSC signal. As the temperature increases, the sample eventually reaches its melting temperature (T_m). The melting process results in an endothermic peak in the DSC curve. The ability to determine transition temperatures and enthalpies makes DSC an invaluable tool in producing phase diagrams for various chemical systems, whether the system is miscible or immiscible. The technique is widely used across a range of applications, both as a routine quality test and as a research tool. The equipment is easy to calibrate, using low melting Indium, for example, and is a rapid and reliable method of thermal analysis. A few notable specific applications of DSC are:

The result of a DSC experiment is a curve of heat flux versus temperature or versus time. There are two different conventions: exothermic reactions in the sample shown with a positive or negative peak; it depends upon the different kind of technology used by the instrumentation to perform the experiment. This curve can be used to calculate enthalpies of transitions. This is done by integrating the peak corresponding to a given transition. It can be shown that the enthalpy of transition can be expressed using the following equation: $\Delta H = KA$, where, ΔH is the enthalpy of transition, K is the calorimetric constant, and A is the area under the curve. The calorimetric constant will vary from instrument to instrument, and can be determined by analyzing a well-characterized sample with known enthalpies of transition.

Most of well known spectroscopic methods of great value in the qualitative and quantitative chemical analyses are based upon our ability to measure energy absorption or emission caused by transition from one energy state to another. The great potential of thermal spectroscopy (redefined DSC) for quantitative analysis was not realized in the past because of the absence of a suitable, fast scanning, the calibration run for the synthetic compounds and the baseline technique used in the area measurement. However, it can be used for the quantitative analysis of constituents present in wide spectrum of materials.

4.6.1 Polymorphism

Many materials can exist in two or more different crystal forms. The chemical reactivity and physical properties of different forms varies frequently one to another. Technological handling requires one of the perfect suitable forms, hence the phenomenon is of great importance in chemistry and conveniently studied by DSC.

For example, CCl_4 was not completely understood until, in Nov. 1966, Rockmon reported an X-ray diffraction study, which revealed the existence of a previously unknown sophisticated modification of CCl_4, in addition to the cubic form which had been thought to be stable from −47.7°C to −22.75°C and the monoclinic form stable below −47.7°C.

Physiologically active substances which are present in minor quantities in foodstuffs can be studied in their pure forms, or in aqueous solutions. Crystalline caffeine has been studied extensively in the anhydrous or monohydrate form or in solution, particularly by calorimetric techniques [18]. Caffeine shows a solid-solid transformation around 140°C and melts at around 235°C, as shown in the calorimetric curve of Fig. 4.22.

DSC may also be used in the study of liquid crystals. As matter transitions between solid and liquid it often goes through a third state, which displays properties of both phases. This anisotropic liquid is known as a liquid crystalline or mesomorphous state. Using DSC, it is possible to observe the small energy changes that occur as matter transitions from a solid to a liquid crystal and from a liquid crystal to an isotropic liquid.

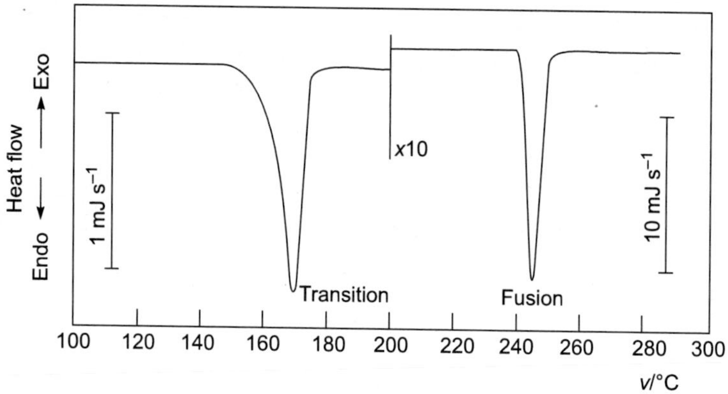

Fig. 4.22 DSC curve of pure β caffeine (Adapted from Ref. 18)

4.6.2 Specific Heat

It is evident that calorimetric techniques are often used to determine specific heats of materials like foods; the methods, and a synthesis of results, have been presented in the literature. The use of a standard such as synthetic sapphire simplifies the determination of specific heats; by measuring the standard, an empty cell, and the substance, the specific heat value C, of the sample can be obtained directly by comparison. For a given temperature,

$$C_2 = C_1 \frac{m_1}{m_2} \frac{(Q_2 - Q_0)}{(Q_1 - Q_0)}$$

where m_1 is the mass of the standard substance, m_2 is the sample mass, C_1, is the specific heat of the standard substance, and Q_0, Q_1 and Q_2, are the required heat quantities for empty cell, the standard substance and the sample, respectively. For example, the values obtained vary between 0.9 J g^{-1} K^{-1} for very dry food products to 4.18 J g^{-1} K^{-1} for water; it is therefore evident that the moisture content of a food has a strong influence on its specific heat value.

4.6.3 Purity Determination

DSC curves may also be used to evaluate purity of drugs and polymer samples. This is possible because the temperature range over which a mixture of compounds melts depends on their relative amounts. This effect is due to a phenomenon known as freezing point depression, which occurs when a foreign solute is added to a solution. (Freezing point depression is what allows salt to de-ice sidewalks and antifreeze to keep your car running in the winter.) Consequently, less pure compounds will exhibit a broadened melting peak that begins at lower temperature than a pure compound. DSC is used widely for examining polymers to check their purity and composition. Melting points and glass transition temperatures for most polymers are available from standard compilations, and the method can show possible polymer degradation by the lowering of the expected melting point, T_m, for example. T_m depends on the molecular weight of the polymer, so lower grades will have lower melting points than expected.

In the pharmaceutical industry, it is necessary to have well-characterized drug compounds in order to define processing parameters. For instance, if it is necessary to deliver a drug in the amorphous form, it is desirable to process the drug at temperatures below those at which crystallization can occur. The above-mentioned transitions are well observed in DSC curve and used in industry regularly.

4.6.4 Analysis of Polymeric Materials

The energy involved in different temperature induced structural transitions is used for purity and relative compositional analysis of the pure and processed polymeric samples. For example, the packaging of paste and cream are done in tube having reverse flexibility against applied mechanical pressure. In general, toothpaste tubes are made of multilayer laminates with variable relative compositions polyethylene, polypropylene along with barrier layer from inorganic materials. The DSC analysis provides variable energy for heat of fusion at different temperature, which can be used for quantification of present polymer in product. The comparison of DSC curve of pure and blende of LDPE and LLDPE in different compositions are shown in Fig. 4.23. [19] This relative energy analysis is used for compositional analysis of polymer in blend and composites.

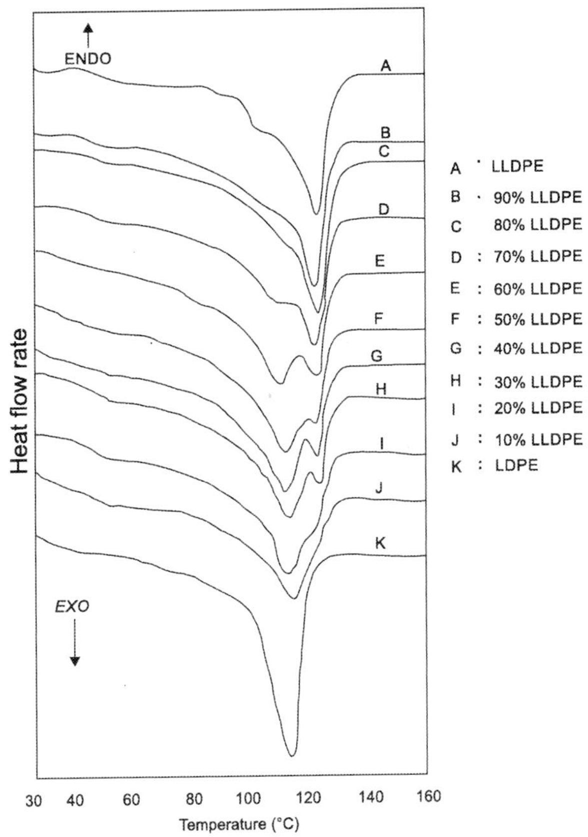

Fig. 4.23 DSC curves of LLDPE, LDPE and their blends

The other hybrid form of polymers are hydrogel, nanocomposite, laminates and melt blends with several advanced applications. DSC also provides important characteristic features about the hybrid form of polymer for their applications in terms of stability and functionality. The presence of heterogeneity during cooling due to irregular crystallisation in hybrid, shown in Fig. 4.24, provides transition for individual crystal at different temperature, which can be used for their quantification of individual component, i.e., three peaks depicted in Fig. 4.24 for different components A, B and C [20].

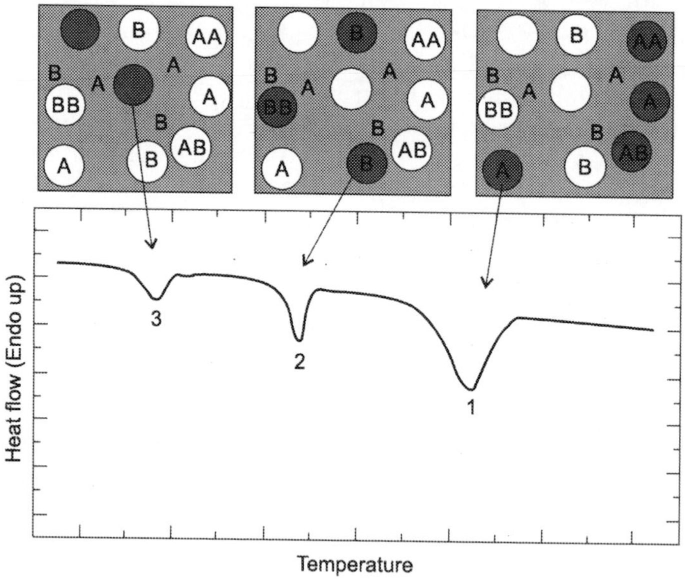

Fig. 4.24 DSC curve of immiscible polymer blend along with structural illustrations [20]

The presence of the contaminants in processed and pristine polymer are also identified by DSC analysis after comparing different thermal transitions like melting points, heat of fusion and energy of crystallisations. In an example the inadequate processing of polymer through extruder may lead to the presence of residual polymer in a particular batch, which did not allow to pass the quality standards for marketing. In this regards, DSC allows to measure the residual polymers present in processed piece after comparing the DSC curve of pure and processed polymer specimens. The re-cooling of melted polymer provides useful information about polymer composition and structure like nucleation efficiency, reversible transitions and thermal fractionation.

4.6.5 Forensic Application

The measurement of energy involved in transformation of biological reaction and components is frequently explored forensic sciences to solve and support evidence to solve the criminal cases. Its uses to screen body organ, explosives, oxidative stability and thermal stability. The body organs of humans are made up of polymer and fibres, which bears their unique thermal signature for their identification in analysed sample. These characterisations of different body organs like hair, skin and nails help the police and forensic expert to catch the criminals and solving the criminal cases. Similarly, the soil consists of complex organic as well as inorganic components and its DSC curve provides information about crime sites. The residual presence of soil on the shoes and clothes is helpful to tag a crime on the basis of DSC analysis. For example, soil analysis helps to study illegal mining, murder at construction sites after beating and the burial cases after murder [21].

Hair is the exo-skeleton of human and comprises specific protein, i.e., keratine as well as possess exclusive microscopic characteristic like shape, colour, and root appearance. These variation of hair helps a forensic experts to determine age, gender, and racial origin to solve the criminal issues like the enthalpy energy of African are lower ~5.2 jg^{-1} than Asian ~6.3 jg^{-1}. DSC curve of hair of caucasian is shown in Fig. 4.25, which is indicating two peaks at 232°C and 243°C. The reasons for origin of peaks are assigned to melting of alpha-helix present in cells or to the presence of amorphous cystine [22].

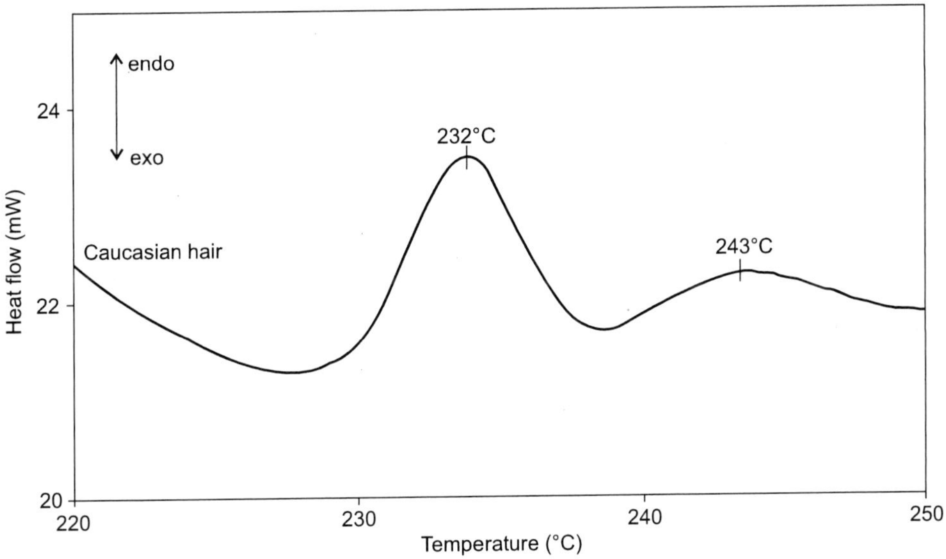

Fig. 4.25 DSC curve of Caucasian hair

The dry, wet, conditioned and treated hairs exhibit their thermochemical behaviours, which help to solve the criminal cases and human conditions. Skin is another widely analysed by forensic experts in terms of composition and morphological changes with time. Although, skin is a complex type of external tissue and its composition changes with age, growth and races in terms of pigments, lipid and cholesterol contents. Thus, the thermal monitoring of these contents is used to support the skin originality during forensic investigations. Monitoring of illicit drugs like steroid are also monitored using DSC methods and it is also used for forensic applications. In this context DSC used to identify the composition drugs like anabolic steroid present in body and excreted fluid after comparing the melting point and enthalpy for purity and composition. The DSC curve of extracted oxymetholone, shown in Fig. 4.26 is used as signature [23].

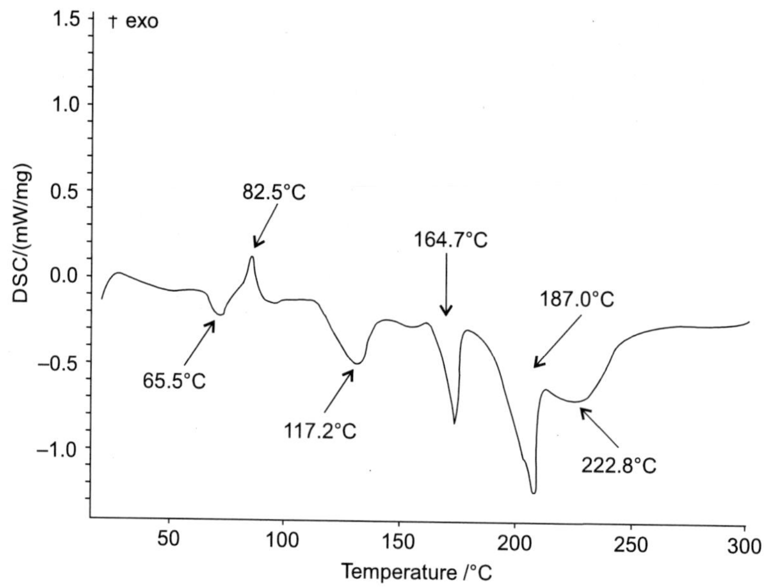

Fig. 4.26 DSC curve of extracted oxymetholone [23]

4.6.6 Food and Nutrition

DSC has been highly an important analytical technique in food and nutrition in order to evaluate the energy involved in thermochemical transition in edible and related components like denaturation of protein, polymorphism in lipid and gelation of starch. The information provided by DSC is also helpful in estimating food adulteration and fraud after comparing the information with available standard ones. The estimation of water contents as free (freezable) or bounded form (un-freezable) in food item is another significant application of DSC. The presence of free and bounded water has great significance in terms of texture, chemical degradation and microbial activity. Determination of glass transition temperature is another significant use of DSC in food items to infer information about diffusion, mobility of different ingredients. DSC curve of beef muscles and white egg is shown in Fig. 4.27, the graph is indicating a deflection peak labelled as G for glass transition temperature. However, the exothermic peak labelled C is revealing crystallisation of vitreous water along with a broad endothermic peak for abstraction of water. The colorimetric data for water removal is also important for sorption of water by dry milk and whey protein [24].

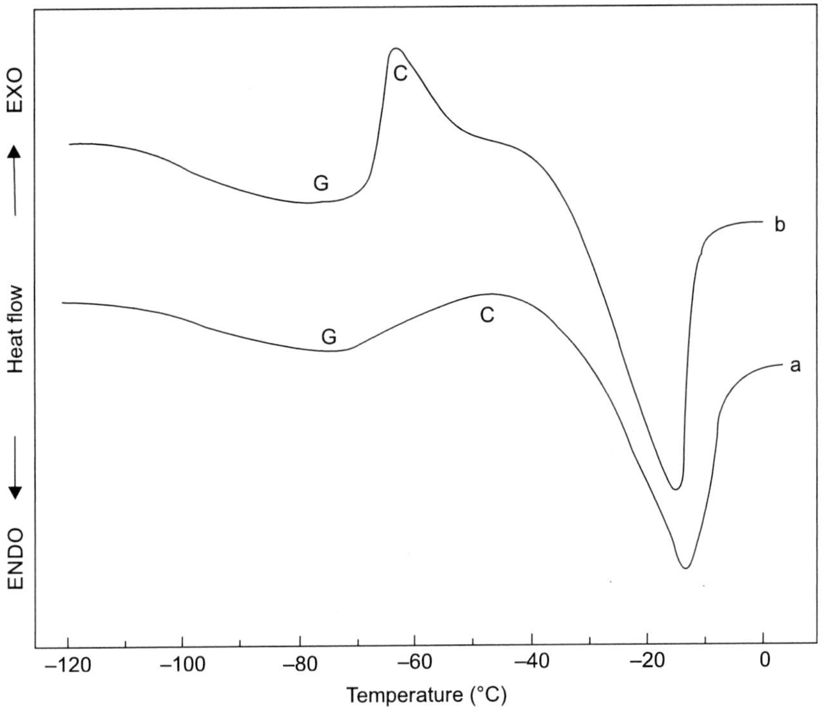

Fig. 4.27 DSC curve of a) beef muscle and b) white eggs [24]

The water sublimation data of sublimation and freeze-drying process is also important for frozen alcohol water mixture. For example, the heat of vaporisation for water and alcohol mixture changes with percentage composition of alcohol for 40% to 60% in alcoholic drink like whisky. The calorimetric data is important to understand the difference in interaction between ethanol and water from whisky due to structural alteration.

The study denaturation of protein is another important application of DSC in food chemistry, which is related to stability of protein in source in the terms of composition and pH. The denaturation of protein develops the texture of foods as well as indicates the effect of cooking of foods. Thus, DSC study of protein in terms of denaturation reveals the effect microbial effect, processing and thermal treatments. The analysis also screen the temperature and pressure for denaturation of protein during processing to assist food chemists and engineers. For example, the high pressure processing and thermal treatments at 100 MPa leads to partial denaturation of protein present in blue crab protein, while the processing over 300 MPs leads to denaturation of myosin at lower temperature [25]. The monitoring of temperature induced transition helps to screen the quality and quantity of protein present in edible items.

The processing of foods is another important step in food sectors, during processing of food different transitions occur during transition. DSC curves of spray-dried and freeze-dried egg whites are shown in Fig. 4.28 (a) and (b). The peaks are indicating significant change in freezing point and enthalpy of meting, i.e., 64.1 j/g in spray drying and 69.6 j/ g in freeze-drying [26].

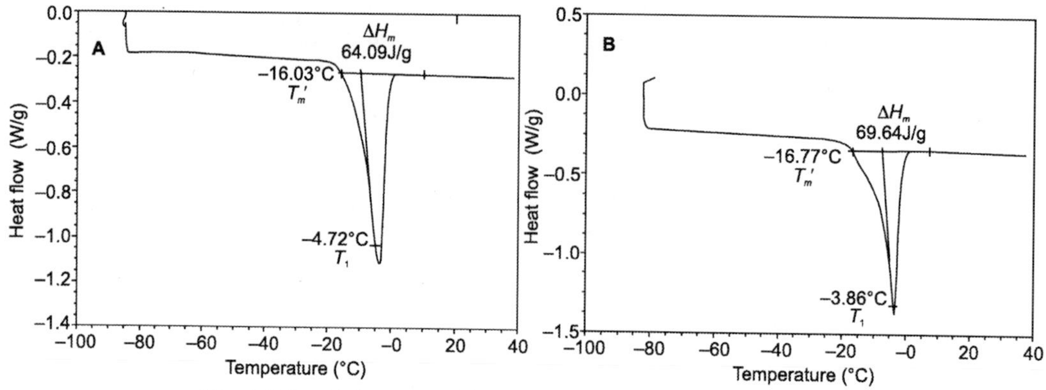

Fig. 4.28 DSC curve of (a) spray-dried (b) Freeze-dried egg

Another application of DSC in food chemistry is to study quality of cooked food, gelation in starch, composition of lipid and protein in beef, interaction of muscles protein with salts and quantitative presence of free and bound water.

4.6.7 Wax and Fats

There are two important criteria for characterization of waxes and fats by DSC.

Melting peak: Melting peak shape of waxes and waxes formulation can be quite complex; they are much less sensitive than fats to variation in the crystallization condition.

Crystallization peak: Crystallization of waxes is easily nucleated with the result that solidification occurs readily with little or no supercooling; in fact, the crystallization peak for most waxes are virtually mirror images of their melting point.

Although reading of DSC curve, applying to these criterion, we can easily make a statement whether the supplied material is wax or fat.

4.6.8 Oxidative Stability

Using differential scanning calorimetry to study the oxidative stability of samples generally requires an airtight sample chamber. Usually, such tests are done isothermally (at constant temperature) by changing the atmosphere of the sample. First, the sample is brought to the desired test temperature under an inert atmosphere, usually nitrogen. Then, oxygen is added to the system. Any oxidation that occurs is observed as a deviation in the baseline. Such analysis can be used to determine the stability and optimum storage conditions for a compound. The details of the application and method is discussed in OIT method.

4.6.9 Magnetic Transitions

A more specific application of DSC employed is the detection of magnetic transitions in the materials from Diamagnetic ⟷ Paramagnetic ⟷ Ferromagnetic. The magnetic transitions are of second order with $\Delta H = 0$ but with a specific heat maximum at the transition temperature, and technique is sensitive to detect these minor variations. A typical DSC curve for a magnetic transition of Ni is shown in Fig. 4.29. Similarly, other transitions can also be measured with DSC.

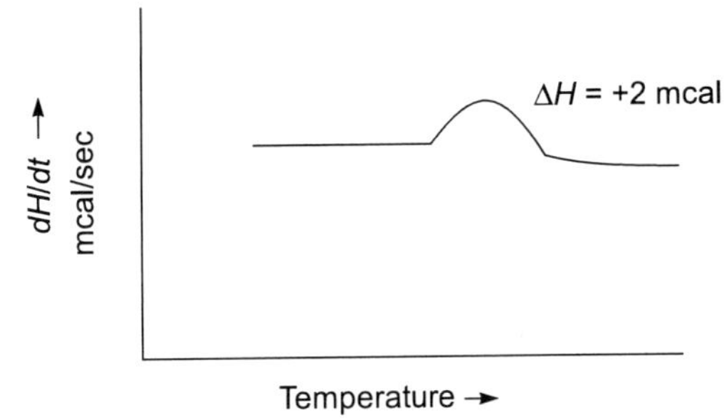

Fig. 4.29 A typical magnetic transition observed in DSC

4.7 NANOCALORIMETER

4.7.1 Introduction

Nanocalorimetry is a newer dimension of thermal analytical techniques to measure the thermal properties of small samples at faster rate. It is used to evaluate the thermal transition associated with very low energy transition such as the metabolism of bacteria (less than 10^6 cal), quantification of reactions and protein binding interactions. In past around 1780, Lavoisier et al. developed the so called ice calorimeter, which determined the heat evolved by a living organism present within it from measurements of the quantity of ice melted. However, the advent of micro-electromechanical systems (MEMS), microchip-based technology and micro-fabricated membrane has effectively advanced the nanocalorimeters for several advanced applications like direct measurements of high-speed, multi-step reactions proceeding on sub-monolayer films, interfacial surface reactions for precise sensing of surface interacting molecule like gas sensing. Although the measurement in small scale is a complex phenomenon, thus, the innovation of nanoscale thermal transition measuring techniques also permits it to integrate other surface analytical techniques like electron microscope, spectroscopy and X-ray diffraction techniques for extended applications [27].

The term 'nanocalorimeter', consistent with the current and accepted definition of microcalorimeter, is used to describe these calorimeters because of the power detected by the instrument rather than the dimensions of the object formed with integrated circuit technology. The main application of these devices to traditional measurements and differential scanning calorimeters (DSC) are measuring thermal transition of small amount with small sample size (mass ≈ ng and volume ≈ nano litre) with high sensitivity of nano joule. The use of small sample mass also effectively reduces the associated risk in laboratory during these measurements. Thus, the advances in calorimetry studies at nano-scaled objects is an important application for the future of modern calorimetry in wider range of applications like sensing of explosives, reaction steps, biological interactions and illicit drugs [28]. Furthermore, to perform "calorimetry at nanoscale" the ultra fast scanning is required due to followings:

(i) Most of the modern materials are used in non-equilibrium states. Thus, to study the actual thermodynamic state of such materials at room temperature, e.g., of a semicrystalline polymer, requires high heating rates to prevent reorganization during the scan. These non-equilibrium states are generated either by rapid cooling or heating of the materials.

(ii) Currently focus of materials in nano-scale systems needs to perform calorimetric measurements using nano-gram samples because the heat flow needed to heat the sample at moderate rates is extremely small, which cannot be detected by conventional calorimeters.

(iii) The combination of high heating and cooling rates with small sub-milligram sample masses open up the application of calorimetry for advance analysis in combinatorial chemistry to optimize the molecular properties.

This ultra fast scanning nanocalorimetry provides important information on thermodynamic properties, kinetics and structural changes in materials during fast thermal treatments. For example, the actual thermodynamic state of semicrystalline polymers can be investigated only at sufficiently high heating rates, ca. 10^3 K/s and even more. Such rates are needed to prevent reorganization during the temperature scan. Thus, cooling rates up to 10^3 K/s are necessary for studies of phase transitions under the same conditions as during processing of polymers in different techniques like injections.

4.7.2 Instrumentation

In precise measuring the heat consumption in thermal event the thermal sensor and heater on a nanoscale membrane are incorporated in excellent isolation along with film heater, sample holder and thermostat. The change in temperature is recorded from integrated thermocouples or by monitoring of changes in the resistivity (temperature coefficient of resistance of the heating element with excellent sensitivity. An optical technique has recently been used for calibration of the heating response in order to eliminate the errors, e.g., extrapolation of lower temperature calibrations to higher temperatures. The basic configuration of a nanocalorimeter is shown in Fig. 4.30 along with schematic illustration in Fig. 4.31.

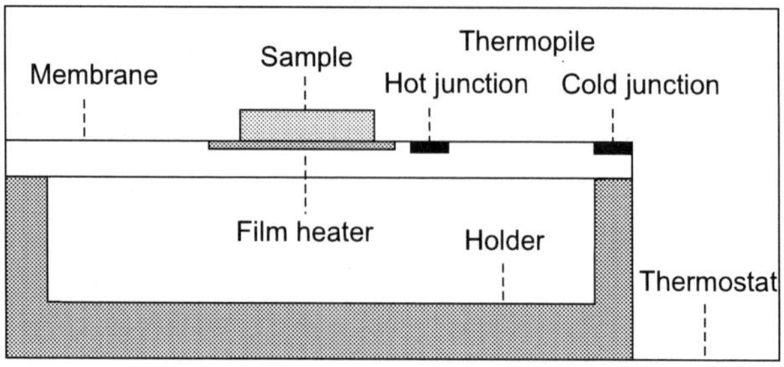

Fig. 4.30 Block diagram of nanocalorimeter

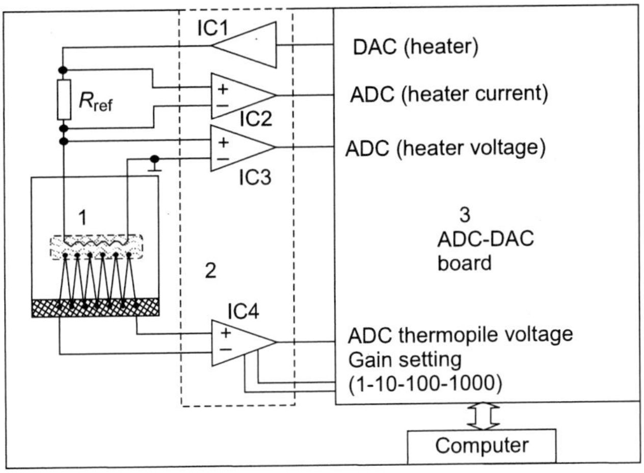

Fig. 4.31 Scheme of the nanocalorimeter [Adapted from Ref. 29]

The fast-cooling scanning calorimetry is proposed by using thin film sensors as shown in Fig. 4.30 and 4.31 electronic details. The semiconductive film-thermopile and a semiconductive resistive film-based heater is used. The thermopile hot junctions are arranged around the heater. The cold junctions are placed at the periphery of the cell, ca. 1 mm from the centre, where the membrane is attached to the holder. Thus, the cold-junction temperature is the temperature of the holder and it is close to the temperature of the thermostat. An additional copper–constantan thermocouple was utilized for the measurement of the holder temperature, which was used as the reference temperature. Furthermore, the resolving power in thermal change in a material depends on the thermal mass of the sensor, mass of analysed sample, determination and minimisation of heat losses during analysis. Hence, several nano-calorimeters are developed with particular specifications for exclusive applications in both adiabatic and non-adiabatic in nature. For example, the determination of melting point of a polymer requires external heating source of heat with different temperature resolution. However, the determination of liquid phase interaction does not require external heating sources. Setaram has developed a portable calorimeter, similar in cross section to that shown in Fig. 4.32, which has a square 10 × 10 mm silicon substrate in which was fabricated 8.5 × 8.5 mm square membrane 50 J. l m thick. This calorimeter bas a sensitivity of ~ 1 V' W-1. A monocrystalline silicon aluminium thermopile with 164 junctions, with an estimated sensitivity of about 100 mV' K -I, and a heater are integrated into the membrane. The nanocalorimeter is mounted in standard ceramic PGA housing.

Where the sample holder is thermally isolated from its surroundings. A 300 μm square membrane 3 μm thick is suspended by a 100 μm wide and 3 μm thick wire above an etched square cavity 2.5 × 2.5 mm. A heater and gold nickel thin film thermocouple are integrated into the membrane. BOTTOM: Cross section along X-X' showing the pyramidal shape of the cavity over which the membrane is suspended.

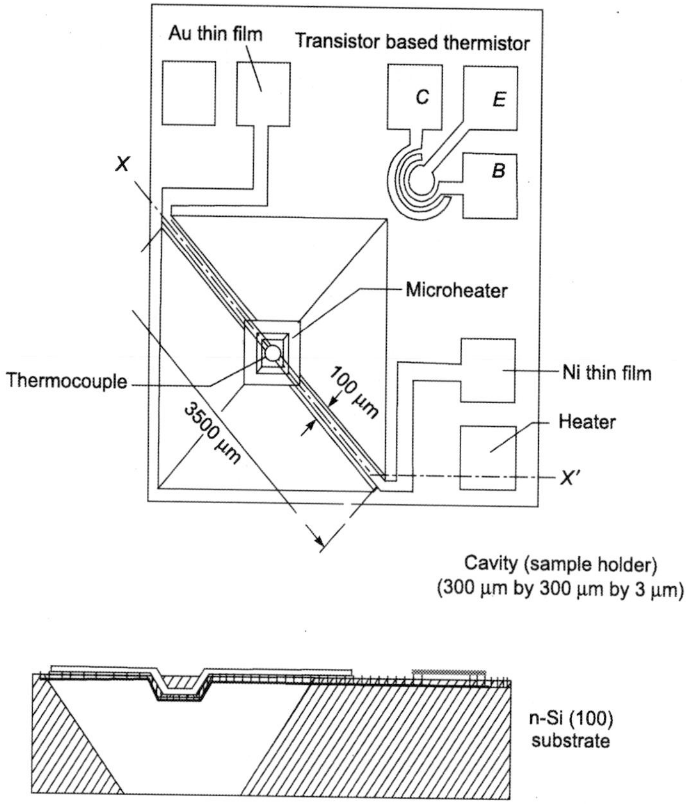

Fig. 4.32 TOP: Schematic of a MEMS nano-calorimeter

Calibration is another important step of this method and needs to be performed before to carry out an analysis. In this regard, the several standards are reported to calibrate the performance for the calorimetry in order to ratify the results in terms of onset temperature, reactions enthalpy and other properties. A list of some established standards is listed in Table 4.6.

TABLE 4.6 List standards for calibration of calorimeter

Material and transition	Transition temperature (°C)	Heat of reactions(j/g)
Gallium (s to l)	29.8	79.9
Benzoic acid (s/l)	123.0	148.0
Tin (s/l)	231.9	60.4
Indium (s/l)	156.6	28.6

4.7.3 Applications

Advanced analytical tools and smaller sample size analysis yield better resolution in calorimetric analysis. For example, comparison of conventional DSC and nanocalorimetry analysis of heat capacity of Ni–Si interfacial reaction is depicted in Fig. 4.33.

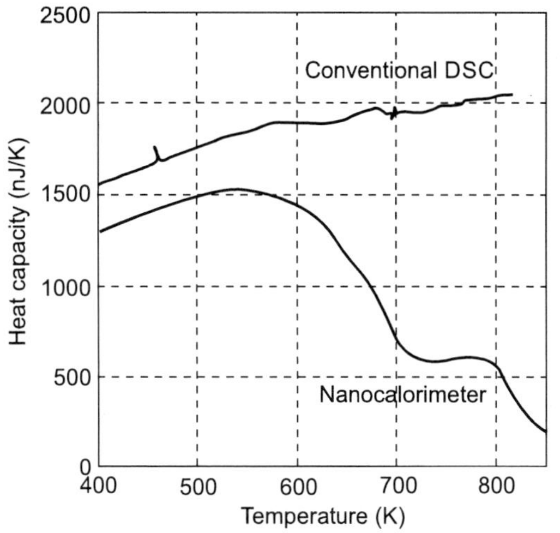

Fig. 4.33 Heat capacities of a Ni–Si interfacial reaction recorded by conventional and nanocalorimetry. [30]

Thus, the advantageous feature of nanocalorimetry has been exploited in several applications, i.e., materials characterization, chemical sensors, pharmaceutical applications, imaging and integrated effect with other techniques.

4.7.4 Materials Characterizations

Advances in small scale calorimetry for nanoscale analysis explore the several dimensions in characterization of in situ materials changes, restructuring effect and surface interacting behaviour like melting induced reorganization, recrystallization, hydration/dehydration kinetics, melting behaviour and nucleation like linear polyethylene standard reference material (number-average molecular weight Mn ca. 100,500 g/mol and the weight-average molecular weight Mw ca. 119,600 g/mol with the ratio Mw/Mn ca. 1.2 with the crystallinity of 60% at room temperature). For study of fast melting and crystallization the sample must be small enough and the sample should stay at the heater during thermal cycling, so the adhesion between the sample and the cell should be sufficiently good. Study of melting and crystallization of linear polyethylene is only little dependent on heating rate but the crystallization temperature shows the expected cooling rate dependency. Polymorphism is another important features in pharmaceutical compounds, metal oxides and polymers for incorporating distinct properties in polymers like tacticity, strength and conductance. Here, this method is used as an alternate methods to evaluate the polymorphism and crystal formations. For example, furandicarboxylate-based polyesters biopolymer, i.e., poly(propylene 2,5-furandicarboxylate) (PPF), has been reported as an alternate for petroleum based terephthalate homologues. The PPF exists in two forms α with dark and larger banded spherulites, while β with brighter and smaller spherulites. The method is used to screen the different phases on the basis of melting behaviour during calorimetric study and formation of metastable states. The single α phase nucleates at lower temperature, i.e., 120°C, while at higher both forms of PPF were coexist [31].

Structural rearrangement, biotransformations and binding behaviour of different biomolecules and polymers like protein and enzymes. These transitions associated with very small energy consumption such as cell metabolic reaction change, which are the challenge for measurement under wet conditions. For example, the protein stability and its interaction with ligands are important for biochemistry and biomedical practices. The biding behaviour of protein develops different structures depending on interacting metal and condition like hemocyanins an copper containing protein responsible for electron transfer during energy production. The rate constant and energy liberation of this protein during copper in fully folded condition is 1.5 times lower than that initial addition, which yields different amount of enthalpy and entropy, thus the screening of energy production helps to estimate the types of interaction occurring in hemocyanins as well as other biochemical transformation [32]. The mechanism of interactions is shown in Fig. 4.34.

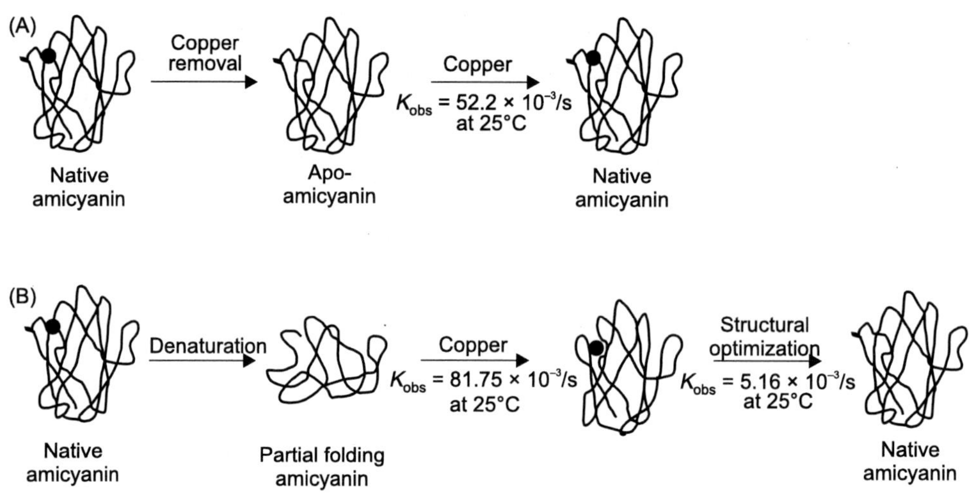

Fig. 4.34 Nanocalorimetric assisted study of protein folding

The methods used to fabricate MEMS are particularly useful for the measurement of the thermal properties of compounds that may be vapour deposited atop, and otherwise essentially identical, nanocalorimeter membranes. The variation of the calorimeters thermal characteristics can be used to evaluate the thermal properties of additional layers. The thermal properties of monocrystalline silicon, polycrystalline silicon as well as the dielectric layers Si_3N_4 and SiO_2, and conductive layers of Al have been determined with nanocalorimeters are listed in Table 4.7.

TABLE 4.7 Seebeck coefficient S, thermal conductivity k, and heat capacity C_p of microsensor materials

Materials	$\frac{k}{W \cdot K^{-1} \cdot m^{-1}}$	$\frac{C_p}{MJ^{-1} \cdot K^{-1} \cdot m^{-3}}$	$\frac{S}{\mu V \cdot K^{-1}}$
Mono silicon	150	1.6	±250 to 750
Poly silicon	20 to 30	1.8	±100 to 300
Aluminium	200	2.5	–2
Silicon nitride	3	2.2	
Silicon	1	1.6	

Thus, the measurement of specific properties are used as signature for characterizing the desired properties and chemical transformations.

4.7.5 Sensors

Presently, sensing science has advanced the analytical chemistry tremendously for various applications after using different induced properties like electrical, optical and thermal properties. Thus, the effective monitoring of thermal properties like enthalpy of reaction and change in temperature has frequently used for sensing of gas, biomolecules present in neighbouring atmospheres after surface chemical reaction and biochemical transformations. For example, the detection of the quantity of both glucose and urea in blood from measurement so the enthalpy of an enzymatic reactions.

To detect glucose concentration the reaction

$$C_6H_{12}O_6 \qquad O \xrightarrow{GOD} C_6H_{12}O_7 + H_2O_2 \tag{1}$$

can be used, for which the enthalpy of reaction ΔH = +79 kJ. mol^{-I}, In Eq. (1) GOD is an acronym for glucose-oxidase (β-D-Glucopyranose aerodehydrogenase), an enzyme that catalyses the glucose-specific reaction and, thus, makes the sensor selective. The sensitivity of the sensor can be enhanced by adding a second enzyme, catalase, which decomposes the hydrogen peroxide according to the reaction Eq. 2

$$H_2O_2 \rightarrow H_2O \qquad \frac{1}{2}O_2 \tag{2}$$

for which $\Delta f H_0 = +100$ kJ mol^{-1}. Thus, the total enthalpy of reaction is about 180 kJ . mol^{-1}. The nanocalorimeter described in literature has a sensitivity of about 50 mV – mol^{-1} – L^{-1}, which can be observed with a minimum glucose concentration detection threshold of 20 μL mol L^{-1}.

Another important application is the use of nanocalorimeters with living cell is, where either the sensitivity of cells to medication or toxins can be determined from measurement of the heat, which can be up to 300 pW per cell, produced during metabolism. In addition, the concentration of nutrition in the fluid surrounding the cells can be determined from the metabolic heat output.

The use of a MEMS, with a silicon membrane, integrated silicon thermopiles, and an enzyme coating that was placed on the opposite side of the membrane to that of the thermopiles, to detect constituents in blood and other liquids. Both of these devices could be used for medical diagnostics techniques including. For example, the determination of glucose concentration in blood sample, which is significant for diabetes. Similarly, when a nanocalorimeter membrane is coated with the enzyme urease the concentration of urea can be estimated employing the enthalpy of reaction ~ ΔH = + 61 kJ . mol^{-1}. The other chemicals which can be identified are penicillin *G*, creatinine, up to micro molar concentration, which are indicators of kidney malfunctioning. The advantageous feature of technique that no energy is released thus enzyme activity and biochemical reactions proceed in natural path.

Similarly, the ultra-sensitive calorimeters can be used for real-time in situ detection of specific gaseous species within a gas mixture, such as detecting protein to all illness based on a chemical concentration within exhaled breath. A modular based sensor system for the detection of gases in gas mixtures , which is also called electronic nose, for which specific gases alter the electrical resistance. Each nanocalorimeter in the matrix has a different catalytic or enzymatic coating so that each gives a unique response to specific gases. This method gives a multidimensional signal allowing the determination of type and concentration of components in a mixture.

4.7.6 Surface and Morphological Study

An apparatus based on atomic force microscopy methods, shown in Fig. 4.35, has been developed to determine the thermal conductivity, thermal diffusivity and heat capacity of a solid by TA instruments (J. 1 T A 2990 microthermal analyser) [33]. In this apparatus a probe, with a tip of radius 50 nm, is lowered close to the sample surface and moved in the *x*, *y* plane while maintaining a fixed distance *z* from the surface. These measurements provide the topography of the surface which must be flat. In this particular device, a laser is used to determine the vertical displacement. The tip is a Wollaston wire, which is coated platinum that acts as both the resistive heater and thermometer. This probe can be used to determine the thermal properties of the sample in 30 × 30 μm square blocks with a spatial resolution < 1 μm, and a total scanning area of (100 by 100) μm. Constant probe heating provides values of the surface thermal conductivity while ac heating yields thermal diffusivity and heat capacity.

The combination of nanocalorimetry with other analytical techniques like scanning electron microscope (SEM), transmission electron microscope (TEM), X-ray diffraction (XRD), mass and Raman spectroscopy has been significantly improves for characterization of materials and materials properties in terms of thermal effects. For example, the scanning electron microscope is widely explored for study of surface morphology, size and chemical composition. However, the integration of SEM with nanocalorimeter allows the visual monitoring of thermal changes like melting and phase transition proceeding during course of heating along with energy measurements. A simple example for structural change in PEO after melting and

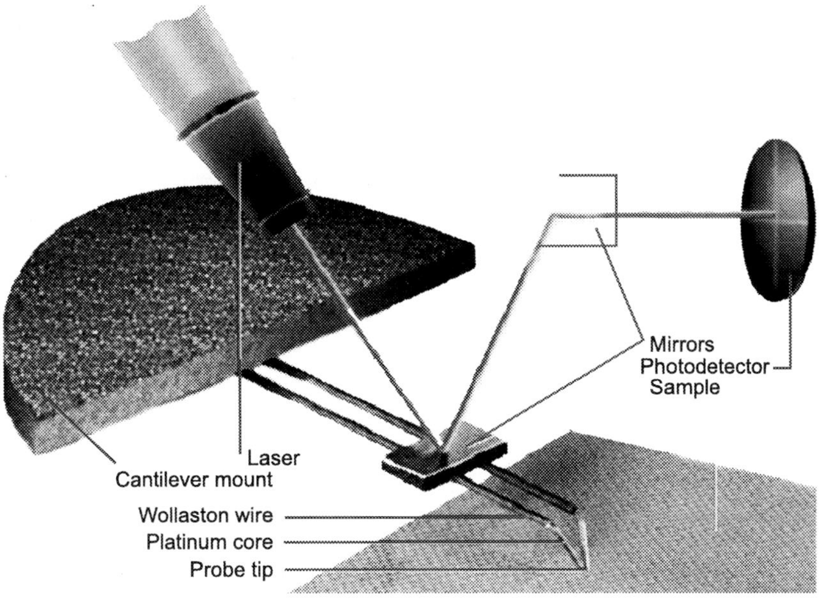

Fig. 4.35 Schematic of the TA instruments microthermal analyser with platinum probe and laser used to determine its vertical displacement.

re-solidification has been investigated by Yi et al. for better information about effect heating on physio-chemical changes along with energy compensations [34]. Similarly, the effect of heating on structural change in materials structure like recrystallization, onset crystallization and crystal deformation have been investigated after coupling the both techniques.

4.8 KINETIC STUDIES

The technique has been successfully employed to determine the kinetic parameters of a chemical reaction involving a substantial enthalpy change. The theoretical treatment is based on fundamental equations of heat balance. While the analytical derivation was simplified by carefully choosing the experimental conditions. The calculation of the kinetic parameter is possible for those thermochemical reactions which could not be identified by TGA.

Generally, a thermochemical reaction is based on Eq. (1);

$$d\alpha/dt = f(\alpha, t) \quad (1)$$

where $d\alpha/dt$ is the rate of reaction, $f(\alpha, t)$ is α, function of amount reacted, and t is the absolute temperature at time T.

It is also known that the rate of heat of a reaction H, is also directly proportional to $d\alpha/dt$ as per Eq. (2),

$$H = \beta\rho\,(d\alpha/dt) \quad (2)$$

where β is the heat per unit volume and ρ is the sample density. Although the kinetic equation used depends on the reaction mechanism, generally one expression which excludes diffusion controlled, fusion and inversion type reaction,

$$d\alpha/dt = A(1 - \alpha)^n e^{-E/RT} \tag{3}$$

where A is the pre-exponential factor, α is the fraction of sample reacted $(0 < x > 1)$ and n is the order of reaction temperature. However, the process is not accepted to all because it is not known to which it applies. Thus, different methods and equations have been developed using both isothermal and nonisothermal DTA and DSC results for calculation of kinetic parameters. First use of DTA in the study of reaction kinetics was made by Murray and White by establishing relationship between the rate of decomposition of Kaolin and temperature (heating rate). Since the weight loss is function of time at constant temperature, the calculated rate constant is similar to thermogravimetry. The important methods are discussed below:

Kissinger Method [15]

He derived Eq. (4), which is observed suitable for kinetic parameters calculation;

$$d[\ln(\beta/T^2_{max})]/d(1/T_{max}) = e^{-E/R} \tag{4}$$

where β is the heating rate. For any value of n,

$$E/RT^2 \max = (An/\beta)(1 - \alpha_{max})^{n-1} e^{-E/RT \max} \tag{5}$$

For $n \neq 0$ or 1

$$n(1 - \alpha_{max})^{n-1} = 1 + (n - 1)(2RT \max/E) \tag{6}$$

Since $(n - 1)(2RT \max/E) < 1$, therefore Eq. (4) may be further approximated by:

$$n(1 - \alpha_{max})^{n-1} = 1 \tag{7}$$

Substituting Eq. (7) in (6) gives an approximate value Eq. (8)

$$dx/dt = (A/\beta)e^{-E/RT}(1 - \alpha) \tag{8}$$

Kissinger concluded that order of reaction (n) was obtained from shape index (S) defined as the absolute value of the ratio of the slopes of the tangents to the curve at inflection points.

$$n = 1.26\ S^{1/2} \tag{9}$$

Thus, in conclusion of Kissinger method,

$$E_a\beta(RT^2_p) = Ae^{-E_a/RT_p}$$

where, A is the frequency factor and β is the heating rate, which is also expressed as $\beta = \dfrac{dT}{dt}$ (dT and dt are temperature and time change, respectively). Taking logarithm of Eq. (4), Eq. (5) resulted:

$$-\ln(\beta T_p^2) = -\ln\left(\frac{AR}{E_a}\right) + \left(\frac{1}{T_p}\right)\left(\frac{E_a}{R}\right)$$

Thus, activation energy, E_a can be determined by plotting $\ln(\beta T_p^2)$ against $\frac{1}{T_p}$, where T_p is the peak temperature.

Flynn-Wall Method

In this method, Eq. (10) can also used to calculate activation energy (E_a):

$$\log\beta = \log\frac{AE_a}{g(\alpha)R} - 2.315 - \frac{0.457E_a}{RT} \tag{10}$$

where, β is the heating rate, A is the frequency factor, E_a activation energy, T is recorded peak temperature and $g(\alpha)$ is the integral function of mass conversion, α. So, the activation energy for different conversion values can be calculated from a plot of log β versus 1000/T.

Piloyan Method

Piloyan, Ryabochikov, and Novikova developed Eq. (11) by using Δt i.e., observed temperature difference between sample and reference materials. The method required a single DTA curve in non-isothermal conditions.

$$\ln\Delta t = C - \ln f(\alpha) - \frac{E_a}{RT} \tag{11}$$

C is constant, T is temperature in K corresponding to a given value of temperature difference (Δt). In DTA analysis, the role of temperature change is prominent. Thus Eq. (11) can written like Eq. (12),

$$\ln\Delta t = C' - \frac{E_a}{RT} \tag{12}$$

where C' is constant.

Borchordt and Daniels reported a more useful method using the concept:

In DTA curve, the enthalpy of reaction the $\Delta H = A$ (where ΔH is the heat of reaction and A is the area of peak. It is assumed that heat evolved is directly proportional to the number of moles reacted (N_0). It follows the relation Eq. (13),

$$-dN/dt = -N_0/A\ (dH/dt) \tag{13}$$

and the rate constant is given by

$$k = (AV/N_0)^{n-1}\ (dH/dt)/(A - \alpha)^n \tag{14}$$

The equation for K may be written after rearrangement:

$$K = \Delta T/(A - \alpha) \tag{15}$$

For first order or

$$K = (AV/N_0)^{n-1}\Delta T/(A - \alpha)^n \quad (16)$$

Maycock reported Eq. (17) using DSC curve, useful to calculate E using linear Arrhenius plot:

$$-E = R\ \ln d_1 - \ln d_2/1/T_1 - 1/T_2 = 4.58 \log (\alpha_1/\alpha_2)/1/T_1 - 1/T_2 \quad (17)$$

where d_1 and d_2 are the rate of heat evolution at T_1 and T_2 respectively.

Davies and Dollimore [15] developed Eq. (18) for kinetic parameters determination of polymorphism kinetics. The method was employed for changing vaterite to calcite and aragonite to calcite. For both transitions, the kinetic equation describing the process is found to be Eq. (18):

$$\frac{\partial x}{\partial t} = \frac{4}{3}k[-\ln(1-\alpha)]^{3/4}(1-\alpha) \quad (18)$$

The values of activation energy were found to be 452 ± 19 kJ mole^{-1} and 208 ± 8 kJ mole^{-1} for aragonite and vaterite transitions respectively, using a procedure which enabled the calculation of kinetic parameters from a rising temperature experiment.

Quantitative analysis of the calcium carbonate polymorphs calcite, aragonite and vaterite has also been studied using complimentary techniques such as X-ray diffraction and infra-red spectroscopy. Since there is no weight change during these transitions, conventional methods of determining the kinetics by TG cannot be employed. A dynamic temperature technique is proposed here which may be adopted to determine the kinetics of any polymorph transition by DTA.

Ozawa Method

The basis of this method is that for a change in heating rate (β) different temperatures correspond to different specific values for fractional mass change (1 – α). Equations (19) and (20) were established by Ozawa for different value reaction products (1 – α):

$$\log\beta = -0.4567\frac{E}{RT_p} - 2.315 + \log\frac{AE}{Rg(a)} \quad (19)$$

$$\log\beta + \frac{0.4567E}{RT_p} = \text{constant} \quad (20)$$

where, E is activation energy, R is gas constant, T_p is peak temperature and $g(\alpha)$ is the functional form of α representing the reaction mechanism.

The above equations (19) and (20) can be used to calculate E at different values of α. Further, the slope of plot between log β and against $\frac{1}{T_p}$ equation yield activation energy of a reaction. Further, if it is assumed that reaction is first order, then the intercept of plot between log β and against $\frac{1}{T_p}$ yields pre-exponential, A factor.

Presently, both isothermal and non-isothermal methods are used but it is assumed the most reliable kinetic methods are isothermal methods. In non-isothermal kinetics, the difference in the temperature dependence of the rates for nucleation and growth is taken into account. This difference in the temperature dependence is a very important point needed to derive non-isothermal kinetics of a reaction, consisting of multiple elementary processes one can consult reference books.

4.9 CALIBRATION

Since DSC is a dynamic method, in principle it consists of measuring the change in temperature associated with physical or chemical changes occurring during the gradual heating of the substance. It is essential that all aspects of the technique are standardized in order to obtain reproducible results. These include pretreatment of specimen, particle size, and packing specimen, dilution of the specimen and nature of the inert diluents. DTA and DSC techniques are required to be calibrated in routine manner for:

(i) Baseline
(ii) Temperature
(iii) Heat of reaction
(iv) Determination instrument constant

The procedure with suitable example is explained in Chapter 3.

SECTION C: THERMOMETRIC TITRATIONS

4.1 INTRODUCTION

This technique is assumed to be an extension of the knowledge derived from TGA, DTA and DSC for deriving analytical information about simple and complex compounds formation and on line chemical reaction processes [5, 35-36]. In this section, we will first introduce thermometric titration (TT), with instrumentation including experimental details and then discuss some analytical applications.

4.2 PRINCIPLE

Similar to the case of thermogravimetry, basic principle of thermometric titrations is based on monitoring the change in temperature with the addition of titrant and determining the end-point from a plot of temperature versus volume of titrant. The titrant is added to an isothermal titrate in an adiabatic titration calorimeter. In most instances there occurs a change in enthalpy concomitantly, yielding a corresponding heat of reaction. As a result these are also called enthalpy titrations and the titration curves are called enthalpograms. It can be exothermic or endothermic. In thermometric titration, change in temperature occurs only when titration is in progress and sample reactant is present. Thus, start and end points of a titration are readily observed and the number of moles titrated is calculated as in regular titration. By determining the heat capacity of the system under study, heat of reaction can be readily determined. In addition, equilibrium constant can be evaluated under appropriate conditions.

On both enthalpograms, the baseline represents temperature-time blanks recorded prior to the start of the actual titration, corresponding to the beginning of addition of titrant, is the end point and the excess reagent line. Thus, the titration branch of the enthalpograms. In cases where $\Delta H < 0$, it has an ascending slope whereas for $\Delta H > 0$, it has a descending slope. The excess reagent branches of the enthalpograms are drawn with ascending slope because the dilution of most titrants is an exothermic process. In order to minimize variations in heat capacity during titration, it is customary to use titrants 50-100 times more concentrated than the unknown titrated. Thus, the volume of titrate solution is maintained virtually constant, but the titrants are diluted appreciably. The heat of dilution can be corrected conveniently by the linear back extrapolation. Under these conditions, the extrapolated ordinate height represents a measure of the change of temperature, ΔT, due to the titration reaction, as well as of integral heat, O, evolved or absorbed, viz.

$$\Delta T = \frac{Q}{K} = -\frac{N \cdot \Delta H}{K} \tag{1}$$

where K denotes the effective heat capacity and N is the number of moles reacted in the titration. From Eq. (1), it is evident that the quantity can be used for obtaining the following information:

Heat of titration reaction, ΔH, whenever a known amount of titrate is used.

Concentration of an unknown 'titration'; whenever ΔH is known (ΔT is proportional to concentration and independent of tirate sample size).

Quantity of an unknown in a known volume of titrate.

Operationally, enthalpograms are thermometric titration curves, where temperature is experimentally observable variable upon addition of titrant. The basic property which determines the shape of enthalpograms is enthalpy change. This accounts for the designation, "thermometric titration" (TT) and "enthalpy titration" used in literature. Other similar terms include calorimetric titration, thermochemical titration, thermal titration, similar to the usage for potentiometric/amperometric techniques.

Another type of thermometric titration includes direct injection enthalpimetry (DIE) which yields a plot of temperature versus the time following injection of the titrant. In this case, no end point is obtained but the magnitude of the temperature change is proportional to the concentration. The speed of analysis is enhanced and processes with equilibrium unfavourable for titration are readily studied by using large excess of one reagent.

4.3 INSTRUMENTATION

Experimental setup of TT consists of a reagent delivery system (motor driven automated burette), an adiabatic reaction cell (Dewar flask with stirring device), an electronic temperature sensing system and an amplifying and data processing system. A typical experimental assembly and adiabatic cell are illustrated in Fig. 4.36. In TT temperature control is most important and depends upon the results of room temperature. However, for precise calorimetric results the titrant and sample must be close to the same temperature as far as possible and this is achieved by a thermostat.

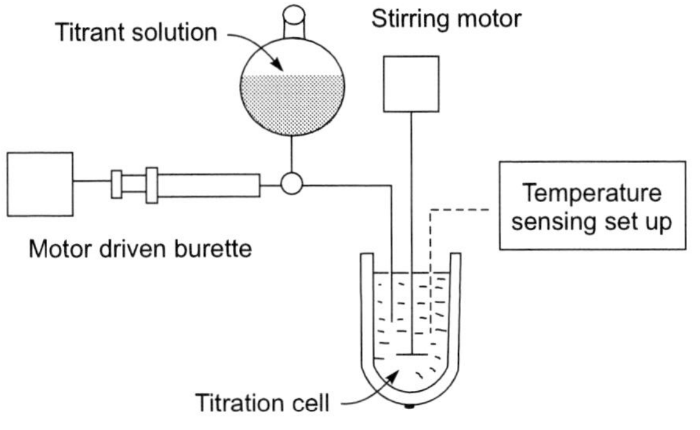

Fig. 4.36 Experimental setup of thermometric titration

The main components of TT experimental setup are discussed below.

4.3.1 Delivery Pump

A constant delivery pump permits the time axis of a strip chart recorder to be used as the volume of titrant axis. Typically, a syringe driven by a synchronous motor which drives the screw is used and solution is delivered at constant rate ranging down to a few μL per min.

In Direct Injection Enthalpometric (DIE), the syringe is rapidly emptied at the start of the experiment to deliver the titrant instantaneously into the sample cell.

4.3.2 Adiabatic Cell

These have widely varying designs ranging from an insulated beaker to a Dewar flask type cell (Fig. 4.28). All the cells are designed to minimize the heat transfer from the cell to the environment, thus maximizing the temperature change observed. When only titration end point is of interest, the simple cell is sufficient. However, if quantities such as heat of reaction, equilibrium constant or kinetic parameters are sought, it becomes essential to use better cells which have thin walls to minimize the heat capacity and maximize the speed of response to temperature change.

4.3.3 Temperature Measuring System

It is the heart of thermometric titrator; generally, either thermocouple or thermistor is used. The thermistor is a temperature-sensitive semiconductor whose resistance obeys the equation (2),

$$R_T = Ae\frac{b}{t} \tag{2}$$

where A and b are constants whose values depend on the nature of the thermistor. A thermistor has the advantage of being small in size, fast response to temperature change and chemical resistance.

4.3.4 Amplifier and Recorder

It is often advantageous to amplify the signal obtained from the thermistor. ADC amplifier is used to obtain good signals for temperature changes of the order of 10^{-4}°C or less. For still lower temperature changes of the order of 10^{-6}°C, an AC Wheatstone bridge with a lock-in amplifier is used. The two most popular data acquisition systems are the strip chart recorder and the digital data storage system.

4.3.5 Calibration

In order to bring the cell quickly to thermal equilibrium, it is essential to use calibration heater. It has two purposes, to determine the heat capacity of the system and to control the temperature in the cell itself. The heat evolved or absorbed is calculated from the temperature change using the relationship.

$$\Delta Q = \Delta T C_p' \tag{3}$$

where C_p' is the heat capacity of the system which is measured as the amount of heat necessary to raise the cell temperature by an unit.

4.4 THEORY TO EXPERIMENT

An idealized thermometric titration curve is represented in Fig. 4.29 showing four regions. Region I (AB) is the baseline which is ideally horizontal but in practice it always has finite slope due to frictional heat added by stirring, resistive heat added by the thermistor and the transfer of heat from the cell to the thermostat. Its heating due to friction and resistance is constant and equal to W. The slope of Region I is represented by

$$\frac{dT}{dt} = C(T_c - T_e) + W \tag{4}$$

In case of Region II (BC) slope is due to similar effects as in Region I and following cases; temperature change generated by the reaction, the heat of dilution of the reactants ΔH_0 and the difference in temperature,

$$\frac{dT}{dt} = -C(T_c - T_e) + W + \left(\frac{-\Delta H}{C_p'}\right) + \left(\frac{dn_p}{dt}\right) + \frac{\Delta H_D}{C_p''} + \Delta T_R \cdot k \tag{5}$$

where between titrant and sample after the start of titration (ΔT_r), where k is a constant. In Region III, equivalence point has been passed and the slope of the curve is described by

$$\frac{dT}{dt} = C(T_c - T_e) + W + \frac{\Delta Hd}{C_p'} + \Delta Tr \tag{6}$$

After this, no more titrant is added and the slope in Region IV (CD) obeys Eq. (4). In real titration, however, a rounding is observed at the equivalence point and this case is dealt with separately.

Equations (4)–(6) can be combined to obtain the heat of reaction ΔH and if the production of product (n_p) is equilibrium, controlled equilibrium constant K of the reaction can be calculated as:

$$-\Delta G = RT \ln K = \Delta H + T \cdot \Delta S \tag{7}$$

4.5 ANALYTICAL CALCULATIONS

Knowing the concentration of either the titrant or the sample, concentration of the unknown may be calculated from the volume added to reach the end point. Heat of reaction is, in general, obtained from Eqs. (4) to (6). However, in practice a graphical method is followed. Heat capacity at the mid-point of the curve is obtained by extrapolating the part of the curve in Region III back to Region II and thus measuring Q from the baseline to the extrapolated line at the mid-point of titration.

4.6 APPLICATIONS OF THERMOMETRIC TITRATION

Since heat of reaction is the most general property of chemical processes, thermometric titrations have a wide range of applicability in quantitative analysis [37]. It includes the determination of the concentration of an unknown substance, the reaction stoichiometry, equilibrium constant and the thermodynamic quantities: ΔG, ΔH and ΔS. These have been applied successfully to all types of titration processes including acid-base (neutralization) reactions, redox reactions, precipitation reactions, and complexometric reactions in media ranging from aqueous at room temperature to molten salt at 350°C. Precision and accuracy of measurements depend largely on the enthalpy of the reaction involved ranging from 0.2%. Lowest limit of concentration that can be successfully titrated is up to 10^{-4} M. The main limitation of the method is that it is non-selective. The nature of information derived from the judicious mathematical analysis of thermometric titration curves is summarized in Table 4.8.

TABLE 4.8 Summary of applications of thermometric titrations

Type of Application	Information Obtained	Procedure
Fundamental	Heat of reaction	From extrapolated ordinate heights
Analytical	Free energy of reaction Entropy of reaction	From curvature in equivalence pt. region.
	Reaction Stoichiometry Concentration of unknown	From the dynamic equation End point determination Direct enthalpimetry

As an illustration, Fig. 4.37 shows potentiometric (pH) and thermometric titration curves for HCI and H_3BO_3 with NaOH in dil. aqueous solution. The potentiometric curves of the two acids (Fig. 4.37) are very different yielding a large end point inflexion for the HCI ($K\alpha$ ~ •) and virtually none for boric acid. However, corresponding enthalpograms (Fig. 4.37) are strikingly similar, because the heats of neutralization of the two acids are comparable, –13.5 kcal/mole for HCl and –10.2 kcal/mole for H_3BO_3. As a result, H_3BO_3 can be determined with better precision and accuracy by thermometric titration. In the enthalpogram for H_3BO_3 and NaOH (Fig. 4.37), *AB* represents a trace of the temperature of the solution before the addition of titrant and *C* is the end point. Line *BC* shows the gradual evolution of heat of reaction. Linear portions of the curves are extrapolated to give the initial and equivalence points and the vertical distance between them (BC) gives temperature difference (ΔT) is used to evaluate enthalpy.

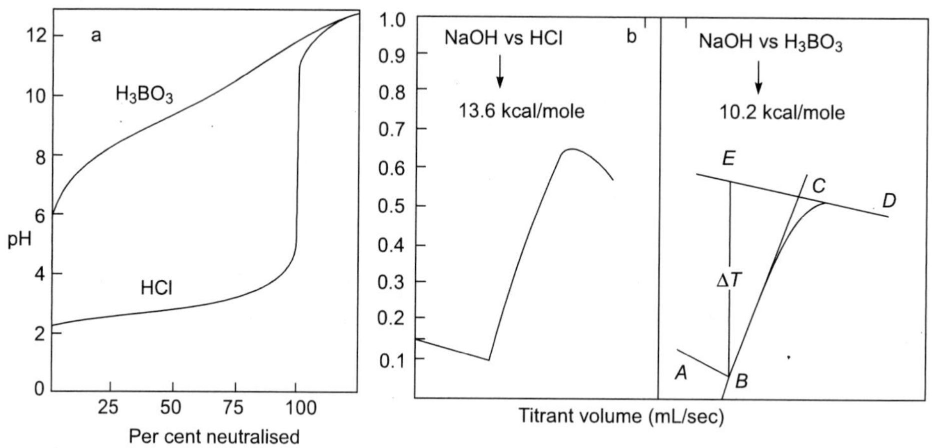

Fig. 4.37 (a) Potentiometric and (b) thermometric titration curves for HCl and boric acid with NaOH

Thermometric titrations possess extraordinary potentiality for determining heats of reaction in a dilute solution rapidly and conveniently so that ΔH values can be set virtually equal to the ideal thermodynamic parameters corresponding to infinite dilution.

Non-aqueous systems are well suited for thermometric titrations though heat of mixing of solvent and dilution pose a problem. The lower specific heat of many organic solvents introduces a favourable sensitivity factor. Under strictly anhydrous conditions, even diphenylamine, urea, acetamide and acetanilide are readily titrated with perchloric acid in glacial acetic acid. Lewis bases such as dioxane, morpholine, pyridine and tetrahydrofuran may be titrated with Lewis acid $SnCl_4$ in CCl_4, benzene and nitrobenzene. TT are very useful in titrating acetic anhydride in acetic acid, sulphuric acid acetylating bath. Acids by dilution with fuming acids are free anhydrides in fuming acids. Analysis of mixtures is possible when the two species have different equilibrium constants and heats of reaction with the titrant. For example, in the titration of mixture of calcium and magnesium with EDTA calcium ($K_f = 10^{11}$) reacts first and exothermally ($\Delta H = -5.7$ kcal/mole) and then magnesium ($K_f = 10^{9.1}$) reacts endothermally ($\Delta H = 5.5$ kcal/mole).

Precipitation and ion-recombination reactions also yield good results in TT. For example, halide reacts with silver or mercury (II), and cations such as Mn(II) with EDTA and oxalate. Titration of silver with halide can be carried out at elevated temperatures in molten state.

Thermometric titration can be recommended where the analyst wishes to simplify the conduction of a variety of titrations by using one sensor for all. For example, a laboratory might conduct routinely acid/base, redox, complexometric, sulphate and chloride titrations.

The titration environment is considered unsuitable for conventional titration sensors. For example, glass membrane pH electrodes must be kept adequately hydrated for proper operation. The use of such electrodes in substantially non-aqueous media as in the determination of trace acids in lipids and lubricating oils can lead to loss of performance as the membrane fouls and dehydrates, and/or if the reference junction is partly or completely blocked.

A thermometric titration methodology which cannot be emulated using other types of titration sensors will deliver superior or results otherwise unobtainable by other techniques. Examples are the determination of fluoride by titration with boric acid, the analysis of orthophosphate by titration with magnesium ions, and the direct titration of aluminium with fluoride ions.

Applications for thermometric titrimetry are drawn from the major groupings, namely:

(a) Acid/base (acidimetry and alkimetry): Determination of fully dissociated acids and bases, titration of weak acids, titration of acid mixtures, non-aqueous acid-base titrations

(b) Redox, titrations with permanganate, dichromate and thiosulphate, titrations with hypochlorite

(c) Precipitation, titrations with silver nitrate, titration of sulphate, titration of aluminium with fluoride and total orthophosphate

$$[Mg^{2+} + NH_4^+ + PO_4^{3-} \longleftrightarrow MgNH_4PO_4\downarrow]$$

(d) Complexometric: Direct EDTA titrations with metal ions are possible when reaction kinetics are fast, for example, copper, calcium and magnesium. However, with slower reaction kinetics such as those exhibited by cobalt and nickel, back-titrations are used.

EXERCISE

1. Describe DTA and DSC in brief.
2. What is the difference between DTA and DTG?
3. Compare DTA and DSC.
4. What is a thermocouple?

5. What is a transducer?
6. Calculate the molar change in enthalpy for melting of Indium metal at 157.6°C. The heat absorbed is 13.6 cal per gram.
7. All the thermal analysis instruments have features in common. Discuss these common features and the way in which the individual techniques differ from each other.
8. Estimate the purity of a material (may be made from the analysis of a DSC melting endotherm). Describe the procedure.
9. What type of standards are needed for calibrating heat and temperature of a DSC?
10. How can a DSC be used for measuring the specific heat capacity of solid samples? Is a particular standard required?
11. What are the essential components of the experimental setup for carrying out thermometric titration?
12. Explain why reaction cell should be adiabatic and continuous stirring required when the reaction is in progress.
13. What is the chief limitation of thermometric titrations?
14. Even after the equivalence point, an increase in temperature is observed. Why?
15. Explain how thermometric titrations are different from classical titrations. Discuss their advantages.
16. Why is it essential to calibrate the thermometric titration unit? How is it done?
17. What are the various heat terms involved in different stages of an enthalpogram? Derive an expression to determine ΔS for a reaction.
18. In a thermometric titration of acid A with a base B, the slopes of three regions were found to be 1.0×10^{-5}, 8.0×10^{-4}, and -0.5×10^{-5} °C/sec, respectively. The overall temperature change was 0-100°C and heat capacity of the cell was determined to be 1.00 Cal/°C. The titration rate of A was 6.0×10^{-8} moles/sec., and the titration of B into pure water gave a slope of 2.0×10^{-5} °C/sec. Calculate the heat of reaction.
19. Calculate molar enthalpy of decomposition of $CaCO_3$. If 10 mg of calcium carbonate is heated in ambient condition and 15.5 J of energy released.

    $CaCO_3(s) \rightarrow CaO(s) + CO_2(g)$ ....
20. Calculate molar enthalpy of decomposition of enthalpy of Indium metal. If 5 mg of Indium is heated in ambient condition and 15.5 J of energy released at 157.6°C.
21. Calculate the percentage of individual polymers in analysed polymer blend of PE and PV sample from the data below:

Compound	Heat of reaction observed (kJ)	Molar heat of fusion (kJ)
PE	28.5	60.5
VA	35.6	77.5

REFERENCES

1. Mackenzie, R. C. (1970). Differential Thermal Analysis. Academic Press,London.
2. Brown, M. E. (2001). Introduction to Thermal Analysis. Kluwer Academic Publisher, London.
3. Wandlandt, W. W. (1986). Thermal Analysis. Wiley, New York.
4. Garn, P. D. (1965). Thermoanalytical Methods of Investigations. Academic Press, New York.
5. Willard, H. H., Merrit Jr., L. L., Dean, J. A. and Settle Jr., F. A. (1986). Instrumental Method of Analysis. Wordsworth Publishing Company, USA.
6. Blazek, A. (1972). Thermal Analysis. Van Nostr and Reinhold, London.
7. Vassallo, D. A. and Hardsen, J. A. (1962). Precise phase transition measurements of organic materials by differential thermal analysis. *Anal. Chem.* 34, 132-135.
8. Mazieres, C. (1964). Micro and semi micro differential thermal analysis (μ.D.T.A.). *Anal. Chem.* 36(3), 602-605.
9. Gunzzle, H. and Williams, A. (2001). Handbook of Analytical Techniques. Wiley–VCH, Weinheim, Vol-2.
10. Ewing, G. W. (1990). Analytical Instrumentation Handbook. Marcel Dekker Inc., New York.
11. Meyer, R. A. (2000). Encyclopaedia of Analytical Chemistry. John Wiley & Sons Ltd., Vol 15.
12. Skoog, D. A., Holler, F. J. and Nieman, T. (1998). Principles of Instrumental Analysis. 5th Edition, New York.
13. Dean, J. A. (1995). The Analytical Chemistry Handbook. New York, McGraw Hill, Inc.
14. Endoh, K. and Suga, H. (1999). Phase diagram of salt-water system determined by TG-DTA. *Thermochim. Acta.* 327, 133-137.
15. a. Kissinger, H. E. (1978). Reaction kinetics in differential thermal analysis. *Anal. Chem.* 29 (157), 1702; b. Davies, P., Dollimore, D. and Heal, G. R. (1978). Polymorph transition kinetics by DTA. *J. Therm. Anal.* 13, 473-487.
16. a. Ozawa, T. (1965). A new method of analysing thermogravimetric data. *Bull. Chem. Soc. Jpn.* 38, 1881-1886; b. Ozawa, T. (2000). Thermal analysis—Review and prospect. *Thermochimica Acta.* 355, 35-42; c. Font, J., Muntasell, J. and Navarro, J. (1986). Influence of deconvolution on kinetic studies by DTA: Part II. Determination of the kinetic parameters. *Thermochim. Acta.* 99, 11-17.
17. Xiao, X., Sichen, D., Seetharaman, S. and Sohn, H. Y. (1997). Determination of kinetic parameters using differential thermal analysis—Application to the decomposition of $CaCO_3$. *Metall. Mater. Trans. B.* 28B, 1157-1164.
18. Raemy, A. and Lambelet, P. (1991). Thermal behaviour of foods. *Thermochimica Acta.* 193, 417-439.
19. Bhardwaj, I.S., Kumar, V. and Palanivelu, K. (1988). Thermal characterization of LDPE and LLDPE blends. *Thermochimica Acta*, 131, 241-246.
20. Müller, A.J. and Michell, R.M. (2016). Differential scanning calorimetry of polymers. In: *Polymer Morphology: Principles, Characterization, and Processing* (pp. 72-99). John Wiley and Sons New Jersey.

21. Lombardi, G. (1985). Thermal analysis in forensic science. *Thermochimica Acta*, 93, 313-315.
22. Popescu, C. and Gummer, C. (2016). DSC of human hair: a tool for claim support or incorrect data analysis? *International Journal of Cosmetic Science*, 38(5), 433-439.
23. Berneira, L. M., de Freitas, S. C., da Silva, C. C., de Mattos Machado, A., de Pereira, C. M. P. and Dos Santos, M. A. Z. (2019). Application of differential scanning calorimetry in the analysis of apprehended formulations of anabolic androgenic steroids. *Forensic Science International*, *296*, 15-21.
24. Quinn, J. R., Raymond, D. P. and Harwalkar, V. R. (1980). Differential scanning calorimetry of meat proteins as affected by processing treatment. *Journal of Food Science*, 45(5), 1146-1149.
25. Velazquez, G., Méndez-Montealvo, M. G., Welti-Chanes, J., Ramírez, J. A. and Martínez-Maldonado, M. A. (2021). Effect of high pressure processing and heat treatment on the gelation properties of blue crab meat proteins. *LWT*, 146, 111389.
26. Zhao, J. H., Liu, L. S. and Sablani, S. S. et al. (2020). Comparison of the thermal transitions of spray-dried and freeze-dried egg whites by differential scanning calorimetry. *Food Bioprocess Technol*, 13, 1329-1343.
27. Yi, F. and LaVan, D. A. (2019). Nanocalorimetry: Exploring materials faster and smaller, *Appl. Phys. Rev.* 6, 031302.
28. Yi, F., Gillen, G., Lawrence, J., Forbes, T. P., Staymates, M. and LaVan, D. A. (2020). Nanocalorimetry of explosives prepared by inkjet printing, *Thermochimica Acta*, 685, 178510.
29. Adamovsky, A., Minakov, A. A. and Schick, C. (2003). Scanning microcalorimetry at high cooling rate, *Thermochimica Acta*, 403, 55–63.
30. Kummamuru, R. K., Rama, L. D. L., Hu, L., Vaudin, M.D., Efremov, M. Y., Green, M.L.,. LaVan, D. A. and Allen, A. H. (2009). Measurement of heat capacity and enthalpy of formation of nickel silicide using nanocalorimetry, *Appl. Phys. Lett.*, 95, 181911.
31. Righetti, M. C., Marchese, P., Vannini, M., Celli, A., Lorenzetti, C., Cavallo, D. and Androsch, R. (2020). Polymorphism and multiple melting behaviour of bio-based poly(propylene 2,5-furandicarboxylate), *Biomacromolecules*, 21, 2622–2634.
32. Jeoung, S., Shin, S. and Choi, M. (2020). Copper-binding energetics of amicyanin in different folding states, *Metallomics*, 12(2), 273-279.
33. Lever, J. T. and Price, D. M. (1998). Using microthermal analysis to characterize the nanoworld, *American Laboratory*, 30, 15.
34. Yi, F., Stevanovic, A., Osborn, W. A., Kolmakov, A. and LaVan, D. A. (2017). Multi-environment nanocalorimeter with electrical contacts for use in the scanning electron microscope, *Mater. Horiz.*, 4(6), 1128-1134.
35. Jordan, J. (1963). Thermometric titrators. *J. Chem. Educ.* 40(1), A5.
36. Barthel, J. and Wachter, R. (1975). Thermometric Titrations. John Wiley, New York.
37. Vaughan, G. A. (1973). Thermometric and Enthalpimetric Titrimetry. Van Nostrand Reinhold Company, London.

5

Oxidation Induction Time/Temperature and Rarely Used Techniques

5.1 OXIDATION INDUCTION TIME/TEMPERATURE (OIT)

5.1.1 Introduction

The method is one of the modifications made for effective use of thermal analysis and consists of exposing the specimen to a gaseous oxidant (oxygen or air) at elevated temperature [1-2]. The schematic representation of technique and influential factors is shown in Fig. 5.1. Generally, two variables are commonly employed:

(a) The dynamic method, which yields a characteristic "oxidation induction temperature.

(b) The static method yields a characteristic "oxidation induction time".

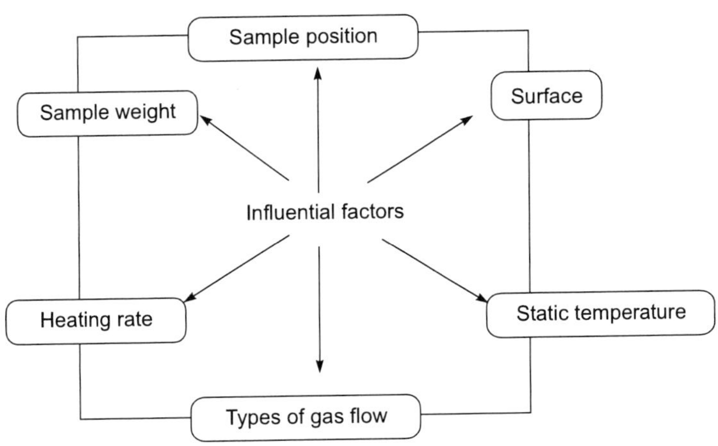

Fig. 5.1 Representation of various factors of OIT

In dynamic method, the specimen is heated in an oxygen-air atmosphere. A shift from the baseline that occurs at a characteristic temperature is a clear indication that oxidation reaction is exothermic. The method has advantages of being less sensitive because the temperature continues to rise. It cannot, therefore, be used to distinguish slight difference in stabilizer effectiveness in polymers. The dynamic method can be used for almost all plastics.

In static method, specimen is heated in an inert atmosphere to a specified temperature above the melting point. This temperature is then kept static. Once equilibrium has been attained, the atmosphere is switched from inert to oxidative (N_2 to air or O_2). The exothermic oxidative reaction usually occurs after a certain time interval. This method has advantages of revealing fine gradation in the degree of stabilization. Static OIT is more sensitive, but more time consuming and is normally used for polyolefin.

5.1.2 Methods for Estimation of Oxidation Induction Time (OIT)

The OIT can be estimated by two ways depending on the requirement: dynamic and static.

5.1.2.1 Dynamic Method

The specimen is heated continuously and the specimen chamber is purged from start with the oxidative gas. Figure 5.2 is typical experimental curve or a semi-crystalline thermoplastic. The first transition is melting and is followed at higher temperature by an exothermic baseline shift (peak) signalling the start of oxidation. The OIT temperature is usually taken to be extrapolated onset temperature, T_{eio}. This is obtained by extending the baseline beyond the start of the exothermic peak to intercept the extrapolated steepest linear slope of the peak. If the exothermic peak has a leading edge or onset temperature difficult to reproduce. It is preferable to use temperature, t_x^{dy}, which is change in heat flux from baseline by a view of x; frequently a value of 0.2 w/g is used.

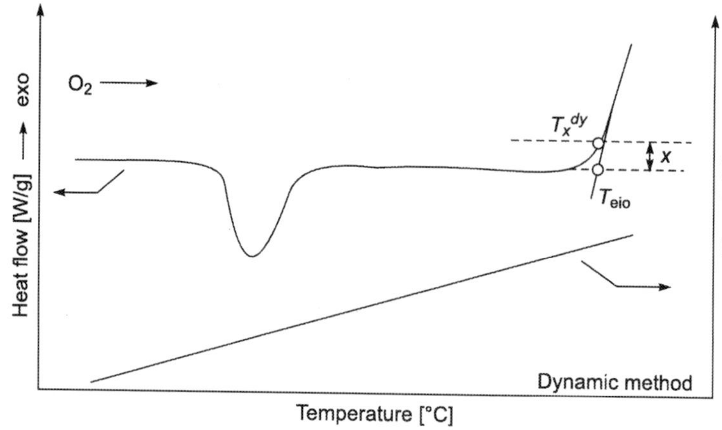

Fig. 5.2 Dynamic OIT curve of polymer typical sample (Adapted from Ref. 1)

5.1.2.2 *Static Method*

In static method, specimen is heated above its melting point where the temperature is held constant. The disturbance in heat flux due to the change in heating rate occurs. The evaluation made from the baseline takes the form of recording the OIT with time. A representative graph is shown in Fig. 5.3, the extrapolated onset time T_{eio} and a constant offset from the baseline, T_x^{stat}, are represented. Further, the T_x^{stat} is preferred where leading edge occurs or onset times, are difficult to determine. The T_{a2}, only shift from the baseline is disturbance in the heat flux core which occurs due to the reduction in heating rate from 10°C/min to isothermal condition.

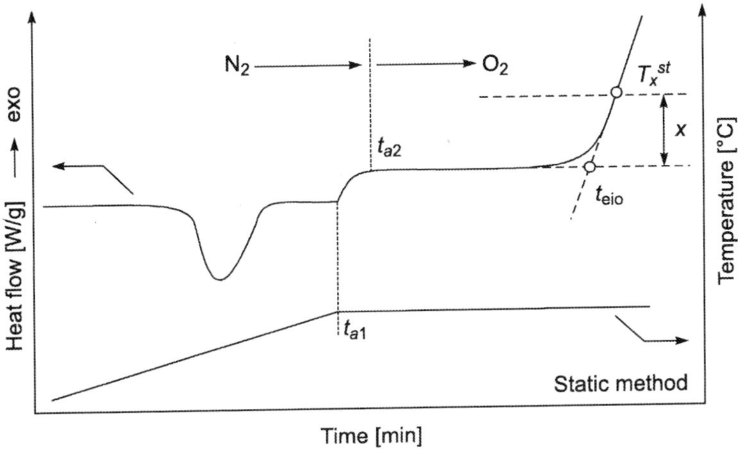

Fig. 5.3 Determination of static OIT (Adapted from Ref. 1)

5.1.3 Applications

OIT is also important to grade the antioxidant nature and aerial oxidation of different fuels and combustible compounds. Therefore, this techniques has been used to quantify the antioxidant compounds like green tea as well as to certify the oxidising conditions of fuels like diesel and petrol [3]. Currently, different oxidants are also added in diesel to modify the oxidation induction time in order to optimise its combustion behaviours. The oxidation induction time of biodiesel prepared from waste cooking oil containing green tea extracted antioxidant is shown in Fig. 5.4. The result reveal significant improvement in OIT value of biodiesel, i.e., 2.88 hr due to addition of green tea extracted antioxidant to 7.11 hr. [4]

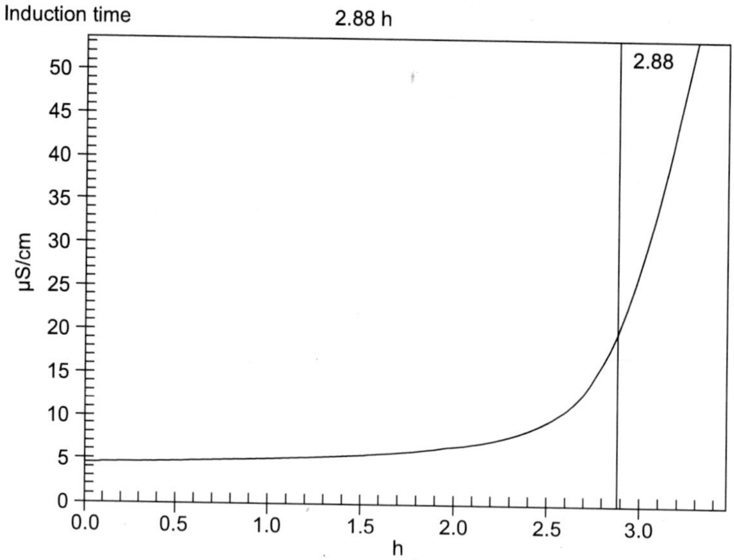

(a)

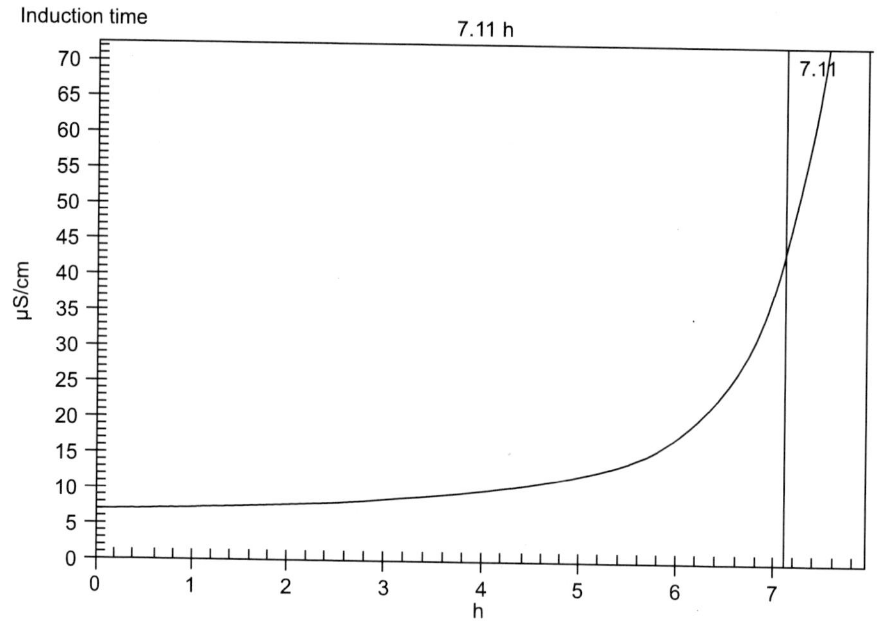

(b)

Fig. 5.4 Comparison of OIT of (a) virgin biodiesel and (b) biodiesel containing green tea extract [4]

OIT technique is also frequently used by food chemists for evaluating the oxidation of processed food as well as impact of additives on edible items like vegetable oils. In general, the edible oils are highly unsaturated and prone to lipid oxidation and additives are added to antioxidant behaviour. Some of the parameters for oxidative unstability are iodine value, peroxide value and free fatty acids. However, DSC based oxidative stability is more hoilistic and useful for grading the product and additive quality. Figure 5.5 shows DSC curve of vegetable oil at fixed temperature and air flow 60 mL/min, which reveals the transition temperature with exothermic peaks and the extrapolation of peaks at 131 min OIT value as an indicator of oxidative stability. [5]

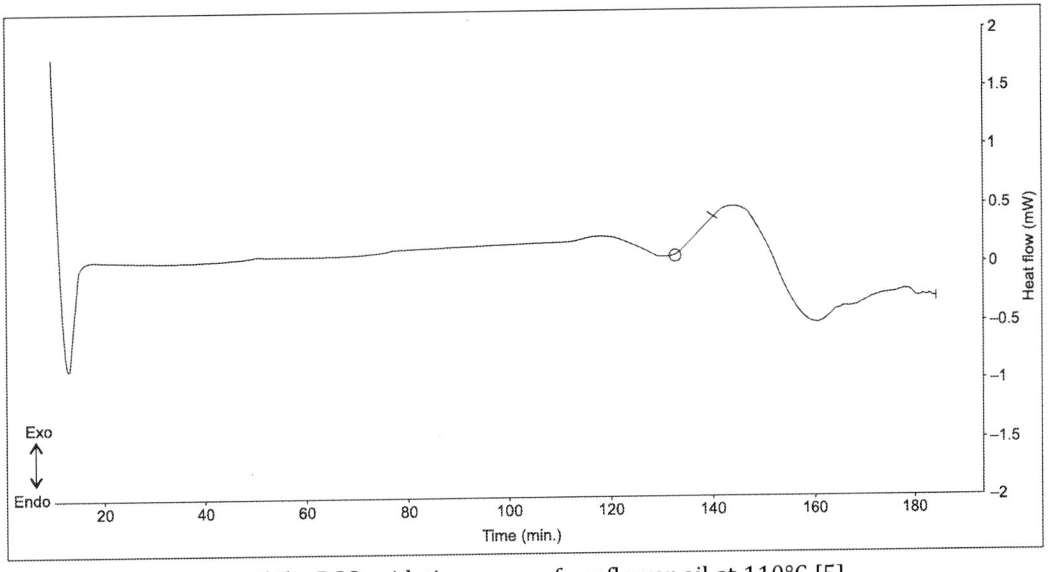

Fig. 5.5 DSC oxidation curve of sunflower oil at 110°C [5]

Different additives like avocado and butylated hydroxyanisole (BHA) are added in sunflower oil to improve its oxidative stability, the OIT trend of pristine and mixed sun flower is shown in Fig. 5.6.

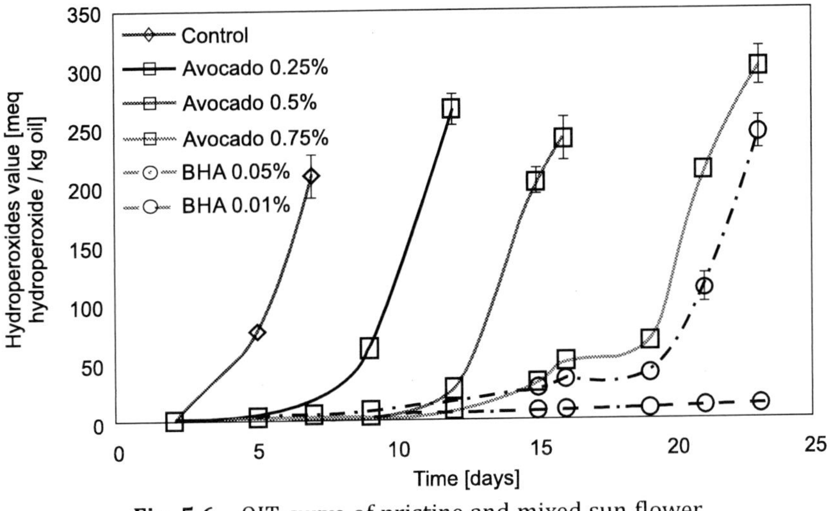

Fig. 5.6 OIT curve of pristine and mixed sun flower

Steps involved in measure of OIT is heating of oil in isothermal cóndition and observe the change of peaks in DSC curve, in generation induced oxidation is exothermic phenomenon and the first peak is monitored for different studied oil for OIT results. The observed impacts on OIT are used to compare the oxidative stability of sun flower oil processed by different process and additives. In processing of edible items like dairy product different active ingredients like herb extracts like vidarikand (*Pueraria tuberosa*), shatavari (*Asparagus racemosus*) and ashwagandha (*Withania somnifera*) extracts (aqueous and ethanolic) are added as antioxidant to improve the shelf life for extended active consumption. This method is very effective to estimate the active impact of additives on the life of oil or butter as well as antioxidants.

Some other areas of application of OIT are:

1. Differentiate fresh and processed juices
2. Comparative stability meat and poultry products.
3. Oxidative stability of rubber and plastic product under use and certify working time.
4. Effectiveness of food processing technology.
5. Evaluate antioxidant activities of active compounds.

Further, the results recorded during OIT study like induction time and peaks a linearity graph can be made to establish to arrange the effectiveness of antioxidants and determine the oxidation activation energy [6]. The use of thermal stabiliser in potential alternate fuel is an important practice to make the oil robust and suitable for better applications. The curve indicating the impact of phenolic, i.e., N,N′-hexane-

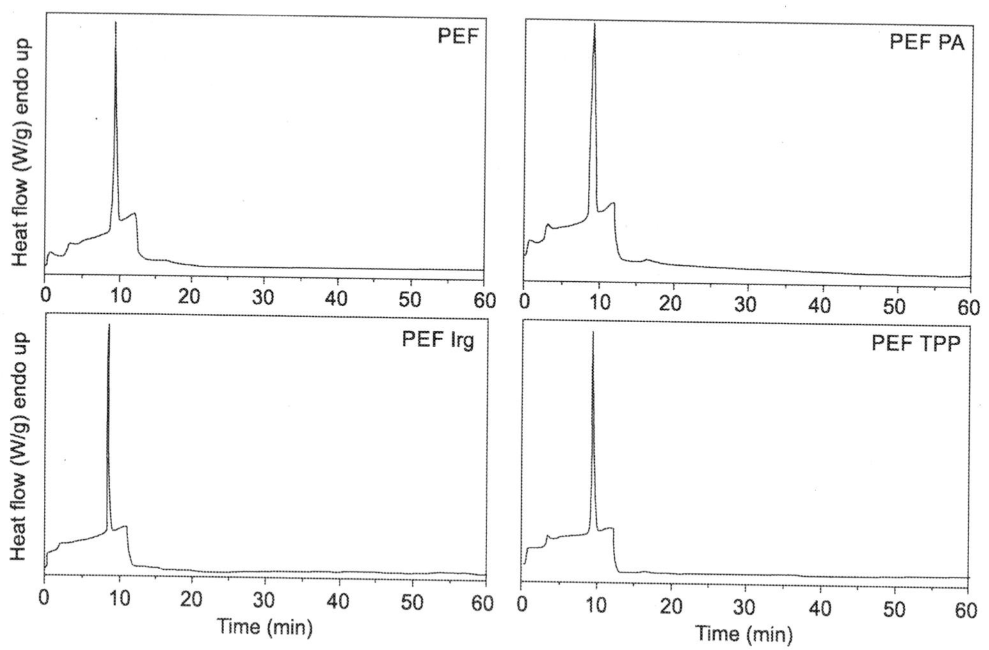

Fig. 5.7 OIT determination DSC thermograms of polyester and additives [7]

1,6-diylbis(3-(3,5-di-tert.-butyl-4-hydroxyphenylpropionamide) Irg and phosphorus based stabiliser (phosphoric acid (PA), triphenyl phosphate (TPP)) on poly(ethylene 2,5-furan dicarboxylate) (PEF) are shown in Fig. 5.7. The curve is indicating that the PEF and PEF-Irg better oxidation initiation than other two, i.e., PEF PA and PEF TPP did not. The OIT value for PEF is 1.45 min, PEF Irg OIT = 1.47, while OIT values for PEF PA and PEF TPP are 2.66 min and 1.74 min because delayed initiation of oxidation due to presence of additives [7].

5.1.4 Influence of Processing Parameters

The OIT can be proved to be useful for analyzing processing conditions of polymers, and how heating destroyed some of the stabilizers intended for processing. Therefore, it is necessary to check the process parts for uniform processing conditions as soon as they have been made. Figure 5.8 illustrates how PS is processed at different conditions a: virgin b: hard c: soft. Both soft and hard PS have lower OIT $T_{0.2\ w/g}$ than virgin materials. The recorded value can be used to analyse the characteristics in processing in terms of hard and soft machine setting.

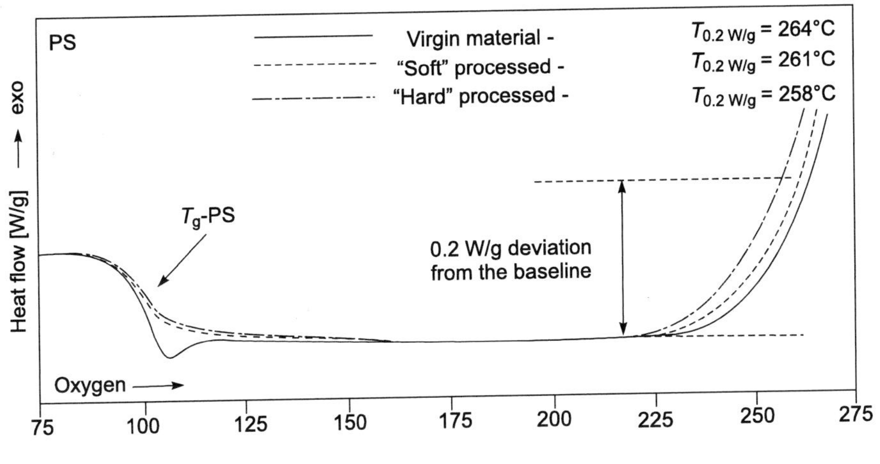

Fig. 5.8 The OIT curve of polymers processed in different conditions (Adapted from Ref. 1)

5.1.5 Media-related Stabilizer Degradation

The contact between the media and the plastic may impair the effectiveness of the stabilizer by dissolving same or all of it out. This effect was studied on specimen taken from different parts of the specimen, such as PP water pipe: inside, outside and centre. The static OIT curve shown in Fig. 5.9 reveals much lower extrapolated onset time for oxidation T_{eio} at the pipe's inner and outer surfaces relative to the centre. It is because stabilizer degraded from the outside. On the inside, the stabilizer may have been extracted by water passing through the pipe, whereas the outer surface of the pipe was exposed to oxidation attack by atmospheric oxygen.

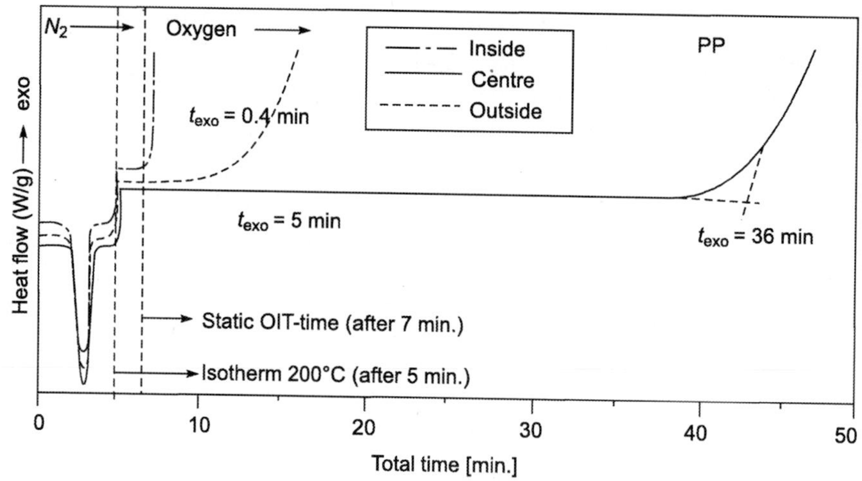

Fig. 5.9 Comparison of OIT curve from different parts of polypropylene pipe sample (Adapted from Ref. 1)

In another example, OIT measurements for the polyoxymethylene performed at 230°C, curve is shown in Fig. 5.10; for a different sample, there is an exothermic peak observed. This observation leads to the conclusion that pre-damaged chains oxidise in the same way as the homopolymers. It is also obvious that there is a tendency for a shift of the endothermic peak to shorter time after different exposures, and the peak is higher as a result of UV-stabiliser absence. The differentiation between these copolymer samples by OIT can be performed.

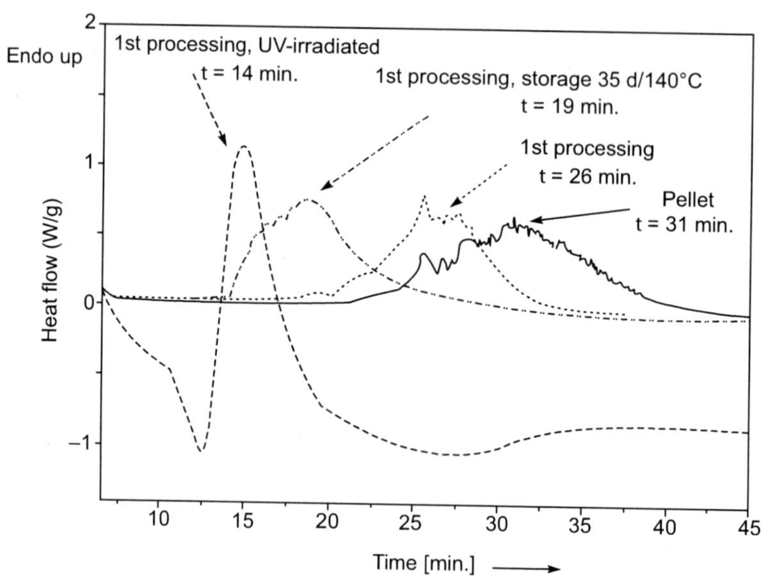

Fig. 5.10 OIT curve of polyoxymethylene (POM) in different ageing states

In general, this method is useful for polymers for evaluating, thermal stability, ageing, presence and effectiveness of stabilizers, and comparison of ageing resistance of different plastics.

5.1.6 Factors Influencing the Results

5.1.6.1 Specimen Preparation

Materials are oxidized by the reaction between oxygen and free radicals. These may be generated by heat, irradiation, mechanical shear and other external factors. Thus, specimen should be prepared as carefully as possible. For comparative measurements, the contact area of the specimen should be as large as possible and they should have same geometry. Generally, careful localized sampling is required. The use of specimen with the same surface is recommended. Figure 5.11 explains the advantages and disadvantages of various specimen geometries for OIT measurement.

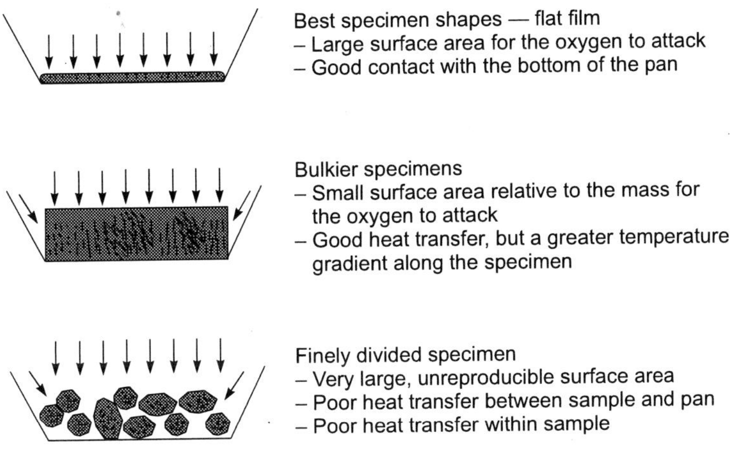

Fig. 5.11 Comparison of different specimen shapes (Adapted from Ref. 1)

5.1.6.2 Specimen Mass

The small specimen of 1-3 mg has a major difference in OIT value, that shows extremely poor reproducibility. The specimen mass above 5 mg gives some reliable result. Generally, specimen mass more than 12 mg is recommended.

5.1.6.3 Pan

All DSC sample pans are usually appropriate. The specimen is left upon to facilitate oxygen access. The specimen should be such that it makes good contact with the bottom of the sample.

5.2 RARELY USED TECHNIQUES

5.2.1 Introduction

These techniques are based on the monitoring of less obvious properties of a material and generally require more specialized equipment.

5.2.2 Thermosonimetry

This technique is also known as thermoacoustimetry and decrepitation techniques. In this technique, the sound wave emitted by a substance or by its reaction product is measured as a function of temperature during heating the materials under controlled temperature for heating or cooling [8]. The sound wave emitted from the thermal stress of the sample during the heating of materials because movement of dislocations, generation and propagation of cracks, nucleation phase and deposition of minerals. Thus, the monitoring of induced sound effect helps to understand the materials properties and phase change like solidification, glass transition as well as solid state interaction among the materials along with ultrasonic mobility in analysed samples. The formation of sound waves during heating is a natural phenomenon in terms of amplitude, intensity, and frequency distribution due to regular oscillation and burst of noise. Further, the heating induced mechanical vibrations are caused by several microscopic processes, i.e., motion of dislocations, chemical reactions, propagation of cracks, nucleation of the new phase, relaxation processes and discontinuous variations of physical parameters like thermal expansion, and compression. Although previously discussed method like DTA detects several above listed changes but the higher sensitivity of TS makes it more effective and informative [9]. The block diagram of thermosonimetry instrument is shown in Fig. 5.12.

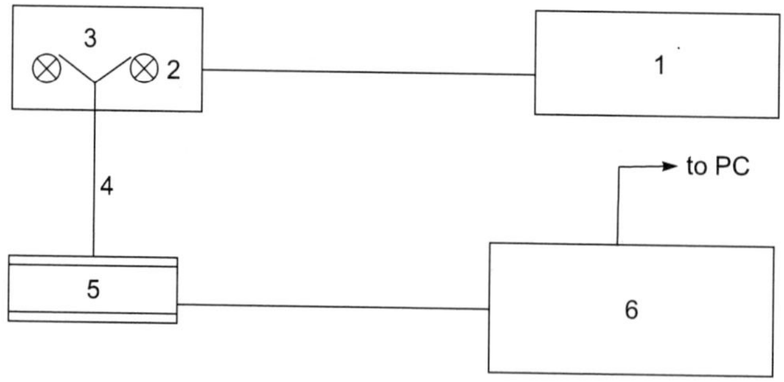

Fig. 5.12 Block diagram of thermosonimeter; 1. Temperature programmer, 2. Heating system, 3. Measuring chamber, 4. Stethoscope, 5. Transducer, 6. Acoustic emission analyser, (Adapted from Ref. 4)

In this analysis, the sample is placed in seismically mounted holder attached with heating system. Heating induced acoustic signals are measured in terms of intensity, amplitude and frequency by attached stethoscope. The basic principle of measurement is disturbance in diaphragm due to impact of sound wave, while the temperature is measured by attached thermocouple. Both the signals are recorded by computer through analyzer for necessary calculation and output. The commercially available TS instruments are developed by Smith and Rach for recording the rates of decrepitation of crystalline mineral and rocks like garnets and mica. The technique has features to analyze inorganic salts and minerals in terms of change temperature induced frequency of sound waves along with intensity.

Applications

The technique is used to study the heat induced mechanical transition associated with physical and chemical change such as lattice change, dehydration, decomposition, phase change, glass transition temperature, softening and melting points. [10] This technique is used to characterize sound related applications such as frost and firing of building materials like ceramic, glass and woods. The lattice change in feldspar is investigated by this technique and graph is shown in Fig. 5.13, the curve is indicating different transitions due to heat induced lattice change and acoustic effect between 270 and 343°C [11].

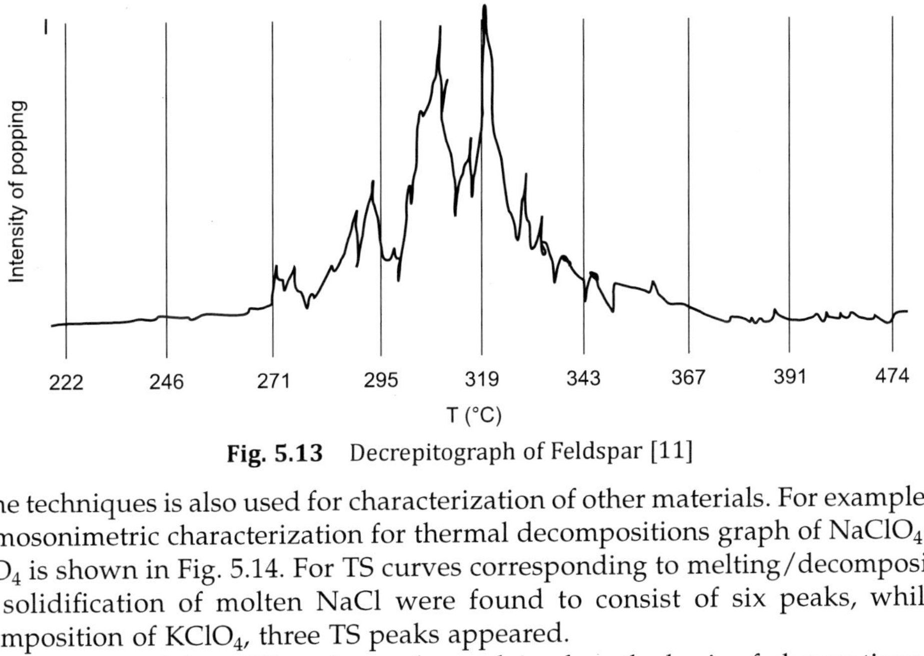

Fig. 5.13 Decrepitograph of Feldspar [11]

The techniques is also used for characterization of other materials. For example, the thermosonimetric characterization for thermal decompositions graph of $NaClO_4$ and $KClO_4$ is shown in Fig. 5.14. For TS curves corresponding to melting/decomposition and solidification of molten NaCl were found to consist of six peaks, while in decomposition of $KClO_4$, three TS peaks appeared.

The presence of these TS peaks can be explained on the basis of observations that the melting of the particles, evolution of bubbles of different sizes, formation of solid products of varied morphologies, vigorous vibration of these solids, and precipitation of NaCl or KCl. The acoustic emission is accompanied by solid-solid transition (polymorphic phase change such as in silver iodide and silver chalcogenides. Figure 5.10 explains the utility TS in comparison of DSC.

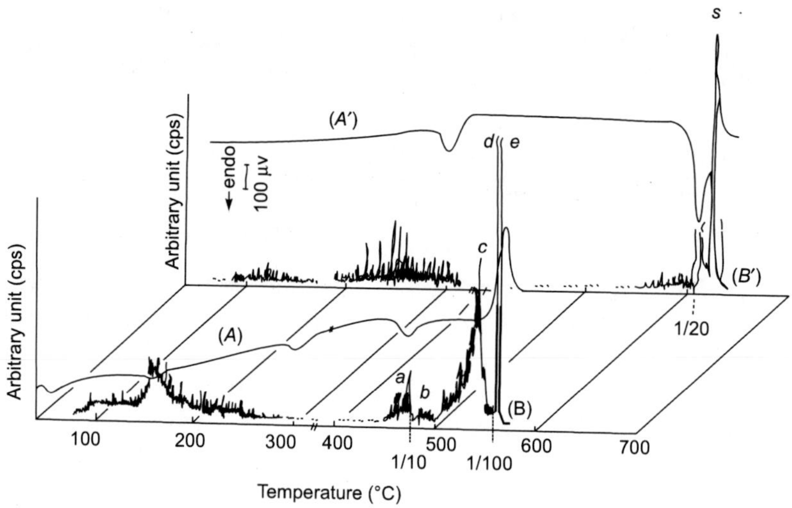

Fig. 5.14 Simultaneous TS-DTA curves of decompositions of $NaClO_4$ and $KClO_4$. (A) and (B), DTA and TS curve of $NaClO_4$; (A') and (B'), DTA and TS curve of $KClO_4$. (Graph adapted from Ref. 10).

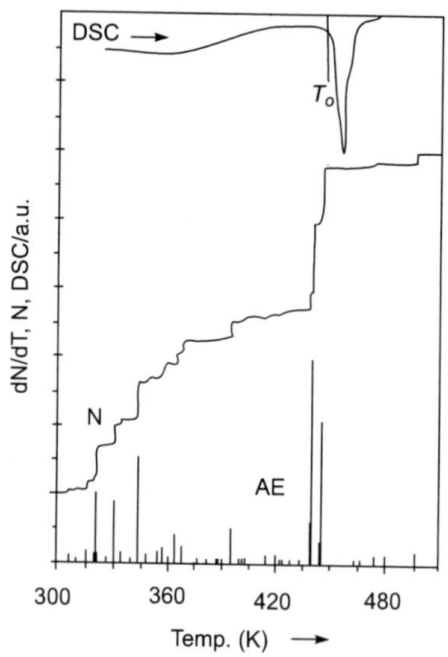

Fig. 5.15 TS–DSC curve of Silver sulphide (N acoustic event). (Graph adapted from Ref. 8)

The technique has wider utility in characterization of different acoustic materials like roof plates, false ceiling, wood and cementous materials after due calibration using standard reference materials. Generally, a machined aluminium rod with defined dimension, i.e., 6.03 cm length and 5 cm diameter was used for calibration in this techniques.

5.2.3 Emanation Thermal Analysis

Emanation thermal analysis abbreviated as ETA, is based on measurement of release of inert (usually radioactive like neon, argon, krypton, xenon, radon)) gas from a substance (or its reaction product) generates as a result of programmed temperature heating. The release of these gases is the indicator for defect states of solid, their changes occurring during heat treatment of the samples. These gases do not react in materials but released in uncontrolled manner through diffusion but depend on surface structure and porosity. Although, the release of these gases changes some mass loss and energy in TG and DTA, however, the use of ETA is a better means to understand the evolution of these gases due to smaller change in mass and energy causes very minor change. Thus, ETA in combination of TG and EGA provides the information about the microstructure based chemical change of the sample with temperature. The method has been also used for monitoring of amorphous and crystalline nature of the samples. [12]. Emanation thermal analysis consists in the measurement of radon release rate from the samples previously labelled and general scheme may be like Eq. (1)

$$^{228}\text{Th} \xrightarrow[T_{1/2} = 1.9\text{y}]{\alpha} {}^{224}\text{Ra} \xrightarrow[T_{1/2} = 3.8\text{d}]{\alpha} {}^{220}\text{Rn} \xrightarrow[T_{1/2} = 55\text{s}]{\alpha} \quad (1)$$

which makes it possible to investigate even rapid changes in the solids and on their surface. [13] The theoretical background for the use of inert gas diffusion as a probe of microstructure changes in solids. The high sensitivity of the emanation thermal analysis (ETA) in the study of solid state processes especially interactions taking place on surfaces and in the near surface layers can be studied. The sensitivity of the method increases towards bulk processes with rise in temperature.

Instrument

A general scheme for ETA instrument is shown in Fig. 5.16 [12].

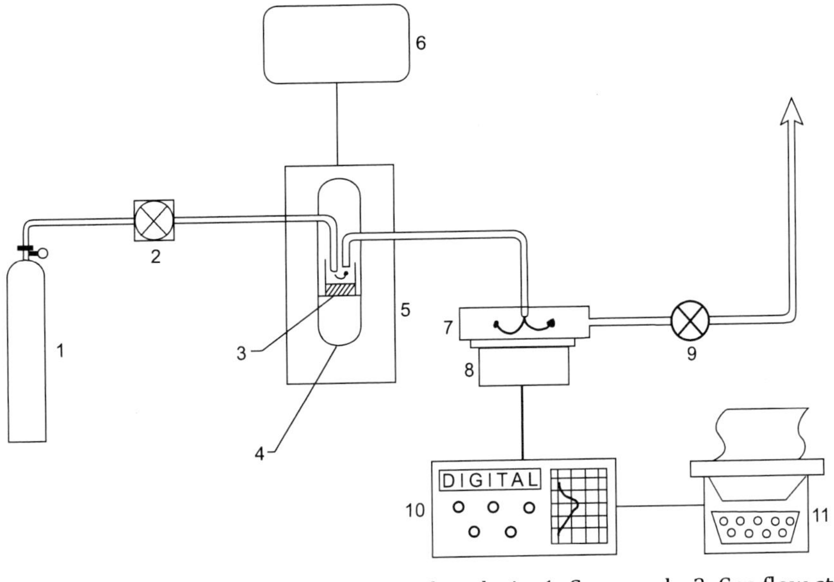

Fig. 5.16 Block diagram of emanation thermal analysis: 1. Gas supply, 2. Gas flow stabilizer and flow rate meter, 3. Labelled sample, 4. Sample holder, 5. Furnace, 6. Temperature controller, 7. Measuring chamber, 8. Radioactive detector, 9. Flow rate meter, 10. Counts meters and 11. Data processor and printer

The basic components are listed in figure caption and their objective is to heat the sample and detect the released radioactive gases with required precision and accuracy. The sensitivity of the method depends on the use of various detectors like scintillator counter, ionization chamber, semiconductor detectors, and G.M. counter. In a typical experiment, the carrier gas is passed through system along with heating and the carrier gas carries the released gas to detector for recognition.

Applications

In general, the shape of ETA curve depends on a number of diffusion paths serving for the gas release. The theoretical considerations summarized above indicate that ETA can be used in the investigation of processes taking place in solids or on their surfaces. Any process in a solid or its phase boundaries leading to changes in either the surface and/or changes in the inert gas diffusion becomes observable from ETA measurements. Different mechanisms of diffusion in the disordered heterogeneous solid were reported as models, namely:

(a) diffusion in two (or more) independent paths (parallel diffusion);

(b) diffusion permitting the reversible exchange of the radon atoms between paths (dissociative diffusion);

(c) diffusion in a solid where the creation of new diffusion paths and/or the annealing of the existing paths is supposed (dissociative diffusion in dynamic regimes);

(d) diffusion in a solid where solid state transitions take place during heating (reactive diffusion); and

(e) the consecutive inert gas diffusion.

This technique is used to detect stability of sample along with quality and quantity of released gases. Thus, like other techniques it is used to study thermally induced solid state process occurring in a solid along their kinetics and thermodynamic parameters. The presence of defects and their movements, channelling effect and annealing effect are also studied by this methods.

ETA method enables to identify the thermally induced different steps in ceramic, semiconductor and their constituents. The heating of kaolinite produces pre-dehydroxylation, dehydroxylation, evolution of structural disorder in metakaolinite, segregation of amorphous silica and formation of precursors of high-temperature phases, crystallization and sintering of both mullite and cristobalite. The results of both samples are quite different and demonstrate the role of the structural disorder of the mineral in the manifestation of these structural changes.

The ETA curve of kaolinite sample is shown in Fig. 5.17.

The indicated peaks in *b* curve are due to *abc* peak of the radon release rate, which reflects the release of absorbed water; a decrease (**gh**) was expected on the ETA curve, an increase of the inert gas (radon) release rate with three bursts (g_1, g_2, and g_3) was observed. On further heating, the increased opened microporosity of the kaolinite collapsed. This process is reflected by the decrease after points, of the radon release rate. Thus, the information demonstrates that the ETA curves can be used as fingerprints of clay samples of different crystallinity and morphology. The information obtained by ETA is usually compared with results DTA, TG and other methods. In contrast to DTA and TG, ETA made it possible to investigate processes that are not accompanied by thermal effects or mass changes, such as of sintering of powdered or gelatinous samples, which has been so far impossible to study by means of common dilatometry.

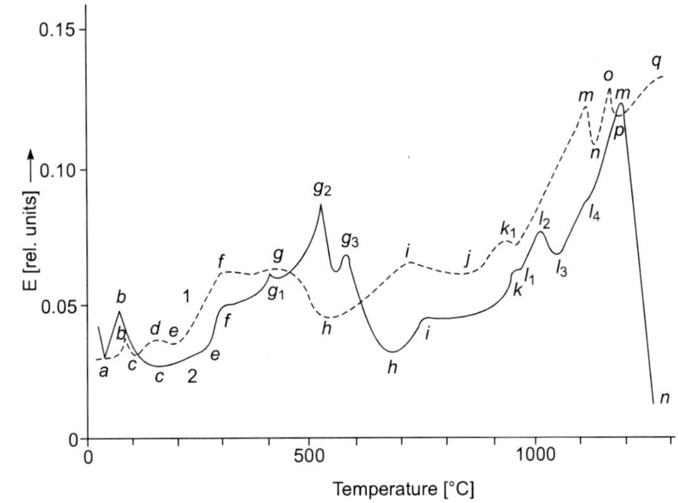

Fig. 5.17 ETA of two kaolinite samples (A) curve 1 and (b) curve 2. (Ref. 14)

The techniques is also used for composite materials and study the working of different materials such as building during its working life. The ETA study for release of radon exhalation or emanation in mortars containing 30 to 90% bauxite has been reported. The results indicated the release profile of radon depends on the porosity and surface area of mortars. The higher porosity yields higher cumulative mesopore volumes which resulted in higher emanation rates but has specific low correlation with radon emanation [15]. In the same line the technique is used to monitor the release various radioactive gases from other sample building materials like cement, rock and soils.

5.2.4 Thermoparticulate Analysis

It is a thermoanalytical technique [16], in which the release of particulate matter or condensation nuclei from a substance (and/or its reaction product(s)) is measured as a function of temperature whilst the substance is subjected to a controlled temperature program. The method can be used to measure the formation of particulate matter after degradation of polymers and other organic compounds as well as initiation of a decomposition process. The extreme sensitivity of instrumentation explored this technique to screen the organic compound as strong source for emission of the particulate matter in the surroundings. A block diagram of this technique with labelled components is shown in Fig. 5.18. In principle of this technique the compounds are passed through a pyrolyzer and oxidized to produce volatile material. The produced volatile matter is further hydrolyzed and precipitated as in case of nitrogen dioxide into ammonium nitrate. The produced ammonium nitrate was further detected by condensation nuclei detector.

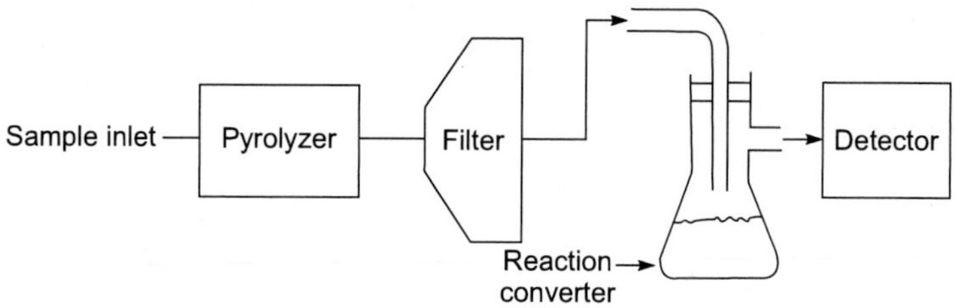

Fig. 5.18 Block diagram of thermoparticulate analyzer

The techniques is useful for estimating organic particulate after thermal decompositions of polymers, metal acetylacetonates, arenesulfonic acids, blocked isocyanates, metal alkyldithiocarbamate and diazonium compounds. This extremely sensitive technique makes possible the detection of particulates emitted from the diazonium compounds during thermal decomposition. The submicron particulates derived in this fashion are readily detectable by their influence on the output current of an ion chamber detector or by their effect on the cloud chamber of a condensation nuclei monitor. A representative TPA curve of p-chlorobenzenediazonium hexafluorophosphate is shown in Fig. 5.19. The curve confirms the strong particulate emission behaviour of compound at 200°C. The basis to produce particulates, the diazonium compounds can be placed in specified groups reflecting their chemical composition, molecular size, and degree of substitution. Vapour phase association of the molecular fragments formed during thermal decomposition might produce the required detectible sized particulate.

This technique is further used to study the condensation process occurring during aerosol formation as well as the nature of emitted particulate particles after integrating gas chromatograph and mass spectrometer. The nature of the emitted of organo-particulates from the blocked isocyanate compounds were carried by Kaczmarek on vapour effluents at thermo-particulation. Initially, the pyrolyzed particulates were collected on a modified divinylbenzene absorbent kept between two filter disks and then analyzed. The technique is useful for estimation of particulate formation ability of coating, tape and other structural parts [17].

5.2.5 Thermophoresis

This technique is also called thermodiffusion, thermomigration or Soret effect, which is based on the effect of a temperature gradient on the non-equilibrium mobility flow of multicomponent (or isotopic) mixtures of mobile particles from hotter region to colder region [18]. Thermal diffusion is regarded "positive" when molecules move from a hot to cold region and "negative" when the reverse is true. Typically, the

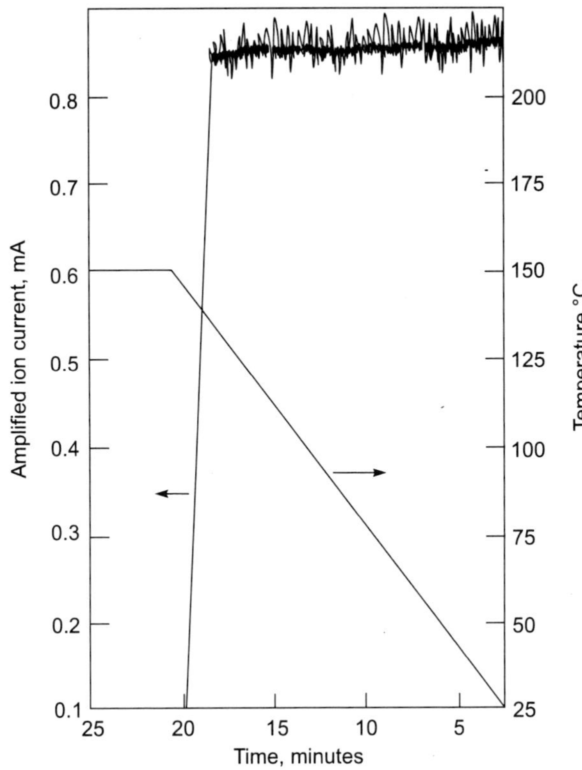

Fig. 5.19 Organoparticulation pattern for p-chloro benzenediazonium hexafluorophosphate

heavier/larger species in a mixture exhibit positive thermophoretic behaviour while the lighter/smaller species exhibit negative behaviour. This effect is relevant to planetary differentiation and is exploited in the manufacturing of optical fibre and uranium enrichment as well as a transport mechanism in fouling of heat exchangers [19]. The quantification of thermophoresis can be made by the particle thermodiffusion coefficient (D_T) and Soret coefficient (ST), which is related by Eq. (1).

$$ST = DT/D \tag{1}$$

It is the ratio of the thermodiffusion coefficient (DT) to the diffusion coefficient D, It effectively indicates the ability of the particle to move in response to a temperature gradient. If D_T is positive, the particle moves towards cooler regions and is termed thermophobic, while if D_T is negative, the particle moves towards warmer regions and is referred to as thermophilic. The factors influencing the sign and magnitude of D_T are the mass and moment of inertia of the particles, length of chain in solution of polymer as well as surfactants. For measuring the DT, the local heating was initiated by laser beam with specified temperature difference gradient to induce the temperature induced molecular flow. A typical block diagram of MST instrument is shown in Fig. 5.20.

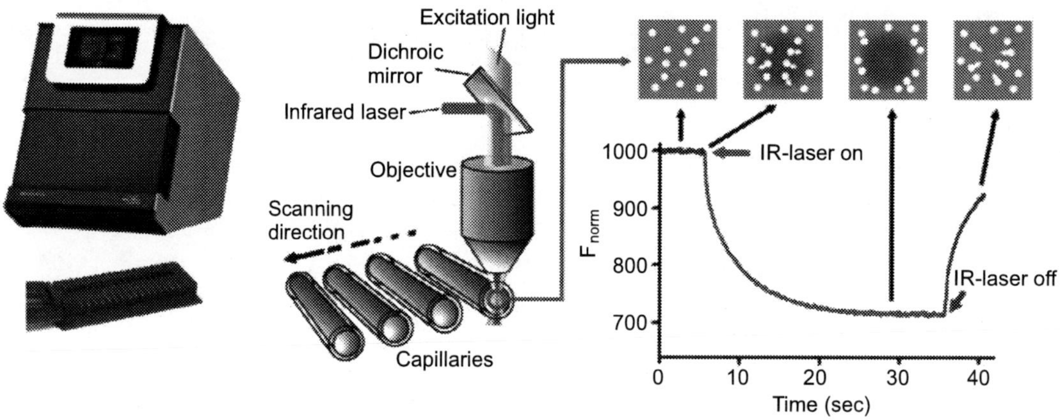

Fig. 5.20 Block diagram of microscale thermophoresis [20]

Thermophoretic effect is more substantial in higher temperature range like the radiant section of a boiler. It is also applicable during the initial stages of deposit formation, i.e., the surface temperature of boiler tubes is around 427 °C as well as the flame temperature is greater than 1427 °C.

Thermophoresis has a number of practical applications in microfluidic technology, microreactors, sensors, tuning of colloidal structure, pumps, separators and mixer manipulation. Impurity ions may move from the cold side of a semiconductor wafer towards the hot side, since the higher temperature makes the transition structure required for atomic jumps more achievable. The diffusive flux may occur in either direction (either up or down the temperature gradient), dependent on the materials involved. Thermophoretic force has been used in commercial precipitators for applications similar to electrostatic precipitators. It is exploited in the manufacturing of optical fibre in vapour deposition processes. The thermal precipitator is widely applied in many aerosol related fields such as dust sampling or even collection of nanoparticles. Thermal precipitation also occurs commonly in many industrial processes. For example, fly ash from coal combustion or incineration can deposit in the heat exchanger. In some cases, such a deposition is undesirable, as the built-up may cause scaling and low energy recovery.

The thermophoretic force acts on a particular particle and the resultant thermophoretic velocity in smaller region is defined as a microscale thermophoresis. It is used to monitor the movements and interaction in different molecules, ions and polymers. The study of interaction also helps to determine the other properties like fluorescent, surface plasma resonance and isothermal calorimetry under the microscopic temperature gradient [21]. This information is further explored in different applications as shown in Fig. 5.21.

Furthermore, the temperature induced mobility has great potential for use in separation of nanoparticles in binary mixture, colloids solution and microfluidic devices. This technique has been also used for advanced biosensing and microsomal ribonucleic acid. Franz (et al.) have used this technique in sensing of inorganic phosphate, an important tool for environment, food quality control and medical diagnosis. The basic principle is the combination of advantages of biological inorganic phosphorous – receptor based on the bacterial phosphate binding protein to the principle of thermophoresis such as the diffusional motion of particles against the temperature gradient to detect sub nano molar range with effective sensing parameters [22].

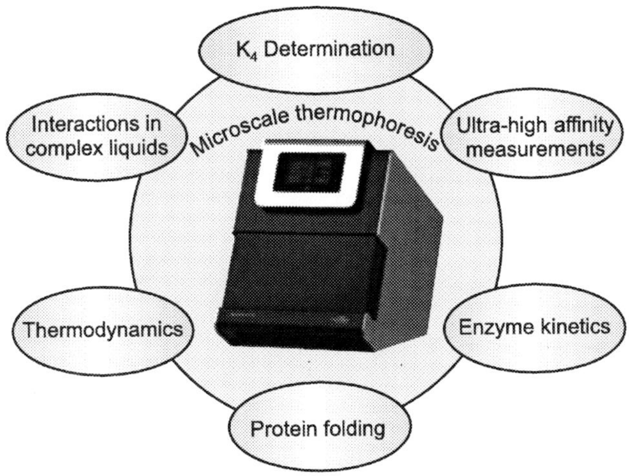

Fig. 5.21 Representative applications of microscale thermophoresis [20]

5.2.6 Thermal Electrical Analysis (TEA)

The technique deals with the measurement of electrical properties as a function of temperature such as dc , ac conductance, resistance, impedance, capacitance and dielectric properties [23]. The measurement may be made on a specific materials in the presence or absence of electric field to standardize the working performance of electrical and electronic materials. The basic principles of this technique are based on thermoelectric effect like "Seebeck effect, Peltier effect and Thomson effect". In this principle the induced heating of a material generates induced potentials and current for different applications like heat generators and temperature monitoring. Currently, the structural and compositional variability has strong impacts on thermal induced electrical behaviour in terms of efficiency and responsiveness. Hence, this technique is also used in characterization of materials with objective to prepare better thermoelectric devices like computer cooler, thermoelectric generator, thermometers, space probe and recharging devices. Another property is electrical insulation, also called dielectric analysis. This property is important for evaluating the dielectric strength of polymer and composites to use as electrical insulating applications. The commercial technique available on the basis of above principle is called dielectric thermal analysis (DETA) of dielectric analysis (DEA), which is manufactured as per ASTM E 2038 or E 2039.

Instrument

In the principle during typical DEA testing, the sample is kept in contact of two electrodes referred to as dielectric sensor. The block diagram of a representative DEA instrument is shown in Fig. 5.22.

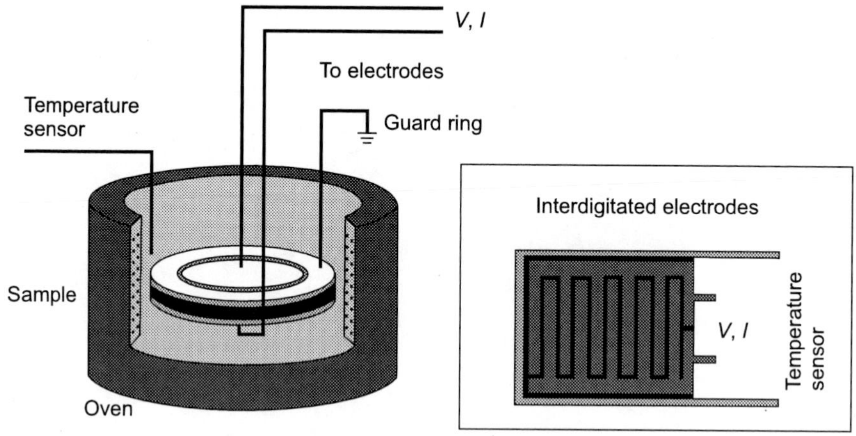

Fig. 5.22 Schematic of thermal dielectric analyzer along with a temperature sensor [24]

Generally, during measurement, the sample is presented as a thin film (~1 or 2 mm thick) and kept between two parallel plates to work as simple electrical capacitor. Further, a grounded electrode as a guard ring is incorporated in order to improve performance from reducing stray electric fields. The temperature of specimen is also monitored with the help of a thermocouple after placing in contact for precise measurements. However, for heating standard oven are used in desired temperature range. Further, an oscillating AC electrical field is applied to a testing sample and induced material's capacitance and conductance are monitored as a function of time and temperature. In general, a part of the electrical charge stored in the material as capacitance while the rest is dissipated through the material sample as conductance. This ratio of capacitance and conductance is a significant property of the materials for its future applications.

The dielectric analysis measures change in properties of the sample as it is subjected to a periodic electric field and in dielectric thermal analysis called DETA sample is subjected to a program temperature heating. Some times technique coupled with other techniques such as TG and DTA, which are already discussed in previous chapters.

Further, during performing an experiment the calibration of instruments and process are also performed. In calibration of temperature was performed by previously discussed methods, while in process a suitable standard with known properties are recorded. However, some other parameters like position of thermocouple, power loss and heat loss are calibrated mostly by engineers with standard protocol developed by manufacture using different models like reduced-order model and FE Model [25].

Applications

The application of technique is found of great use in measuring resistivity of materials such as polymer and trend of variation with temperature. However, the conductance also changes with materials like curing, ionization and cross-linking, etc. Thus, the method has been immensely used in materials characterization like structural transitions, crosslinking, cure, aging and decomposition. Further, the formation of conducting carbon chain by cross-linking or cyclization process can be easily observed beautifully with this technique. One more important application of the technique to study the appearance liquid phase during heating can be detected by the technique. It is also possible to study the chemisorption or formation of defects.

A comparison of DEA and rheometer data during the cure of an epoxy-graphite composite system is shown in Fig. 5.23, which indicates that for the first 150 min the mechanical viscosity and ion viscosity curves nearly superimpose, demonstrating the clear correlation between the two properties. However, at the 175°C final hold, the epoxy resin goes through gelation and the mechanical viscosity can no longer be measured. The ion viscosity signal continues to follow the entire reaction, even as the material cures into a rigid glass.

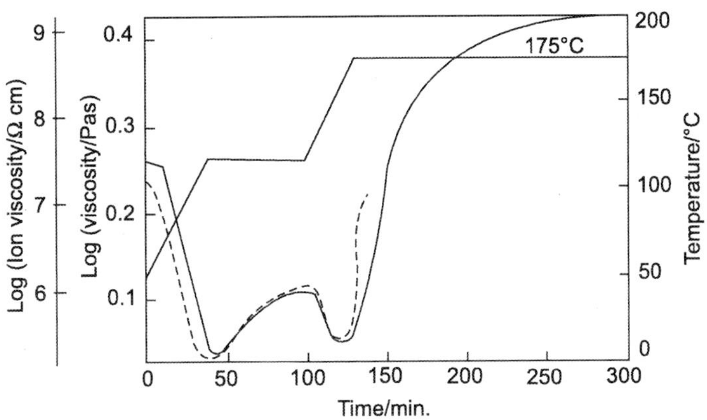

Fig. 5.23 The DEA – Rheometer result of epoxy. [26]

This techniques is used to check the effect of filler and temperature on the dielectric behaviour of polymer and polymer composites. This method is also used to analyze the biological sample like nail, hair and skin. Generally, the tissue of human varies in composition like keratin in nail and hair, hence depending on presence of keratin the dielectric polymer of these human exoskeleton changes and thus the techniques is also found suitable for measurement of the tissue properties. For example, Olszewski and Marzec monitored the dielectric parameters of the human fingernail to analyse the keratin degradation with temperatures [27]. The dielectric loss of wet and dry fingernails is shown in Fig. 5.24.

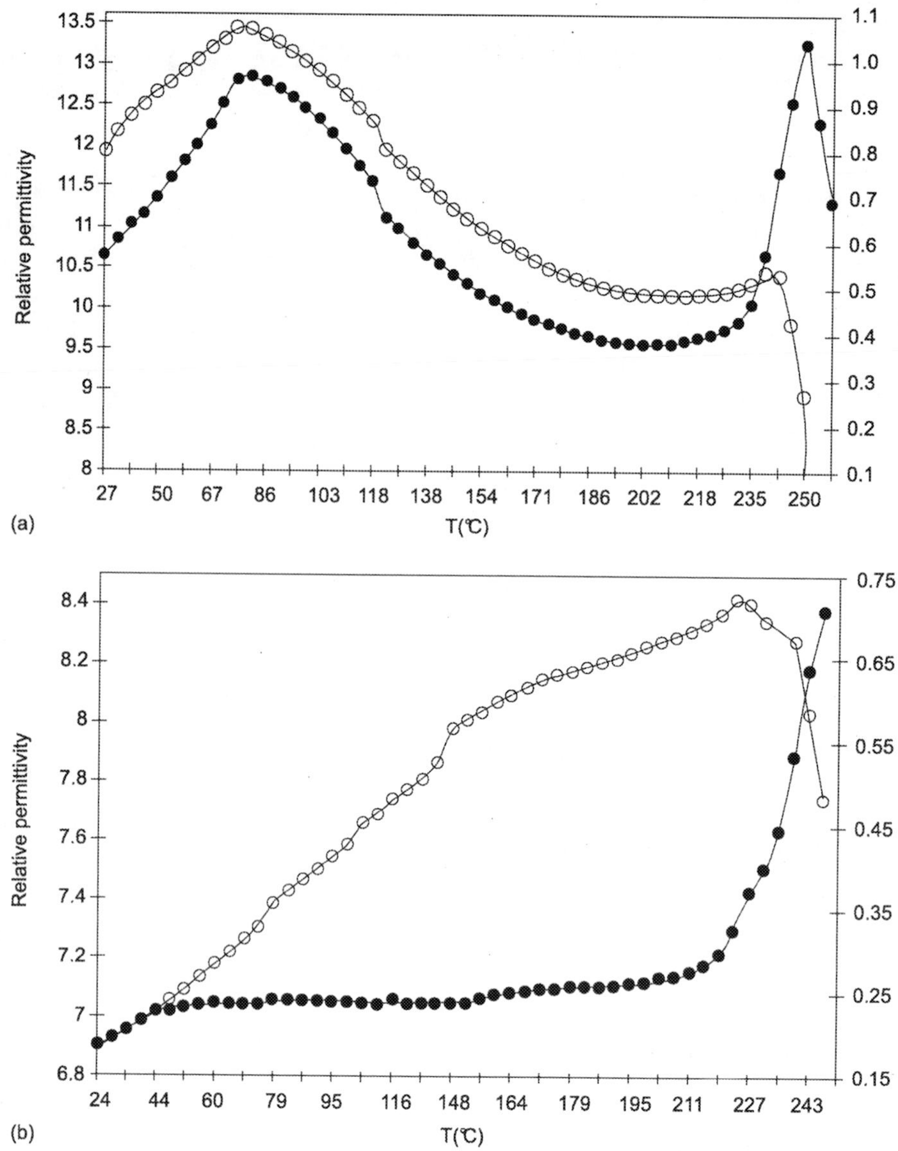

Fig. 5.24 Trend of variation of dielectric loss and relative permittivity of (a) dry and (b) wet nail with temperature [27]

The graph indicates significant effect of dry and wet conditions on degradation of keratine, however, the transition around 200°C indicates the degradation of keratin in both dry and wet nails. However, the loss at 800°C indicates the loss of moisture in wet nail but later change in indicating the conducting due to degradation of keratin.

The technique is also used to measure the composition of a polymer composite, since dielectric varies on the basis of polarity present in composites. This polarity is directly related to added fillers in an example graphite is well known conducting fillers and used in different polymers during processing and formation of composites. The presence of graphite controls the polarity and also yields modified frequency dependent dielectric properties. The change in amplitude and shift in phase in relation to the mobility of the ions and alignment of the dipoles. Thus, dipoles in the material will attempt to align with the electric field and ions (present as impurities) will move towards the electrode of opposite polarity. The dielectric properties of permittivity (ε′) and loss factor (ε″) are then calculated from this measured amplitude and phase change A frequency dependent trends in dielectric constant for pure PNIPAM and PNIPAM and 8% graphite composite is depicted in Fig. 5.25 (a,b).

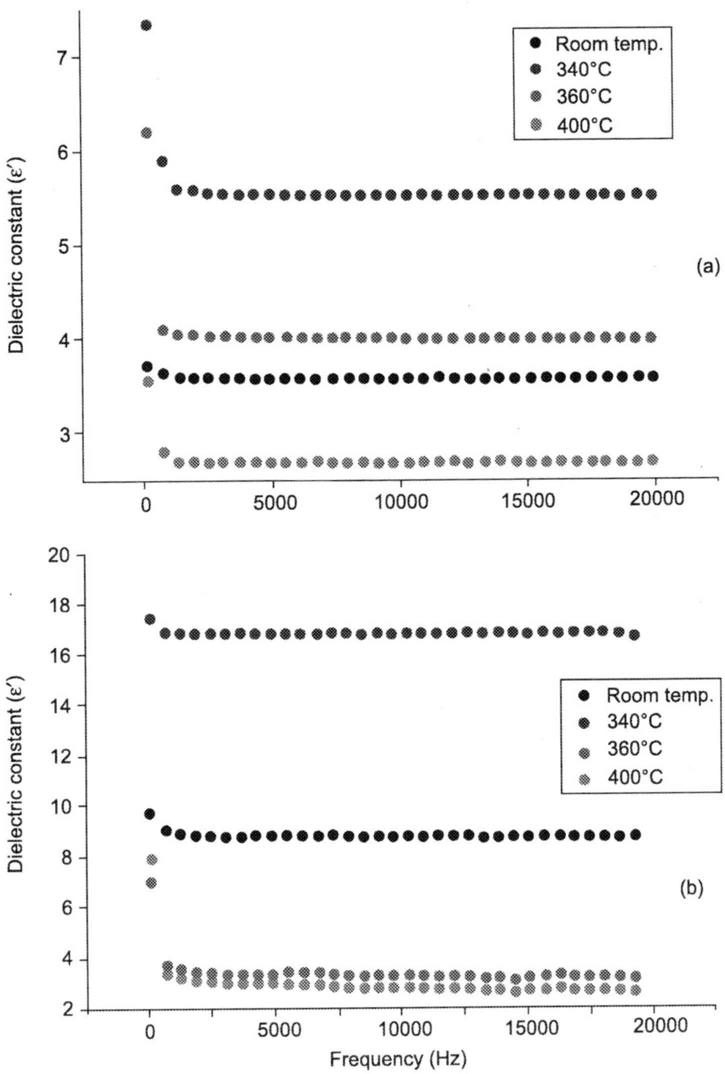

Fig. 5.25 Trend in change of frequency dependent dielectric constants (a) Pure PNIPAM and (b) Composite of graphite and PNIPAM [28]

The graph indicates the significant change in dielectric constant of PNIPAM after due to the presence of graphite at room temperature itself, i.e., 3.60 for pristine PNIPAM and 8.95 PNIPAM and graphite composite. Further, in general, the value of dielectric constant is changing at lower frequency but becomes constant at higher frequency. The values of dielectric constant also change with different temperatures due to degradation of polymer with temperature and formation different matrix. Thus, the monitoring of dielectric properties has been used to study the degradation and structural change in polymers after addition of fillers and formation of composites as a secondary information.

In the line of previously discussed techniques like TG, the thermal electrical analysis is also used to estimate the thermodynamic parameters like activation energy for structural transition like glass transition. For activation calculation using this techniques a two-parameter model was reported [29]. Further, an Arrhenius expression was used to describe the temperature dependence of the dielectric relaxation time. For example, the technique was used to determine the activation energy for the glass transition temperature of an uncured epoxy-amine model resin (tetraglycidyl-4,4'-diaminodiphenylmethane with 25 parts per hundred of 4,4'-diaminodiphenylsulfone hardener) for high performance polymeric composites. The reported result was comparable to results calculated by other methods. Thus, the method has potential to be used in characterization of different aspects of materials like structural deformation, thermodynamic properties along with evaluation of dielectric properties.

5.2.7 Thermal Reflectance Spectroscopy (TRS)

TRS technique is based on the measurement of reflected radiations from a sample subjected to controlled heating [2]. The reflected radiation may be in ultravoilet, visible or infra-red region, which reveals the different types transitions and information about analysed sample. The studies in reflection spectroscopy were also carried out at ambient temperature as well as in subambient temperature, i.e., in the range of 100 to 300°C. Currently, TRS is a very powerful and versatile technique for characterization of nanomaterial, quantum dots, bulk semiconductors, transistors, semiconductor lasers, and verification of growth and processing of devices. It is used to study band structure, electronic transition, photoluminescence and dielectric function of semiconductors. The techniques is also important to study the life and heat induced failure, performance and reliability of a semiconducting devices. The heat induced reflection in visible regions is also extensively explored for the characterization of different complex compounds like $[Co(NH_3)_6H_2O]$ as substitute of liquid absorption and single crystal spectroscopy.

The measurements of the TRS spectra at small temperature increment are carried out for temperature induced thermal transition in more precise manner. The analysis can be also carried out by fixed as well as variable wavelength. The curve recorded at fixed wavelength is called **Isolambdic curve**. The curve reveals the temperature at which the transition begins and ends, also permits the investigation of a single transition, mass loss, enthalpic effect do not interface with the measurement. The

TRS technique is also useful for evaluating thermal stability of a compound and sample structural change function at temperature like other thermal techniques. This technique compliments other thermal techniques like TG , DTA and high temperature XRD.

The technique was initially invented by Wendlandt for characterization of different compounds with the help observed diffuse reflection spectra in visible region [30]. It works on two principles measuring either temperature induced reflection or transmission as represented in Fig. 5.26.

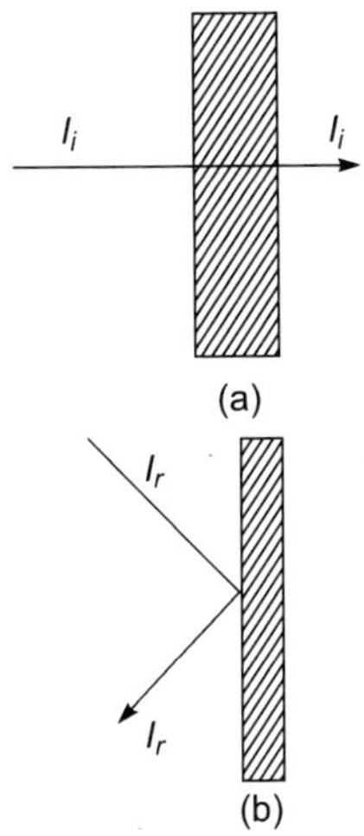

Fig. 5.26 Representation of reflectance and transmittance spectroscopy: (a) Transmittance mode; (b) Reflectance mode.

The use of precise heating and detecting system in synchronism of electronic coordination has advanced the use of several advanced applications like curing, structural transitions and catalytic effects.

The octahedral → tetrahedral structure transition of bis(pyridine)cobalt(II) chloride, $Co(py)_2C1_2$, has been the subject of a number of investigations [31] The thermal transition was found to begin at about 100°C and was completed at about 135° C. The colour change reported for the transition was from violet (octahedral) to a dark blue (tetrahedral) colour, which was stated that the change was non-reversible on cooling to room temperature. The HTRS curves of $Co(py)_2Cl_2$ at 20°C and 120°C, are given in Fig. 5.27. At 20°C, reflectance minima were observed in the curve at 520 and 620 nm, with shoulders at 500 and 550 nm, respectively. The compound reflected rather strongly in the 350—450 nm region and at 580 nm. The blue tetrahedral form, at 120°C, absorbed very strongly in the 500 to 700 nm region with shoulders at 425, 480 and 510 nm, respectively.

Similarly, the technique is considered useful to evaluate the thermal effect of many compounds such as of $CuSO_4$ $5H_2O$ which shows different pattern even in open and closed conditions. The phase thermochromism in Ag_2 $[HgI_4]$ is due to an order-disorder transition which involves in ($\beta \rightarrow \alpha$) phases. The other important uses are the deaquation of $CoCl_2$ $6H_2O$. The result reveals a rather interesting series of structural changes, both in the liquid and solid states.

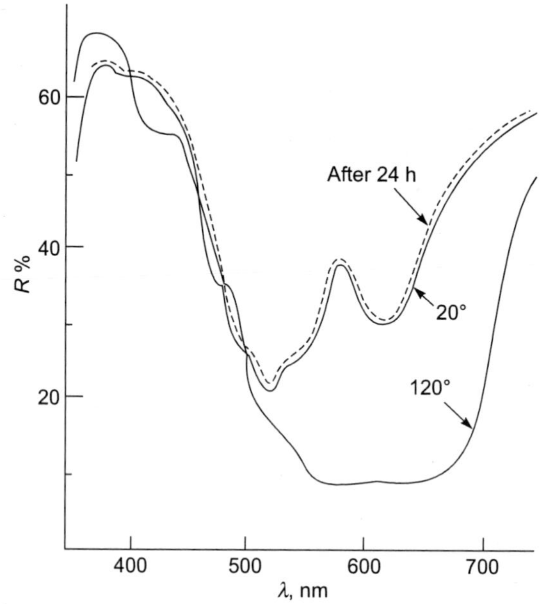

Fig. 5.27 Reflectance Spectra of $Co(py)_2Cl_2$ at various temperatures (Ref. 31).

Like other techniques, the calibration of thermoreflectance spectra is performed for acquiring reliable spectra for the study of not fully characterized optical properties. For this purpose some absolute values of TR coefficient from the literature are used. Further, to ensure requisite accuracy the temperature of the sample also needs to be stabilized with precise monitoring. The precise temperature control in temperature of the device is performed by a reliable temperature controller as well as a water cooled thermoelectric element.

5.2.8 Thermal Activity Monitor (TAM)

The technique, used for all chemical, physical and biological processes, results in either small heat production or heat consumption like a microcalorimeter. [32] It is a versatile technique for studying this thermal activity of chemical and biochemical interaction in terms of heat, heat flow and heat capacity. TAM offers unmatched sensitivity, long-term stability and high measuring capacity to study the materials and pharmaceutical interactions. The modular design, coupled with a wide range of accessories and auxiliary equipment, offers unrivalled flexibility. It is a completely non-destructive and non-invasive techniques for testing of different types of sample in solid, liquid, gas and molten phases. It seldom requires any prior samples treatment nor does it limit analysis to a physical state of the sample. This is a direct and continuous measurement technique for monitoring a process under progress in real time and static condition. Unlike other analytical techniques, TAM give "snapshots" of data for continuous progress of a chemical of physical process.

The basic instrumentation components of TAM are thermostat, calorimeter, sample handling apparatus, computer and software accessories. The static and scanning thermostats are employed for maintaining different temperature with precise variability like temperature accuracy of +/− 1°C. In this setup generally the oil based liquid bath systems are used with continuously circulated heat sink medium to prevent any thermal events. The simple structure of thermostat, TMA instrument and basic principle are shown in Fig. 5.28.

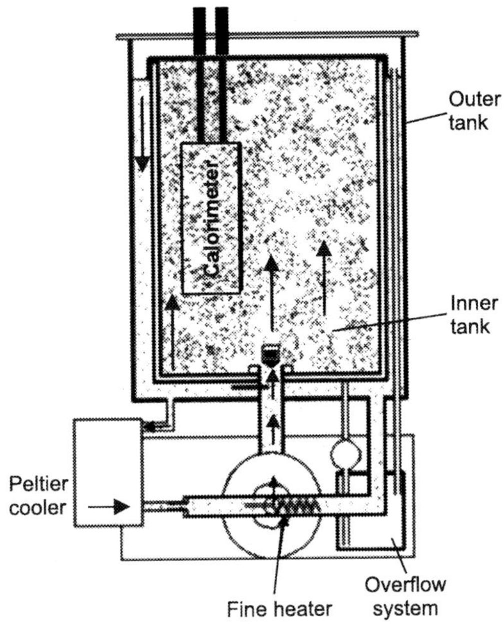

(a)

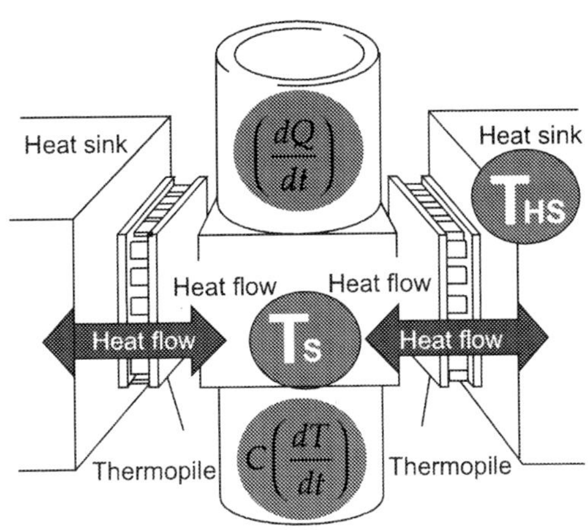

(b)

Fig. 5.28 Simple block diagram TAM (a) thermostat, (b) instrument and basic principle

The attachment of multi-calorimeter and flexible sample holder advances the use of TAM for measuring the multiple reaction at a time. The calibration of this technique is performed by incorporating the reference calorimeters. Both sample holder and calibrating reference calorimeters are shown in Fig. 5.29.

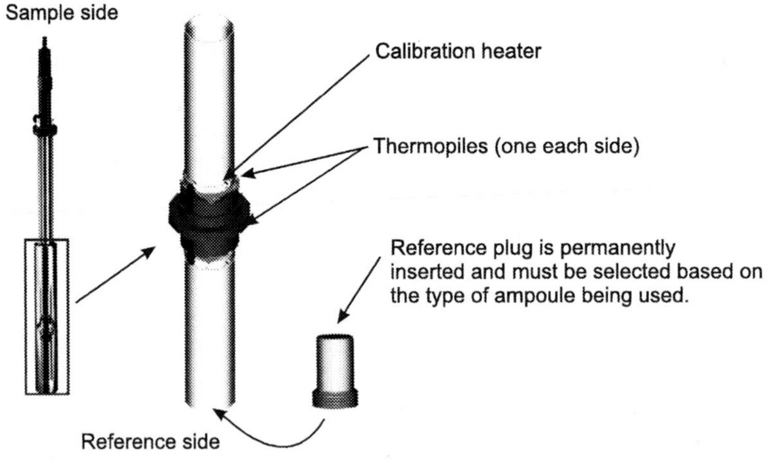

Fig. 5.29 Representative sample holder and reference calorimeter

This technique permits the acquisition of some values related to the microbial degradation of glucose in presence of an inorganic salt (copper sulphate). This causes an inhibitory effect on the degradation due to its behaviour as a pollutant agent in soil, obstructing a normal microbiological growth. The collected observations from TAM and the consequent fit to a kinetic mechanism can establish a methodology to determine the quantitative participation of a given pollutant in a soil microbial system. Thus, this quantitative information on soils can give considerable contribution to understanding important features related to environmental sciences, for example, the thermal hazards analysis of various products, i.e., including benzoic acid for, benzene, and phenol generated during autocatalytic reaction of benzoyl peroxide in both isothermal and non-isothermal kinetic models. The study infers to predict the time, temperature and condition to generate different decomposition products of benzoyl peroxide. The finding is useful to predict the compatibility of different compounds and their autocatalytic behviour like change in decomposition temperatures [33].

EXERCISE

1. Discuss oxidation induction time along with its significance
2. List the analytical factors of oxidation induction time.
3. Explain the principle and basic components of thermosonimetry.
4. Explain the applications of emanation thermal analysis.
5. Elaborate the acoustic surface phenomenon along with its significance.
6. Compare DSC and TS curves of a representative sample.
7. Discuss the principle and application of thermoparticulate analysis.

8. Explain the information derived by reflectance spectra of $Co(py)_2Cl_2$ given below:

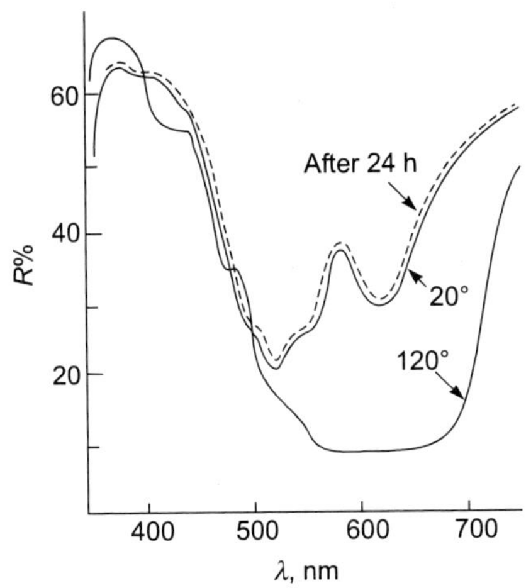

9. Write short notes on
   (a) Microcalorimetry
   (b) Thermochromism
   (c) Isolambdic curve
   (d) Thermo-voltaic detection
   (e) Soret effect

REFERENCES

1. Ehrenstein, G. W., Riedel, G. and Trawiel, P. (2004). Thermal Analysis of Plastics: Theory and Practice. Hanser Publisher, Munich.
2. Wandlandt, W. W. (1986). Thermal Analysis. John Wiley & Sons, Inc., New York, USA.
3. Fearon, P. K., Bigger, S. W. and Billingham, N. C. (2004). DSC combined with chemiluminescence for studying polymer oxidation. *J. Therm. Anal.* 76, 75-83.
4. Bharti, R. and Singh, B. (2020). Green tea (*Camellia assamica*) extract as an antioxidant additive to enhance the oxidation stability of biodiesel synthesized from waste cooking oil. *Fuel*, 262, 116658.
5. Almoselhy, R. (2021). Comparative study of vegetable oils oxidative stability using DSC and Rancimat methods. *Egyptian Journal of Chemistry*, 64(1), 299-312.
6. Pereira, T. A. and Das, N. P. (1990). The effects of flavonoids on the thermal autoxidation of palm oil and other vegetable oils determined by differential scanning calorimetry, *Thermochimica Acta*, 165, 129-137.
7. Terzopoulou, Z., Wahbi, M., Kasmi, N., Papageorgiou, G. Z. and Bikiaris, D. N. (2020). Effect of additives on the thermal and thermo-oxidative stability of poly (ethylene furanoate) biobased polyester. *Thermochimica Acta*, 686, 178549.

8. Klimesz, B., Gorecki, T. and Gorecki, C. (2001). Phase transitions in Ag-based solid electrolytes as detected by thermosonimetry, *Thermochim. Acta*, 374, 145-149.
9. Lønvik, K. (1987). Thermosonimetry, *Thermochim. Acta*, 110, 253-264.
10. Shimada, S. (1993). Thermosonimetry and thermomicroscopy of the decomposition of $NaClO_4$ and $KClO_4$, *J. Therm. Anal.*, 40, 1063-1068.
11. Lach, V. (1987). Some applications of the decrepitation technique and thermosonimetry in research on minerals and building materials, *Thermochim, Acta*, 110, 265-267.
12. Balek, V. (1991). Thermosonimetry and thermomicroscopy of the decomposition of $NaClO_4$ and $KClO_4$, *Thermochim. Acta*, 192, 1-17.
13. Beckman, N. and Balek, V. (2002). Emanation thermal analysis and its application potential, *J. Therm. Anal. Calorim.*, 67, 49-61.
14. Balek, V. and Murat, M. (1996). The emanation thermal analysis of kaolinite clay minerals *Thermochimica Acta*, 282/283, 385-397.
15. Gijbels, K., Krivenko, P., Kovalchuk, O., Pasko, A., Schreurs, S., Pontikes, Y. and Schroeyers, W. (2020). The influence of porosity on radon emanation in alkali-activated mortars containing high volume bauxite residue, *Constr. Build Mater.*, 230, 116982.
16. US patent No. 3,410,662.
17. Phillips, D. C., Smith, J. D. B., Meier, J. F. and Kaczmarek, T. D. (1978). Organic particulate analysis of isocyanate compounds, *Microchem J.*, 23(2), 165-177
18. Hubbard, A. T. and Somasundaran, P. (2006). Encyclopedia of Surface and Colloid Science, 2nd Edition, CRC Press.
19. Piazza, R. (2008). Thermophoresis: moving particles with thermal gradients, *Soft Matter*, 4(9), 1740-1744.
20. Jerabek-Willemsen, M., André, T., Wanner, R., Roth, H. M., Duhr, S., Baaske, P. and Breitsprecher, D. (2014). MicroScale thermophoresis: Interaction analysis and beyond, *J. Mol. Struct.*, 1077, 101-113.
21. Asmari, M., Ratih, R., Alhazmi, H. A. and El Deeb, S. (2018). Thermophoresis for characterizing biomolecular interaction, *Methods*, 146, 107-119.
22. Franz, P., Gassl, V., Topf, A., Eckelmann, L., Iorga, B. and Tsiavaliaris, G. (2020). A thermophoresis-based biosensor for real-time detection of inorganic phosphate during enzymatic reactions, *Biosens. Bioelectron*. 169, 112616.
23. Brown, M. E. (2001). Introduction to Thermal Analysis, Kluwer Academic Publisher, London.
24. Gaisford, S., Kett, V. and Haines, P. (Eds) (2016). Principles of Thermal Analysis and Calorimetry, 2nd Edition, Royal Society of Chemistry, UK.
25. Ayat, S., Wrobel, R., Goss, J. and Drury, D. (2016). Experimental calibration in thermal analysis of PM electrical machines. In 2016 IEEE Energy Conversion Congress and Exposition (ECCE). 1-8.

26. Netzsch -DEA_E_1106_en_24,
27. Marzec, E. and Olszewski, J. (2010). Dielectric analysis of the thermal processes of human nail, *Thermochim. Acta*, 506, 73–76.
28. Biryan, F., Abubakar, A. M. and Demirelli, K. (2018). Product analysis, electrical and dielectric properties depending on thermal influence of poly(*N*-isopropyl acrylamide)/graphite-filled composite, *Thermochim. Acta*, 669, 66-79.
29. Nass, K. A. and Seferis, J. C. (1990). Activation energy determination from dielectric thermal analysis, *Thermochim. Acta*, 170, 19-33.
30. Wendlandt, W. W., Franke, P. H. and Smith, J. P. (1963). High temperature diffuse reflectance spectroscopy, *Anal. Chem.*, 35 (1), 105-107.
31. Wendlandt, W. W. (1969). The reversibility of the octahedral → tetrahedral transition in bis(pyridine)-cobalt(III) chloride $Co(py)_2Cl_2$, *J. Thermal Anal.*, 1, 469-472.
32. Airoldi, C. and Critter, S. A. M. (1996). The inhibitor effect of copper sulphate on microbial glucose degradation in red latosol soil, *Thermochim. Acta*, 288, 73.
33. Liu, S. H., Hou, H. Y. and Shu, C. M. (2015). Thermal hazard evaluation of the autocatalytic reaction of benzoyl peroxide using DSC and TAM III, *Thermochim. Acta*, 605, 68-76.

6

Thermal Mechanometry

6.1 INTRODUCTION

In this technique, the measurement of a change of a dimension or mechanical property of the sample is carried out, while it is subjected to the heating, cooling, temperature modulation with a set of frequency and amplitude or holding the temperature constant [1-4]. The sequence of temperature may be predetermined (temperature programmed) or sample controlled.

Thermomechanometry includes several variations as listed in Table 6.1, and it is classified depending on the way force is applied.

TABLE 6.1 Classification of thermomechanometry

Methods (Techniques)	Abbreviations	Conditions or property
Thermodiatometry (Thermodilatometric Aalysis)	TD	Negligible force (Dimension)
Static force Thermomechanometry (Static Force Thermomechanical Analysis)	Sf-TM or sf-TMA	Static force (Dimension)
Dynamic force Thermomechanometry (Dynamic Force Thermomechanical Analysis)	Df-TM or df-TMA or DMA	Dynamic force (Dimension)
Modulated force Thermomechanometry (Modulated Force Thermomechanical Analysis)	mf-TM or mf-TMA or DMA	Modulated force (Dimension)

(i) When constant or static forces are applied, it is called thermal mechanical analyser (TMA) or static force thermal analyzer (sf-TM).

(ii) When the applied force is zero, it is called thermodilatometry (TD).

(iii) Applied force when changed as per the case of a typical stress-strain analysis or alteration of the variables with time is called dynamic mechanical analyser (DMA).

(iv) When force is changed with frequency and amplitude. The term modulated is a special variant, used for consistent modulated temperature differential scanning calorimeter (mt-DSC), also known as modulated force thermomechanometry (mf-TM).

6.1.1 Principle

The mechanical analyzers are based on Hooke's law. Force applied to solids causes deformation; to understand this phenomenon, it is necessary to define strain measures and the relationship between stress and strain. If an axial force (σx) perturbation is applied, it causes strain (Ex) such as change in dimensionless ratio. It was found that more engineering materials, a linear relationship between stress and strain is held according to Eq. (1). This is called Hooke's law:

$$\sigma x = EEx \tag{1}$$

where E is the material constant called Young's modulus. For an isotropic elastic material, two material constants are needed to characterize the stress-strain relationship; and Young's modulus, which is the slope of the (σx) versus (Ex). The material responds to an applied stress which is dependent very much on the rates and duration of loading. This, indirectly contrasts the behaviour of the elastic solids at low temperature and small deformation, whereas at high temperature they behave in viscous manner. Therefore, termed viscoelastic because they exhibit both viscous and elastic types of behaviour. Consequently, the mechanical behaviour is dominated by viscoelastic phenomenon in contrast to materials such as metal and glass, where atomic adjustment under stress is merely localized and limited but polymers show viscoelastic behaviour in a striking way. For example, silicon putty flows when it is left alone but bounces like a rubber ball when loaded quickly. The work on viscoelasticity concluded that these materials exhibited a 'delayed elasticity'. Which is manifested in the observation that the impression of a stress resulted in an instantaneous strain continued to increase more slowly with time. It is this delay between cause and effect that is fundamental to the observed viscoelastic response. The three major examples of this hysteresis effect are: creep, stress relaxation and dynamic response, which produce a strain oscillating with same frequencies but out of the phase with stress. This dimensional stability is an important consideration in choosing a material for a product;. which is affected by different parameters such as temperature and time.

6.1.2 Effect of Temperature on Mechanical Properties

The effect of temperature on mechanical properties can be understood by the experiment of measuring Young's modulus (E) and different temperatures and most often, the graph over area. This plot is carrying four regions of viscoelastic behaviour at low temperature, glassy region becoming hard and brittle. The glassy modulus E_1 is slowly decreasing as function of temperature, thermal energy is insufficient to surmount the potential barriers for motion of segments of the molecules and the chain is frozen and fixed in position with the increasing temperature. The amplitude of vibration motion becomes greater and eventually, the thermal energy barrier becomes roughly comparable to the potential energy barrier of the segment. In this temperature range, short range diffusion motion begins, segments are free to jump from one lattice to another and brittle glass becomes resilient leather and phenomenon is accompanied by a modulus decrease in several orders. As the temperature is further increased, the modulus reaches to plateau region. This rubbery plateau is characterized by a modulus, E_2. In this temperature range, long-range cooperative motion of the chains would result in motions restricted by the presence of local interaction between the chains. As the temperature is further increased, the modulus will remain constant for a cross-linked rubber up to a temperature where chemical degradation begins to occur.

6.1.3 Time-Temperature Superposition

There is general equivalency between time (frequency) and temperature. For instance, a polymer that displays rubbery characteristics under a given set of testing conditions can show glass behaviour by either reducing the temperature or increasing the testing rate or frequency. It is reported that in viscoelastic materials, time and temperature are equivalent, meaning the data at one temperature can be superimposed upon the data taken at a different temperature merely by shifting the curve. The curve obtained at each is shifted by an amount called shift factor (aT). The shift factor is a function of temperature only and decreases with increasing temperature and unity at reference temperature, T_s. The aT could be given by an equation called WLF equation (Eq. 2):

$$\log aT = -C_1(T - T_s)/C_2 + (T - T_s) \tag{2}$$

where C_1 and C_2 are constant and T_s is the reference temperature.

6.2 THERMODILATOMETER

6.2.1 Introduction

The measurement of thermal expansion of solids and liquids is a classical physical technique known as dilatometry. When special emphasis is put on recording such dimensional change as a function of temperature, during a controlled temperature program, the technique is referred to as thermo-dilatometry [1,2] and is one of the special techniques derived from thermo-mechanometry. For solids, we can distinguish between linear and volume expansion, and instruments for measuring the length of a solid sample as a function of temperature are the most common.

Most solids expand on heating with some exceptions such as vitreous silica and ZnS at low temperature.

The change in linear dimension, L, with T is given by Eq. (3):

$$L_2 = L_1(1 + \int_{T_1}^{T_2} \alpha dT) \tag{3}$$

where L_1 is the length at T_1 and L_2 at T_2 and α is the coefficient of linear expansion. Over a small temperature interval, ΔT is approximately constant for many materials.

So dilation as Eq. (4),

$$L_2 - L_1 = \Delta L = L_1 \alpha \Delta T \quad \text{or} \quad \Delta L/L_1 = \alpha \Delta T \tag{4}$$

The value of α is related to the structure and the type of bonding in the solid. In general, strong bonding results in low value of α, so the order of expansion coefficient is usually covalent and ionic materials < metal < molecularly bonded materials. In some substances, α may vary with direction (anisotropy) behaviour. Amongst the materials that may show the anisotropy are the single crystals, drawn fibres, drawn blown films, fibre composites, liquid crystals and some metallic glasses. In some anisotropic solids, for example, calcite, the coefficient of expansion in the direction of one of the principal axis may be negative.

6.2.2 Instrumentation

Measurement of thermal expansion may be classed as either absolute or difference measurement, the basic component and technical details are shown in Fig. 6.1(a) and (b).

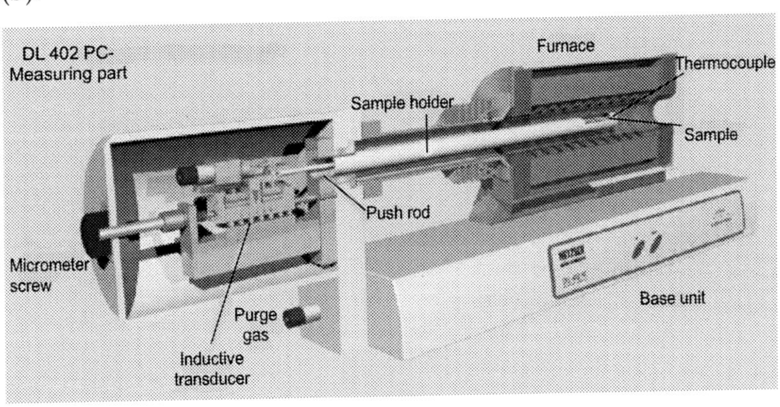

(a)

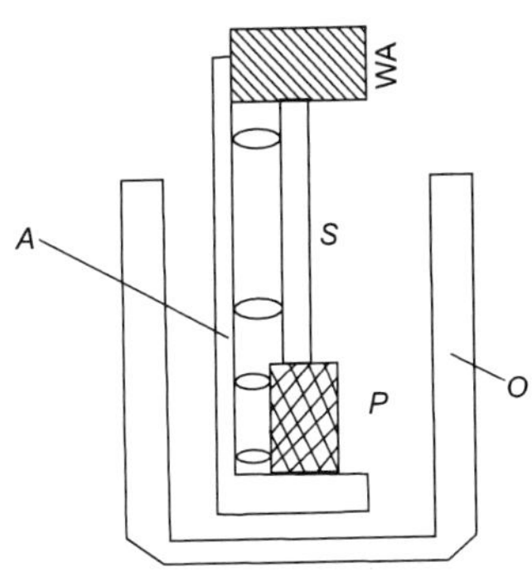

(b)

Fig. 6.1 (a) The measuring principle of thermo dilatometer: (WA = Weight applied, S = sample, P = sample platform, A = Arm) (b) The technical details of Netzsch DL-402 PC dilatometer (Reproduced by permission from M/s Netzsch)

The dimension change of a sample can be measured either absolute or difference with reference (thermally inert) material. The methods are illustrated in Fig. 6.2(a) and (b).

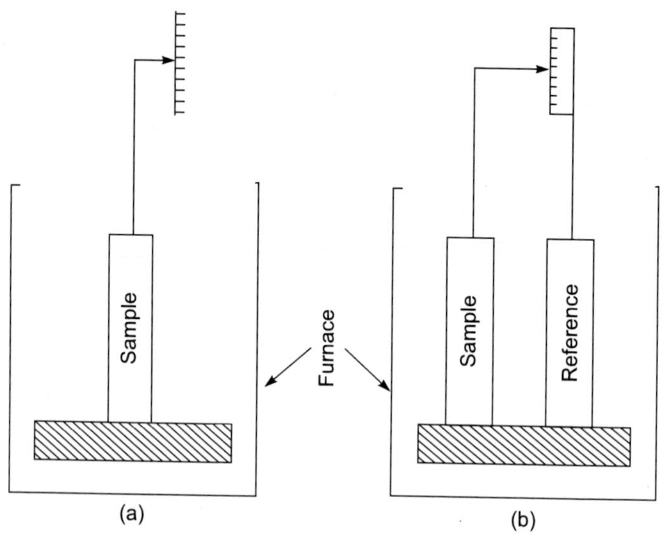

Fig. 6.2 (a) Absolute measurement and (b) difference measurement

The pointer and the scale are outside the furnace and the movements of the sample and reference are transmitted by identical push rod. Measurement may be made in either horizontal or vertical position. In horizontal measurement, arrangement has to be made for suitable spring loading, so that any contraction can be detected. In vertical measurement, the weight of the push rod and measuring system has to be counterbalanced to avoid compression or deformation of the sample.

Calibration of the system uses a solid sample of known coefficient of expansion; for example, aluminium metal under conditions very similar to those used for the actual sample, is necessary to compensate for many dimension changes in the instrument as heating takes place.

Principle: Measurement of sample movement obviously has to be more sophisticated than the pointer and scale used in the schematic diagram. The transducer is usually a linear variable differential transformer (LVDT), the principle of which is illustrated in Fig 6.3.

Movement of the core within produces an output with signs dependent upon the direction of the movement, and magnitude giving the movement of the sample. Application of the movement is claimed to be as much as 20×10^3. Calibration of LVDT is carried out, initially using a micrometer incorporated in the measuring system. Safety devices to prevent damage caused by melting of the sample in the apparatus, are often included. These are activated by rapid sample contraction and to the furnace is cut off immediately. Expansion of a standard material can be used as a temperature indicator. Isothermal measurements may be made to study creep or recrystallization processes in the sample, change in the volume of liquid samples, contained in a suitable cylinder, can be converted into equivalent linear movement by means of closing of fitting piston attached to push rod. A bleed valve on the cylinder ensures that no air is trapped in the liquid.

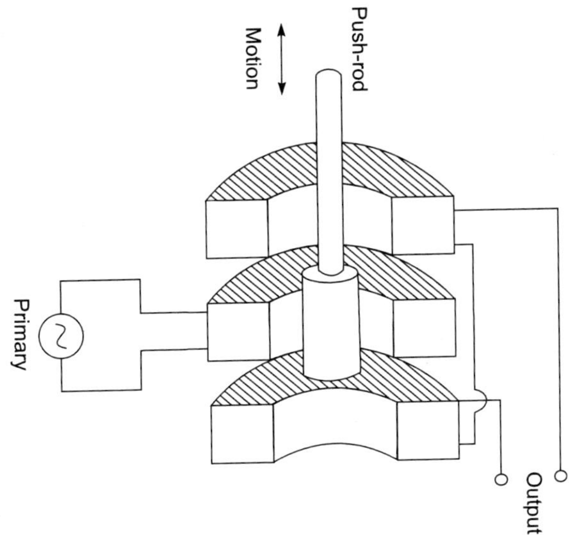

Fig. 6.3 Linear Variable Differential Transformer (LVDT) (Ref. 2)

A schematic diagram of a linear dilatometer based on the principle of Michelson interferometer is shown in Fig. 6.4.

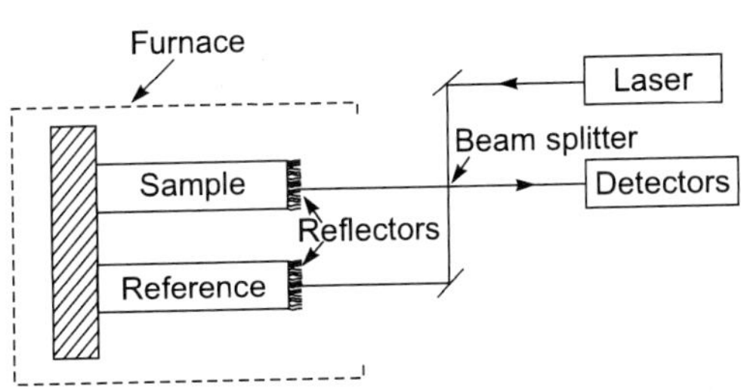

Fig. 6.4 Laser dilatometer based on interferometer

The laser beam splits after a 90° reflection into two separate beams. The two beams are directed at the reflector attached to the ends of the sample and reference materials. After total reflection, the beams are recombined and the interference patterns are examined using photodiode detectors and converted to measurement of expansion or shrinkage. These dilation measurements are accurate to one quarter of wavelength of the laser and induce no mechanical connection with the sample, so very rapid change of the dimension can be examined. The temperature range over which such optical system can be used is much greater than for a mechanical system, which will prove useful in examining expansion coefficient of superconducting material at liquid helium temperature.

Since the LVDT is sensitive to electrical disturbances, thus the use of an optoelectronic transducer is advocated with advantageous features used in place of a LVDT shown in Fig. 6.5.

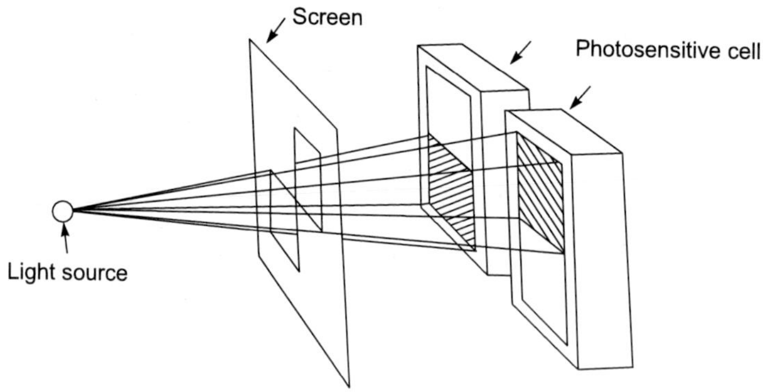

Fig. 6.5 Representation of optoelectronic transducer

6.2.3 Optoelectronic Transducer

The first optical dilatometer was invented by Abbe and Fizeau in the second half of 19th century. This design has a reflected beam of monochromatic light and the measurement of the displacement is carried out by counting the interference fringes between the forward going beam and the reflected beam. After the Abbe invention, many improvements were achieved on the original design and there are now many models available in the market, which use modern optics and designs.

Over the last five decades, interest has grown in the use of thermomechanical technologies for characterising materials in different fields of science and engineering. In particular, the use of optical methods has been playing a very important role in the field of ceramics and have been used to measure the coefficient of thermal expansion of bodies and glazes in order to ensure an optimal match of their thermomechanical behaviour.

An optical dilatometer is a device for measuring changes in the dimensions of a specimen. A monochromatic light source, such as a laser, illuminates the specimen. Some of the light is reflected by the specimen and interferes with the incoming light, creating optical interference fringes. As the specimen contracts or expands, there is a proportional movement of the interference fringes, which can be measured using a camera system. The measurement resolution is determined by the wavelength of the light, and is typically 0.5 μm for blue light. Optical dilatometers are used to measure thermal expansion.

The optical dilatometer has, in fact, replaced the traditional dilatometer when it comes to measure dimensional changes of materials, as a function of temperature, and no contact between specimen and instrument is required. This has a particular advantage for simulation of thermodilatometric, thermoconductometric and thermomagnetometric measurement as LVDT are sensitive to electrical perturbation. A stabilized light source illustrated two photosensitive cells through two rectangular holes of a screen, which are attached to the push rod of the dilatometer movement of the screen; it alters the relative amount of light received by the two cells. Dimensional variations down to 10^{-8} m could be measured and speed of response was about ten times that of a LVDT.

6.2.4 Interpretation of Results

Values of the coefficient of expansion, α, may be determined from the slope of a curve L against T because

$$\Delta L/\Delta T = dl/dt = L_1\alpha \tag{5}$$

Alternatively, the time derivative of the dilation is as

$$(dL/dl)(dt/dT) = L_1\alpha \tag{6}$$

So,

$$dL/dt = L_1\alpha\beta \tag{7}$$

where, $\beta = dT/dt$ is heating rate.

Thus, material coefficient of expansion, α', is defined by

$$\alpha' = (L_T - L_{20})/(L_{20}(T_{-20})) \tag{8}$$

where, L_T and L_{20} are the lengths of the sample at T°C and 20 °C, respectively. Generally, it is discontinuous in the curve of L against T, rather than actual values, the values of α, which are main interest in thermodilatometry, as these indicate occurrence of thermal events in the sample. Thus, by appropriate calibration, a first order transition would appear as a peak in the curve and the area under which it is used. These discontinuities are explained in applications.

6.2.5 Applications

The classical dilatometry is used to detect volume or length changes caused by phase transitions of various types [5-7]. The common types of phase transitions that determine are the solid 1 to liquid 1, solid 1 to liquid 2 and solid to gas. The technique can also be used to detect shrinkage and sintering of a sample on heating at elevated temperature. The linear coefficient of thermal expansion of a sample can also be determined. The other use includes the determination of the glass transition (T_g), softening temperatures, heat distortion temperature (HDT) and so on. The result obtained from thermodilatometer (TD) is suitable for selecting of suitable materials for use, e.g., brake lining, chemical structural change, sintering behaviour, etc. Generally, a low coefficient of expansion, coupled with other qualities such as wear resistance and a suitable coefficient of friction, is desirable for brake lining. Figure 6.6 shows thermal expansion of a representative brake lining material; the continuous expanding material is preferred. Thus, the TD technique is helpful in selecting a material for end use.

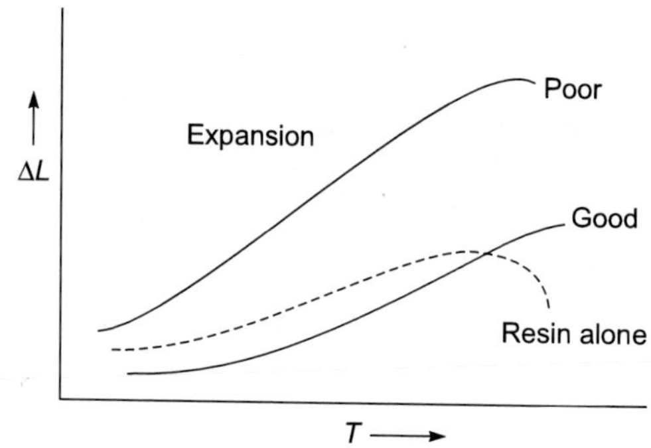

Fig. 6.6 Thermal expansion of brake lining materials

The TD studies for the change in structure from octahedral to tetrahedral in cobalt (II) pyridine coordination compound [Co$(py)_2Cl_2$] is shown in Fig. 6.7 with the abrupt transition at ~125°C.

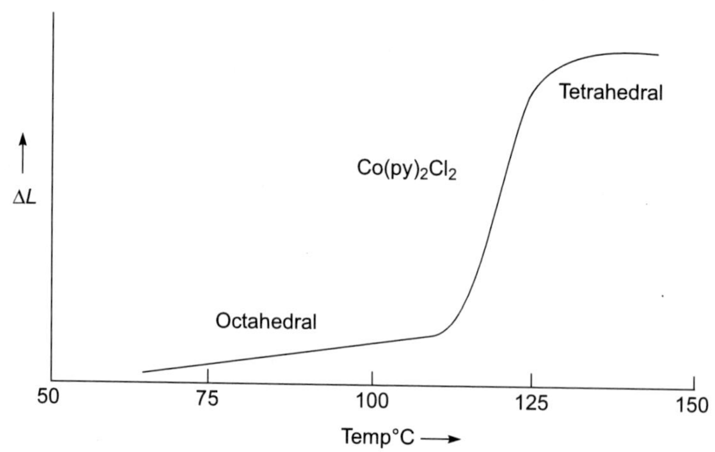

Fig 6.7 TD of [Co $(Py)_2Cl_2$] (Adapted from Ref. 2)

Similarly, TD study can be used for the kinetics of sintering of alloy-like powder such as (Fe-Cr-Co) used in making high energy magnets, the sintering behaviour of clays such as kaolin. It loses structural water at about 550°C and converts to another form called metakaolin and above 960°C, this converts to a spinel structure and then finally above 1100°C to mullite. The reactions are accompanied by the shrinkage shown in Fig. 6.8. The other important applications of TD are: effect of additives on shrinkage of cellular concrete, the defect in non-stochiometric oxides, the properties of ceramics and alloys, etc.

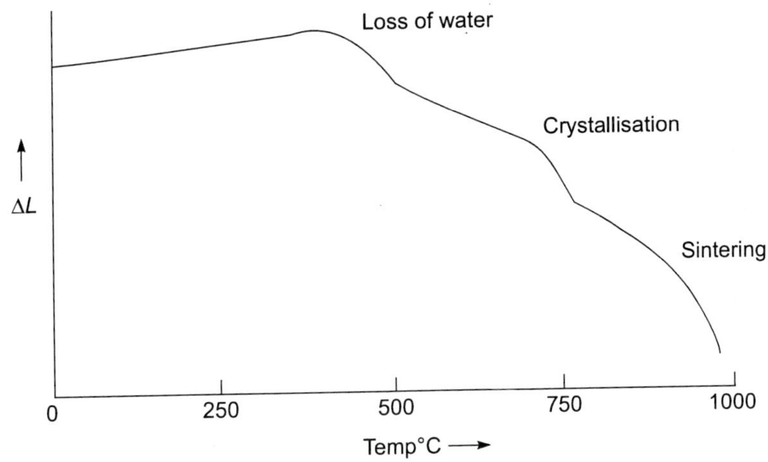

Fig. 6.8 TD study of kaolin (Adapted from Ref. 2)

6.2.6 Liquid Dilatometry

The TD technique is also used for testing liquid samples. Considering the dimensions of the elemental cube as l_1, l_2, l_3 and their relative finite changes as Δl_1, Δl_2, Δl_3, the volume change (ΔV_r) will be:

$$\Delta V_r = V_{\text{final}} - V_{\text{initial}} = (l_1 + \Delta l_1)(l_2 + \Delta l_2)(l_3 + \Delta l_3) - (l_1 l_2 l_3) \tag{9}$$

On dividing by the total unit volume of the element, Eq. (9) can be expressed in terms of the strain components, as follows leads to:

$$\frac{\Delta V^r}{V^r} = \Delta\varepsilon^r = \varepsilon_1 + \varepsilon_2 + \varepsilon_3 + \varepsilon_1\varepsilon_2 + \varepsilon_2\varepsilon_3 + \varepsilon_1\varepsilon_3 + \varepsilon_1\varepsilon_2\varepsilon_3 \tag{10}$$

$$= \frac{1}{2}\left(-1 + \sqrt{\left(1 + \frac{3}{4}\Delta V^r\right)}\right) \tag{11}$$

where, $\Delta\varepsilon_r$ and ΔV_r are the total residual strain and the volume variation, respectively. Thus, the prediction of the residual strains requires the monitoring of the volume changes with cure reaction. Thus, liquid dilatometry has been used successfully to measure chemical shrinkage at different isothermal temperatures in the defined temperature range (120-160°C) using the dilatometer equipment. (Readers will find a complete description of cited dilatometer and data analysis in literature cited). Some preliminary experiments were also performed on room temperature on curing of resins. These kinds of experiments allow to separate the effects of temperature changes and eventually the geometric constraints of different equipment from those due to the chemical reaction.

6.3 THERMOMECHANICAL ANALYSIS (TMA)

6.3.1 Introduction

The technique is based on measurement of mechanical properties as a function of static or dynamic temperature under static force, also called static force thermomechanometry (Sf-TMA). In this technique, dimensional changes in a sample are measured, while the sample is heated, cooled, or studied at a static force. It is a part of a thermal analysis group of techniques aimed more towards the chemical behaviour than physical properties of materials. It was developed for determination of mechanical properties of materials, such as load bearing when heated, penetration, bending, etc. It uses a small sample and usually comes with a variety of shaped tips to do the above. At some point it was an obvious extension to measure expansion coefficients as well, after all the device has a sensitive transducer. The use of TMA is reasonably possible with plastics where the expansion coefficient is in the 100-300 × 10^{-6}/C range, so that a few mm thick sample will expand quite a bit. However, indiscriminate extension of its use to include essentially any material is inappropriate. For example, metal will expand generally about 1/10 of plastics, and some ceramics only 1/100. This means that a TMA is not sensitive equally to all types of materials, such as if a ±1% accuracy for a plastic sample can have a ±100% inaccuracy for a ceramic sample. Therefore, it can neither be calibrated out nor answered with higher amplification of the signal.

6.3.2 Principle and Instrumentation

A block diagram of TMA is given in Fig. 6.9. The sample sits on a support within the furnace. Resting upon it is a probe to sense changes in length, which are measured by a sensitive position transducer, normally a Linear Variable Displacement Transducer (LVDT) labelled as displacement transducer. The probe and support are made from a material such as quartz glass (vitreous silica), which has a low, reproducible, and accurately known coefficient of thermal expansion, and also has low thermal conductivity, which helps to isolate the sensitive transducer from the changing temperatures in the furnace. A thermocouple near the sample indicates its temperature. There is usually provision for establishing a flowing gas atmosphere through the instrument, to prevent oxidation; for example, and also to assist in heat transfer to the sample. Helium is effective in this respect. The load may be applied by static weights, as above, or by a force motor. This latter method gives the advantage that the applied load can be programmed to allow a greater range of experiments. The instrument is calibrated for position measurements by heating a sample whose expansion coefficient is accurately known, such as aluminium, in the form of small machined blocks or cylinders. Sample sizes are commonly around 5-10 mm in height and width. It is important to prepare samples with clean, flat and parallel faces to avoid artifacts in the recorded curves.

A TMA nearly always employs an LVDT (linear variable differential transformer) or a capacitive transducer, both of high sensitivity, short stroke type. These devices have a narrow range within which they are linear to some degree. Both types are

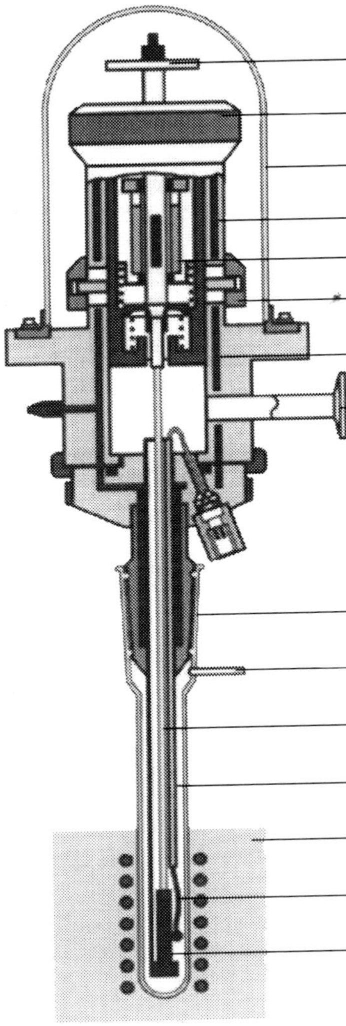

Fig. 6.9 TMA diagram of Netzsch TMA-402 (Reproduced with permission from Netzsch)

prone to drift caused by amplifiers and thermal effects, and therefore they require frequent recalibration. Dilatometers usually use longer stroke LVDTs which are somewhat easier to calibrate and are more stable, or in the case of some dilatometers, absolute digital displacement transducers are employed which do not require periodic calibrations. One major difference between a TMA and a dilatometer is the size of the sample. Normally, dilatometers employ 25 to 50 mm long samples as opposed to the 1 to 2 mm for TMA, thus the amount of expansion for a dilatometer sample is twenty-five times more. This contributes substantially to the measurability and reliability of the data and therefore to the overall accuracy of the measured coefficient.

6.3.3 Types of Measurement

The TMA measurement can be carried out in different modes employing different types of sample holders shown in Fig. 6.10:

(i) Compression
(ii) Penetration
(iii) 3-point bending
(iv) Film extension
(v) Fibre extension.

Two main types of experiments are possible:

(a) measurement of dilation with temperature at a fixed load; or
(b) measurement of dilation with load at a fixed temperature.

TMA curve involves measurement of the change of a dimension of the sample as the load on the sample is increased also normally referred to as stress-strain curve,

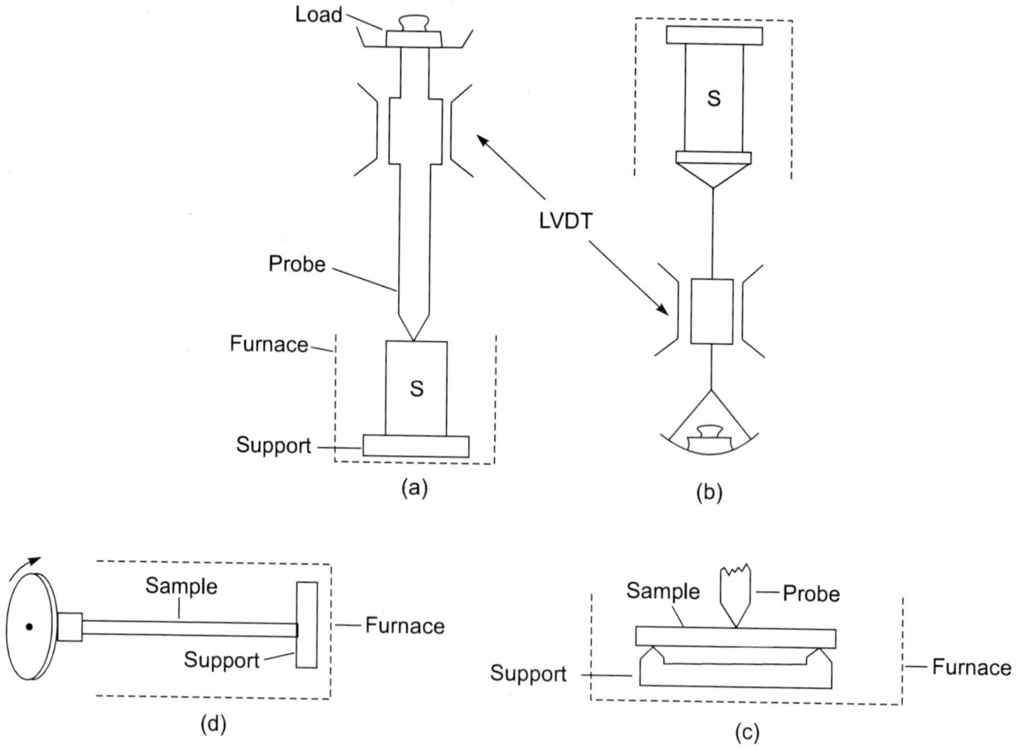

Fig. 6.10 Pictorial representation of different sample holders (modes) of TMA: (a) Penetration (b) Extension (c) Flexure (d) Torsional measurements (Ref. 2)

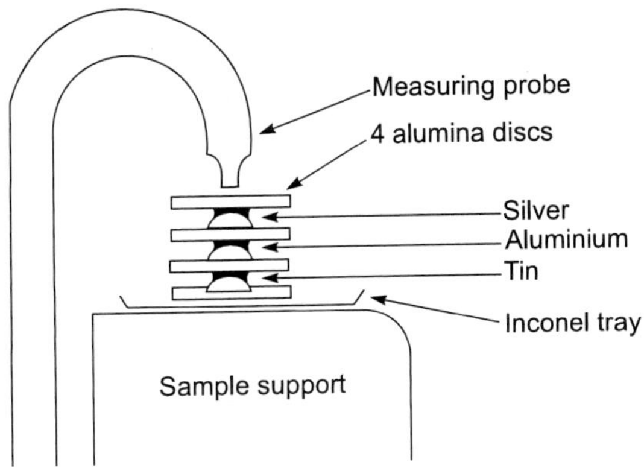

Fig. 6.11 Experimental setup used for TMA calibration and different types of tips (Ref. 2)

while the temperature is held constant. This is effectively the determination of stress-strain curve of the sample. The load is related to the applied stress (force/area) and the change in dimension is related to the strain (elongation/original length, $\Delta l/l$). Different types of materials, different types of stress-strain curves and, of course, the nature of the stress-strain curve for the given material will change as the material at which the measurements are made is changed. The extreme behaviour observed is elastic solids, which obey the Hooke's law, i.e., the applied stress is proportional to the resultant strain (but is independent of the rate strain), and the liquid, which obeys, Newton's law, i.e., the applied stress is proportional to the rate of strain (not the strain itself). Both laws only hold for small values of the strain or rate of strain.

Tensile Stress

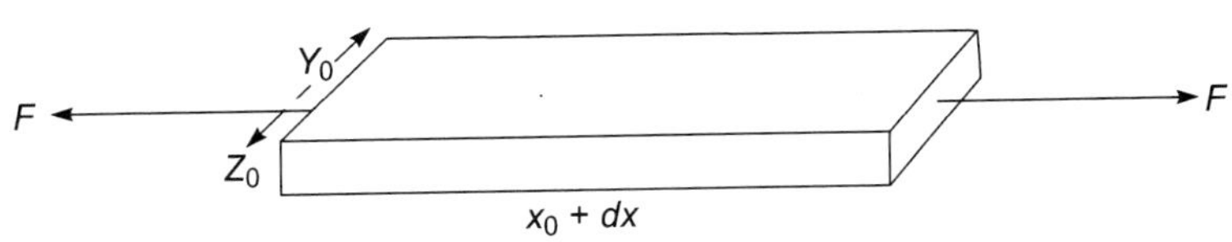

Fig. 6.12

For a sample under tension, the tensile stress, σ = force/area = F/Y_0Z_0 and the tensile strain, $\ddot{E}$ = elongation/original length (dx/X_0), where dx is elongation and X_0 is original length.

According to Hooke's law: $\sigma \alpha \varepsilon$ or $\sigma = \ddot{E}\varepsilon$; where $\ddot{E}$ is Young's modulus (force/area). Thus, for an isotropic body (i.e., properties independent of direction) the change in length and in width are related by Poisson's ratio: $V_p = (d_y/y_0)(X_0/dx)$. The values range from 0.2 to 0.5.

Shear Stress

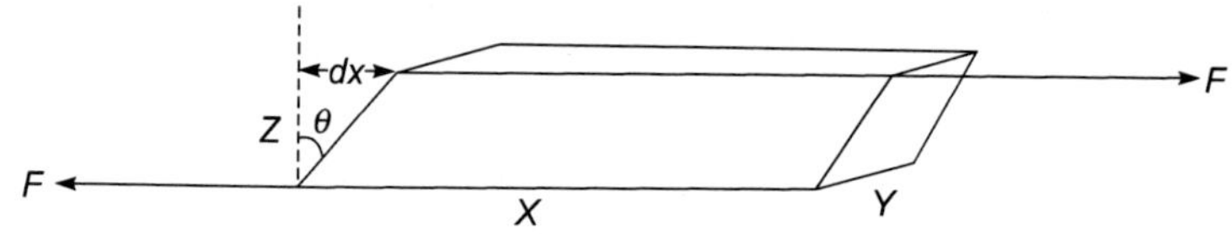

Fig. 6.13

For a sample subjected to a shearing stress, $\sigma s = F/xy = F/A$ and the shear strain, $\varepsilon_s = dz/z = \tan\theta$. Again the Hooke's law is $\sigma s = G\varepsilon_s$, where G is shear modulus.

$$G = \varepsilon_s/\sigma s = F/A \tan\theta \text{ and for small } \theta, \tan\theta = \theta. \text{ So, } G = F/A\,\theta.$$

Compression

When a pressure, P is applied to a body of original volume, V_o, causing volume change, ΔV, the stress $\sigma_B = P$ and the strain $\varepsilon_B = -\Delta V/V_o$.

As per Hooke's law $\sigma_B = B\varepsilon_B$, where B is the bulk modulus, $B = \sigma_B/\varepsilon_B = -PV_0/\Delta V$ which is related to the compressibility of the sample and defined as $1/B = -\Delta V/V_o$.

So, from the three moduli; $\ddot{E}$, G and B only two are independent. They are related via Poisson's ratio:

$$\ddot{E} = 3B(1 - 2V_p) = 2(1 + V_p)G.$$

The above discussion proves the theoretical utility of TMA in material characterizations and can be a versatile tool for assessing the viscoelastic properties of a material. The materials studied are usually rigid or nearly rigid solids, as implied by most of the experimental arrangements shown above. Liquids can be studied in the dilatometer-type accessory, which can also be used to measure the volume changes in irregularly shaped samples, or powders, which are then submerged in an inert liquid such as silicone oil. Powders can also be studied as a layer with a loose fitting lid on top, or when pressed into a pellet. Solid-state transitions can be followed in this way.

The major application areas of TMA are in the polymer field. A list of the main types of experiments and the information obtainable is given in Table 6.2.

Many other techniques exist for studying the important and informative regions of the viscoelastic properties such as glass transition (T_g), but TMA offers advantages for certain types of study. The indentation and penetration probes for instance, can follow transitions in very thin films, such as lacquer coatings on metals. Although quantitative mechanical properties (modulus) can, in principle, be derived from TMA measurements, this may be difficult in practice, and Dynamic Mechanical Analysis may be preferred. TMA is better suited to comparative measurements on a range of materials, and for measurements of transition temperatures and expansion coefficients on relatively small samples, in a conveniently short time. As well as expansion data, which can be determined with high accuracy, TMA is seen to be a sensitive method for determining the glass transition temperature of such materials.

TABLE 6.2 Application areas of TMA

Method	Mode	Measured quantity	As a function of	Information obtained
Bulk sample	Flat probe/ light load	Expansion	Temperature	Coefficient of expansion and T_g
Divided sample	Dilatometer	Volumetric changes	Temperature	Coefficient of expansion and T_g
Thin film	Penetration probe/ Significant load	Depth of penetration	Force	Modulus, cross-link density
			Time	Creep behaviour, cure behaviour
			Temperature	Softening (T_g), melting
Film or fibre	Tension accessory	Uniaxial extension or shrinkage	Force	Modulus, cross-link density
			Time	Creep behaviour, cure behaviour
			Temperature	T_g, melting, cure behaviour, Prepn. history
Fluid	Parallel plates	Distance	Time	Viscosity, gelation
			Temperature	Melting, viscosity, gelation
Bulk or supported	Flexure accessory	Bending	Time	Creep behaviour
			Temperature	Softening (T_g), melting

The change in heat capacity at T_g for a well-cured resin is small, and conventional DSC gives generally poor results in this region. MT-DSC is far better (and indeed a valuable complementary technique), but the experimental time is much greater.

TMA curve shows the dimensional changes around T_g when too great a load is used. The sample slumps after entering its softer state, making measurements doubtful. Curve was recorded under a negligible load, and allows T_g to be measured as the intersection of the portions of the curve above and below T_g.

Thermomechanical analysis is a range of advanced techniques for characterizing the temperature behaviour of materials. It is an indispensable tool for solving materials related problems in both research and quality control. It is particularly suited to polymer materials but a wide range of materials can be examined.

6.3.4 Applications

TMA accurately measures dimensional changes in a material when the material is subjected to a temperature program under static force. The information gathered by technique bears a wide range of applications for stabilizing the thermomechanical standards of the different types of materials. Some important types are discussed here.

Figure 6.14 shows the significance of stress-strain curve (TMS curve) to get significant information. The curve illustrates that as the temperature increases, the modulus of elasticity E (slope of curve) decreases as per Fig. 6.14(a), however the yield strength decreases and maximum elongation generally increases.

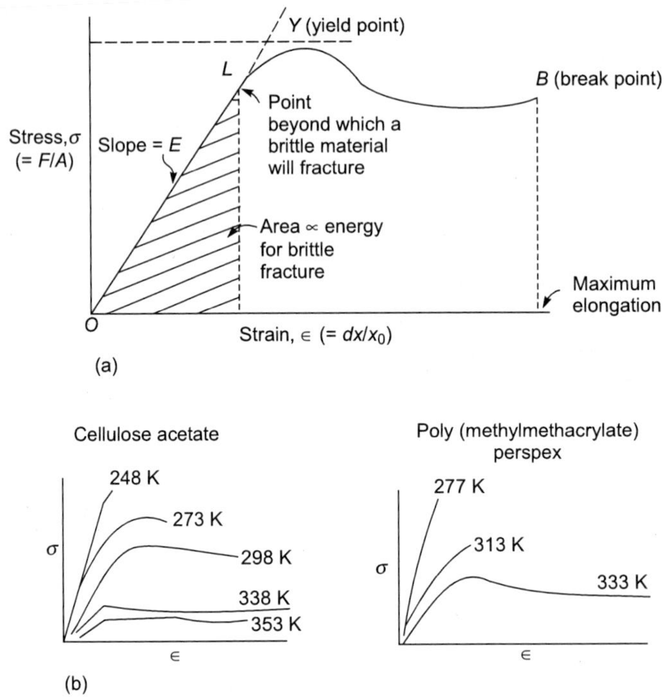

Fig. 6.14 (a) Typical stress-strain curve of a typical viscoelastic sample and (b) Effect of temperature on stress-stress curve: cellulose and poly(methylmethacrylate) perspex (Ref. 2)

TMA can be used to evaluate the thermal expansion coefficient. The thermal expansion coefficients of the partially cured resin samples could be measured using a thermal mechanical analyser, which gives an output proportional to the displacement caused by the temperature variations. Thermomechanical tests have been performed according to the principles stated in the final draft of ISO11359. Specimens sized 3 × 3 × 15 mm were cut and shaped from partially cured resin plates. After placing the specimen in the cell and equilibrating the temperature at 30°C the linear dimension of the sample was monitored up to 310°C, at a heating rate of 2°C/min. From the TMA diagrams (as in Fig. 6.15) [8], the curves of the linear strain ε_l versus temperature can easily be obtained according to the following equation:

$$\varepsilon_l(T,\alpha) = \frac{(l - l_0)}{l} = \gamma^{\text{CTE}}(\alpha)\Delta T \tag{1}$$

where ε_l is the thermal linear strain, l and l_0 are the final and initial sample lengths, respectively; $\gamma^{\text{CTE}}(\alpha)$ and ΔT are the coefficients of linear thermal expansion and the temperature variation. The graph shows the dependence of mechanical behaviour with degree of curing till the glass transition.

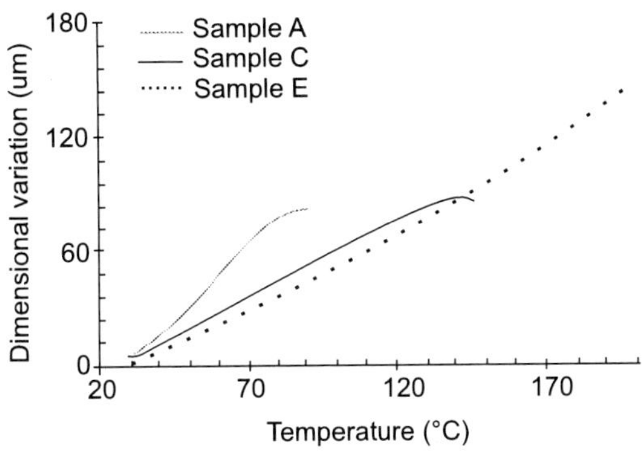

Fig. 6.15 Thermo-mechanical curves of data obtained for partially cured samples. For samples A, C and E, the conversion was 0.68, 0.80 and 0.96, respectively (Graph adapted from Ref. 8)

For a cured epoxy system, the output from TMA will be the coefficient of thermal expansion (CTE) and glass transition temperature. The CTE of the epoxy is a very important parameter because a fine gold wire is embedded in the epoxy compound and a high CTE may cause the wire to break prematurely when the electronic component is subjected to repeated temperature cycles. The deflection point between the different coefficients of thermal expansion can be defined as the glass transition temperature as represented in Fig. 6.16, a typical example of epoxy. TMA is also used to determine the softening point of plastic components and the melting point of the solder. Similarly, the TMA can also be used to study the artifacts, polymorphism, and degree of chemical reactions.

TMA and TD measurements are sometimes carried out on the same sample with same apparatus. The extension and penetration measurement of the neoprene rubber is given in Fig. 6.17. The coefficient of linear thermal expansion (α) can be determined from the slope of the expansion curve. While both glass transition (T_g) and melting temperature (T_m) show up clearly. Another example is the use of penetration measurement for testing of polyethylene coated paper (Fig. 6.18). The structural transition and melting point show up clearly even though the coating was very thin.

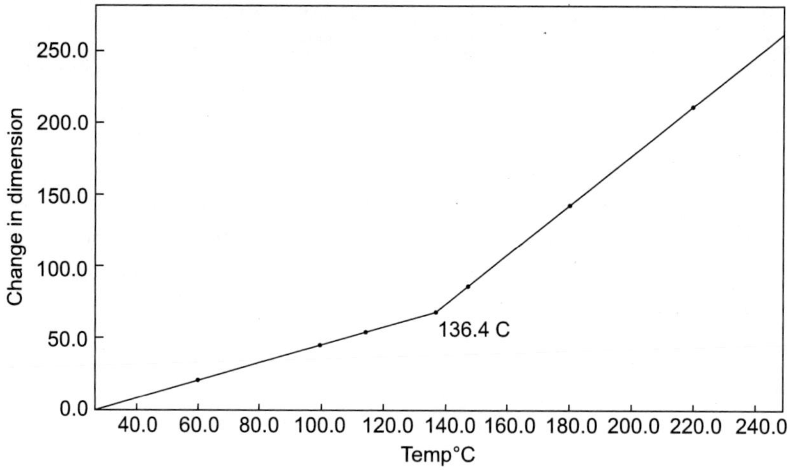

Fig. 6.16 A typical TMA curve of epoxy showing the glass transition temperature (Graph adapted from free material from Google website)

Similar measurement has been carried out on paint film and coatings. The behaviour of polymer film and fibres on heating under load is of great practical importance.

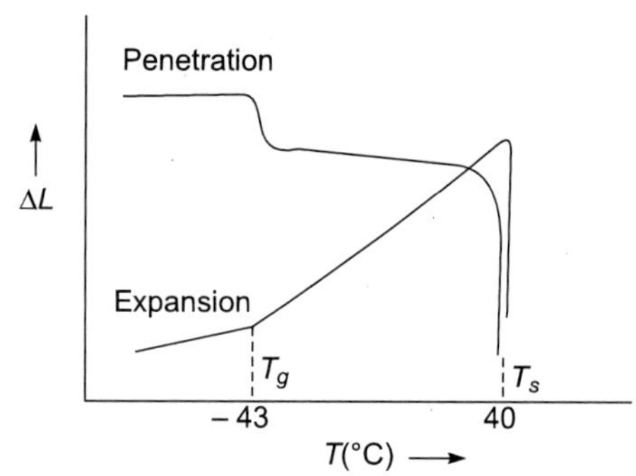

Fig. 6.17 Penetration measurement of neoprene rubber (Graph adapted from Ref. 2)

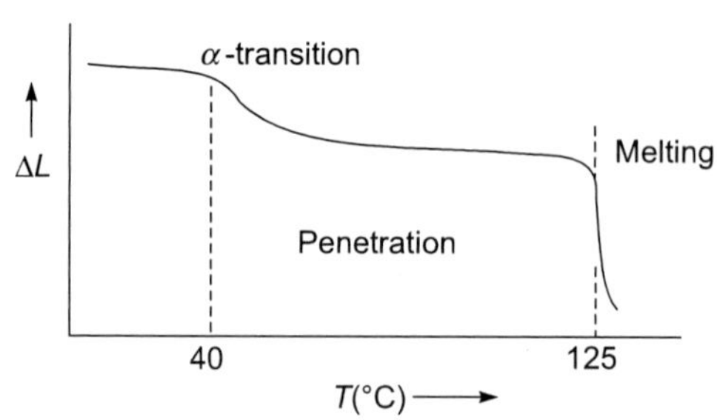

Fig. 6.18 Penetration measurement of polyethylene coated paper (Graph adapted from Ref. 2)

Another interesting application of TMA is a study of the thermoanalytical properties of the polymers used as a denture base. The glass transition temperatures of such polymers are important because there are considerable variations in temperatures in the mouth, as well as during the manufacture of dentures. The TMA measurement used to monitor the effect of various modifications of the polymerization of methylmethacrylate on T_g of the resultant materials. The factors expected to increase the T_g of polymer are: i) The presence of group in backbone which increases the energy required for rotation, e.g., unsaturation and or long side chain. ii) Secondary bonding between chains, e.g., hydrogen bonding. iii) Cross-linking between chains. iv) Molecular mass v) Copolymerization vi) Plasticization.

Wax

It is a class of hydrocarbons bearing its unique properties and can be easily casted. In wax, both the processes expansion during melting and contraction during solidification proceed. The property of change in volume of wax of a low viscosity is used for casting and is reversible with an hysteresis during cooling. A representative diagram is shown in Fig. 6.19.

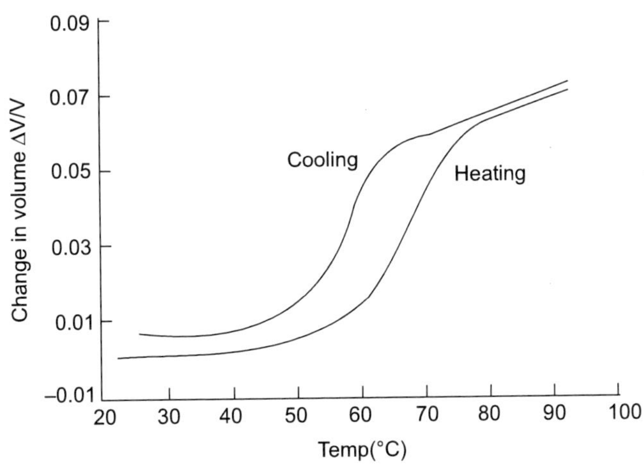

Fig. 6.19 TMA diagram of a wax sample

6.3.5 Crystallization Behaviour

The crystallization behaviour of polyether ether ketone (PEEK) is studied by TMA [9]. In general, the two stage kinetics cannot be clear by DSC. However, the stage is studied by Chen et al. The result is shown in Fig. 6.20, where P indicates primary crystallization and S is secondary crystallization.

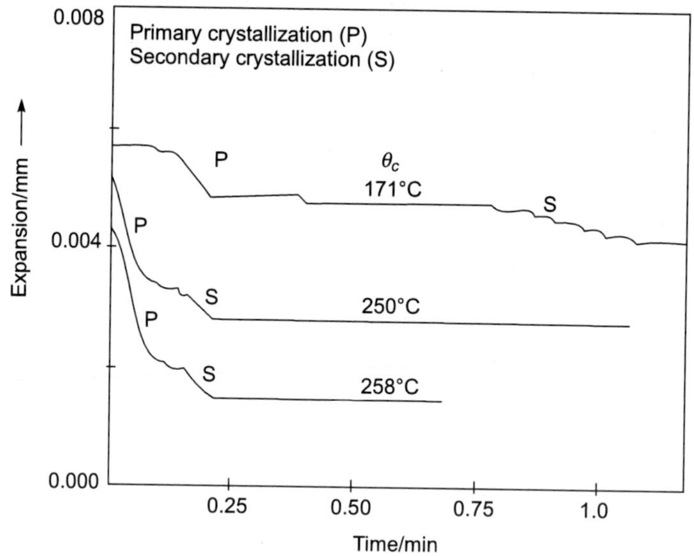

Fig. 6.20 TMA record of PEEK during isothermal crystallization at different temperatures (Adapted from Ref. 9)

In the crystallization curves, the primary crystallization is denoted by *P*, and the secondary crystallization is denoted by *S*. The distinction between these two crystallization stages becomes less clear with increasing temperature. In a polymer primary crystallization has normally been attributed to the formation of molecules (spherulites), and secondary crystallization has been attributed to crystallization taking place inside the molecules (spherulites).

$$\Delta v/V$$

In another development, the technique used was micro drop-coated sensor. The mechanical properties as function to temperature are used as sensor. In principle, the different drops and geometries have been developed with regards to influence the power consumption, the time response, and mechanical behaviour. For this purpose, SnO_2 micro drop is used as a sensing layer.

6.3.6 Modulated Temperature Thermal Mechanical Analysis

The modulated temperature TMA is used to study the behaviour under tension of acrylic fibre with different degrees of orientation. The fibre of all acrylics shrank to different extents on heating but, on deconvolution of the data to separate the contribution from orientation relaxation, they all gave similar values for the coefficient of thermal expansion. This technique provides a method for separating reversible processes such as thermal expansion from irreversible processes such as the deformational changes arising from orientational relaxations in solids. Depending upon the nature of the sample, measurement can be made under tension (film and fibre) or (compression rods or blocks). Data treatment is similar to that used in modulated temperature DSC.

6.3.7 Calibration

There are a number of approaches to temperature calibration. One method uses small flat pieces of pure metals, e.g., indium, sandwiched between thin discs of quartz glass. Melting of the metal sandwich under the probe with a moderate load results in a sharp displacement of the probe. With care, a multi-layer sandwich can be built up using "fillings" of different metals, so that a multi-point calibration can be obtained in one experiment.

6.4 DYNAMIC MECHANICAL ANALYZER (DMA)

6.4.1 Introduction

Dynamic Mechanical Analyzer (DMA) [2,10] is also called modulated force thermomechanometry (mf-TM). In this technique, the sample is subjected to a periodically varying stress (usually sinusoidal of angular frequency, w). The response of the sample to this treatment is measured and provides information about the stiffness of the material (quantified by its elastic moduli) and its ability to dissipate energy (measured by its damping). For viscoelastic materials, the strain resulting from the periodic stress will also be periodic, but will be out of phase with the applied stress owing to the energy dispersion on heat or damping in the sample as shown in Fig. 6.21(a). However, δ is the phase angle between the stress and the strain. Thus, dynamic mechanical analysis (DMA) yields information about the mechanical properties of a specimen placed in minor, usually sinusoidal, oscillation as a function of time and temperature by subjecting it to a small, usually sinusoidal, oscillating force.

If an elastic sample is vibrated over a range of frequencies and the amplitude of vibration is measured, the 'resonance frequency' is that which produces maximum in a plot of amplitude against frequencies. Young's modulus (of elasticity) È, is related to the square of the resonance frequencies, vr:

$\grave{E} = cL^4 \rho vr^2/d^2$, where c is constant, L is the sample length between the clamps, d is thickness and ρ is the density. If the sample is allowed to oscillate freely, it will do so at the natural resonance frequency with decreasing amplitude owing to damping. This form of oscillation is explained in Fig. 6.21(b).

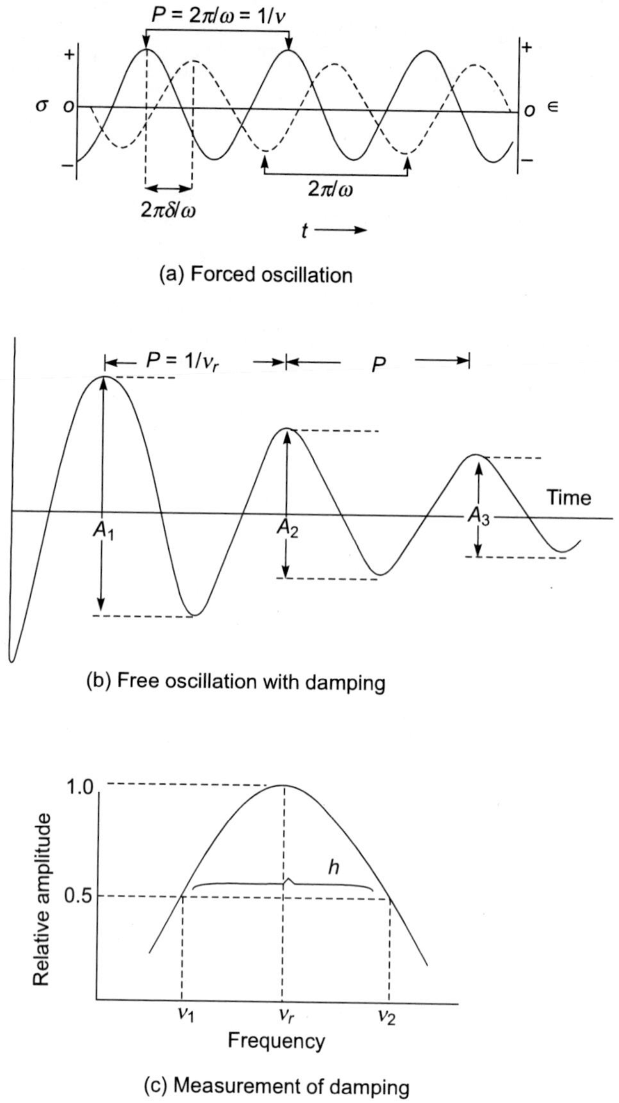

Fig. 6.21 Different types of measurement on DMA (Ref. 2)

The period (P) of the oscillation is constant ($= 1/vr$) and the amplitude (measured from a minimum to the preceding maximum) decays exponentially. The rate of decay measures the amount damping. Damping is usually expressed as the logarithm decrement per cycle, like:

$$\Delta = \log 10\,(A_1/A) = \log\,10\,(A_2/A_3),\ \text{etc.}$$

Units are used in decibels (dB), i.e., $10X\ \log_{10}(A_1/A_2)$, if the amplitude halves, $\log 10\,(A_1/A_2) = \log\,2 = 0.30$. Damping is then 3.0 dB.

Thus, for an elastic material, the modulus E is simply the ratio between the stress and the resulting strain, but for a viscoelastic material, the modulus is complex quantity,

$$E^* = E' + iE''$$

where E' is the storage modulus or in-phase component and E'' is the loss modulus or out-phase component. The ratio E'/E'' is the tangent of the phase angle δ.

Therefore, damping may be determined from the width curve of relative amplitude against frequency as shown in Fig. 6.21(c), $h = (\nu_2 - \nu_1/\nu_1)$ or by measuring the driving force required to maintain a constant amplitude of vibration at the resonance frequency (forced vibration). Damping depends on the physical state of the sample, e.g., at $T > T_g$, polymer dissipates most of the energy supplied and damping is high. At $T < T_g$, the material stores the energy and damping is low. Therefore, the complex modulus form is composed of the storage modulus E' (real part) and the loss modulus E'' (imaginary part). The formulae used for calculating different parameters: complex modulus E^*, storage modulus E', loss modulus E'' and loss factor tan δ are given below. These are dynamic elastic characteristics and are material-specific; their magnitude depends critically on the frequency as well as the measuring conditions and history of the specimen.

$$|E^*| = \frac{\sigma_A}{\varepsilon_A}$$

$$|E^*| = \sqrt{[E'(\omega)]^2 + [E''(\omega)]^2}$$

$$E'(\omega) = |E^*| \cdot \cos\delta$$

$$E''(\omega) = |E^*| \cdot \sin\delta$$

$$\tan\delta = \frac{E''(\omega)}{E'(\omega)}$$

Further, some important information related to modulus characteristics is given in Table 6.3.

TABLE 6.3 Characteristics of different types of modulii

Modulus	Characteristics
Complex modulus	Stiffness of the material (ratio of the stress amplitude to the strain amplitude).
Storage modulus	Proportional to the energy stored during a loading cycle.
Loss modulus	Energy dissipated during a loading cycle.
Phase angle	Phase difference between the dynamic stress and the dynamic strain.
Loss factor	Ratio of loss modulus to storage modulus.

6.4.2 Instrumentation

The representative diagram of a DMA instrument is shown in Figs. 6.22-6.25. The basic components of DMA apparatus are: i) Drive motor, which supplies the periodic stress to the sample, ii) Displacement sensor measures the amplitude and frequency of the response of the sample, iii) Sample clamping system and iv) Furnace and its associated temperature control system. The wide range stress, and optical encoder used instead of linear variable differential transformer (LVDT) are sensitive features of the instrument for better resolution. The optical encoder measures the displacement from light diffraction pattern produced when one grating moves relative to another stationary grating with its attachment. The DMA covers a wide temperature range from –150°C to 600°C. The analyzed samples are usually sheet, rod, film, fibre tested by suitable sample clamp assembly. Some DMA instruments also incorporate a non-contact, direct drive motor to provide the stress over a wide dynamic range from 0.0001 to 18 N. It also uses an optical encoder, instead of traditional linear variable differential transformer (LVDT). The typical resolution of an LVDT is 1 part in 200,000, so to obtain 1 mm resolution, the maximum displacement would be 200 μm. The optical encoder measures displacement from light diffraction pattern produced, when one grating moves relative to other stationary gratings. The resolution is 1 nm over the whole displacement range of 25 nm, corresponding to 1 part in 25,000,000. This increased resolution improves the precision of the measured parameters and allows very stiff materials with very small oscillation amplitude. The cooling attachment in the DMA improves the temperature range of material analysis. However, the clamping system should have high stiffness, low mass and is easy to load and adjust. Some clamping systems available with most DMA instruments are

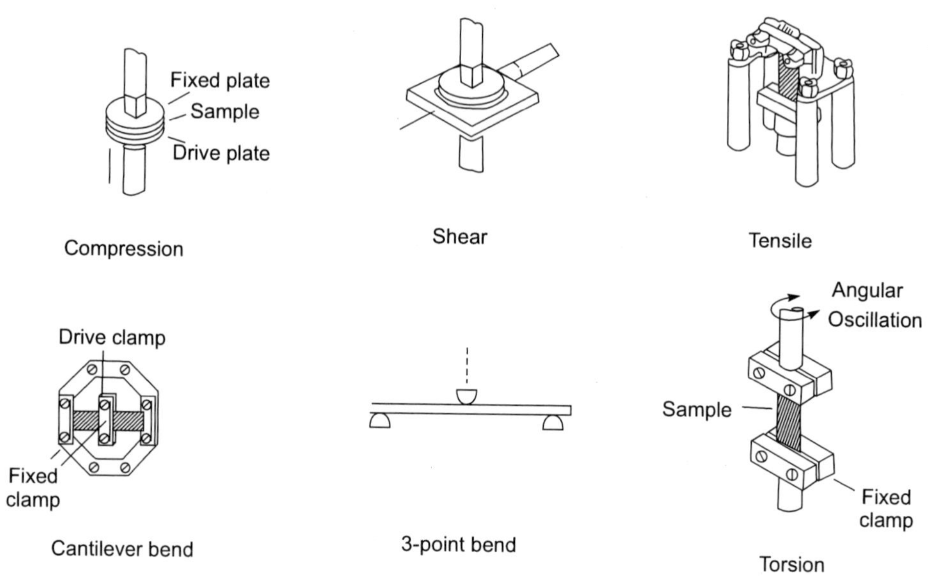

Fig. 6.22 The instrumentation of DMA (Ref. 2)

shown in Fig. 6.23. The specific sample, which is usually in film or fibre form, rigid, thin film can be studied by using specific sample holders. This technique is used for examining mechanical properties of paint as thin film, steel strip. Liquid sample can be supported on film or absorbed into paper or braids can also be analyzed. But, the contribution of the substrate is alone subtracted to give contribution of coating.

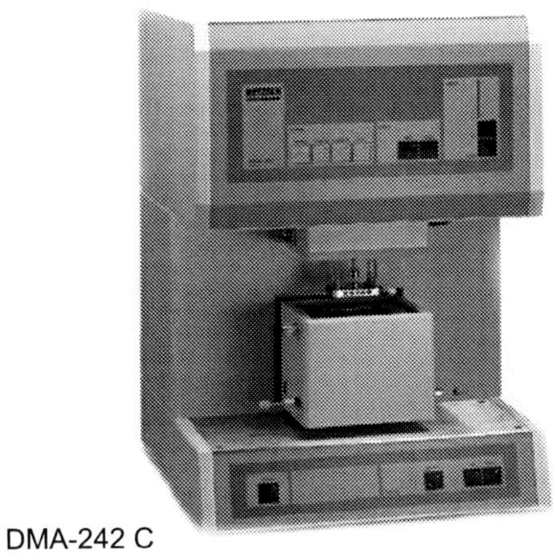

Fig. 6.23 Outer Diagram of DMA-242 C Netzsch

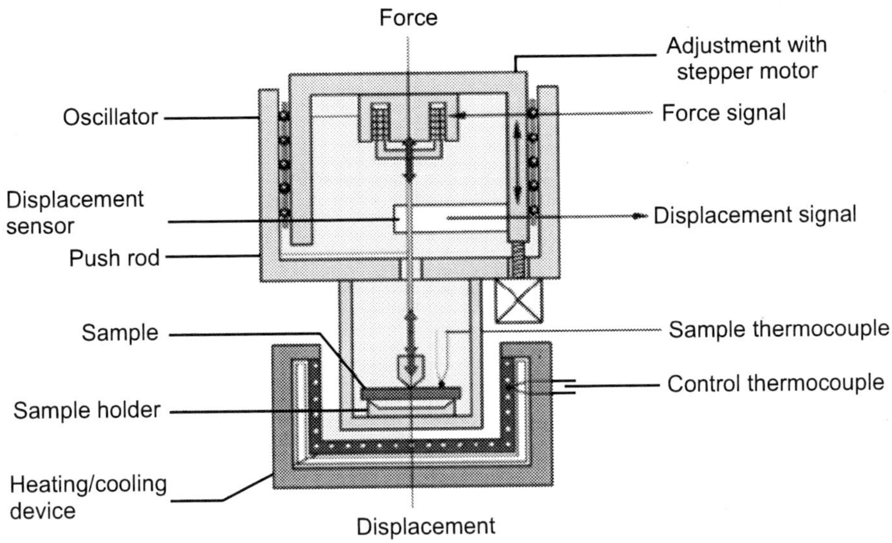

Fig. 6.24 The working of dynamic mechanical analyser (Reproduced with the permission from Netzsch)

6.4.3 Applications of DMA

This is the technique of applying a stress or strain or temperature to a sample and analyzing response to obtain the phase angle and deformation data. Two approaches are used in DMA: (a) forced frequency, where the signal is applied to a set frequency, and (b) free resonance, where material is perturbed and allowed to exhibit free resonance decay. In both the approaches, the technique is very sensitive to the motion of the polymer chain and it is a powerful tool for measuring transitions. It is estimated that DMA is 100 times more sensitive to the glass transition measurement of a polymeric sample than differential scanning calorimeter (DSC). It also resolves other more localized transitions better than DSC. In addition, the technique allows rapid scanning of materials modules and viscosity as a function of temperature or of frequency. DMA may also be referred to as dynamic mechanical thermal analyzer (DMTA).

The change in Young's modulus indicates change in rigidity related to the strength of the sample. Damping measurement gives practical information about the glass transition, change in crystallinity, the occurrence of cross-linking, and also shows the features of polymer chains. The information obtained by DMA has been used in many practical areas such as studies of vibration dissipation, impact resistance and noise abatement. The typical DMA result of two polyethylene samples, linear and branched are shown in Fig. 6.25(a) and (b).

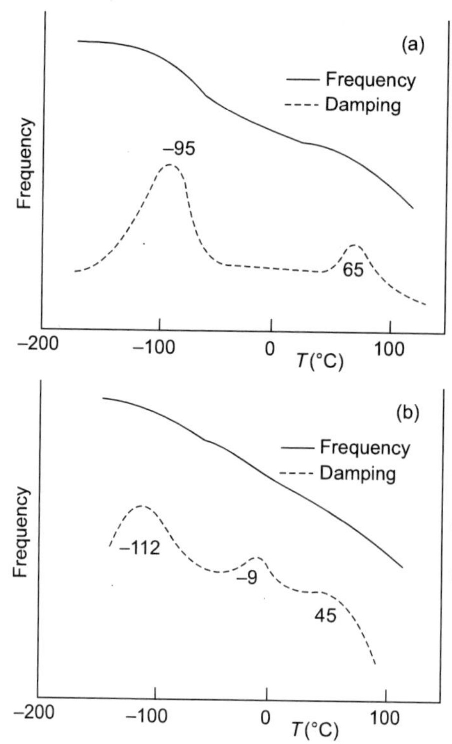

Fig 6.25 A typical DMA curve polyethylene sample (a: linear, b: branched) (Ref. 2)

The damping curve for linear polyethylene (a) shows peaks at –95°C and 65°C. The lower peak has been attributed to long chain $-(CH_2)-$ in crankshaft relaxation in the amorphous phase and the higher peak is because of similar motion in the crystalline phase. Further, damping curve for branched polyethylene (b), has features at –112°C, –9°C and 45°C. Then at –112°C and 45°C peaks are explained as above, while the –9°C peak is attributed to $-(CH_2)-$ relaxation in amorphous phase.

The thermal behaviour of styrene-butadiene rubber (SBR) (its various formulations are used in tyre manufacturing) as illustrated in Fig. 6.26). Different SBR with different isomers and additives are used. A high cis-butadiene content lowers the glass transition temperature, T_g (to as much as –110°C compared to 50°C) giving greater flexibility at low temperature. The addition of carbon black (Fig. 6.26(c)) increases the modulus of elasticity. T_g is also slightly increased. The change in height of DMA damping curve peaks are used as a measure of the extent of vulcanization or additives of elastomer blends.

Further, thermal behaviour of styrene butadiene rubber is illustrated in Fig. 6.26(c) with different styrene – butadiene ratios or isomers or different additives. It is clear that high cis-butadiene content lowers the glass transition temperature T_g, to as much as –110°C compared to –50°C, giving greater flexibility at low temperature. The addition of carbon content increases the modulus of elasticity and T_g is also slightly increased. The complex damping curve at low temperature indicates polymer–carbon black interaction and may lead to adverse properties, e.g., heat build-up. The change in the height of DMA damping curve is used for measure of extent of vulcanization in elastomers blend.

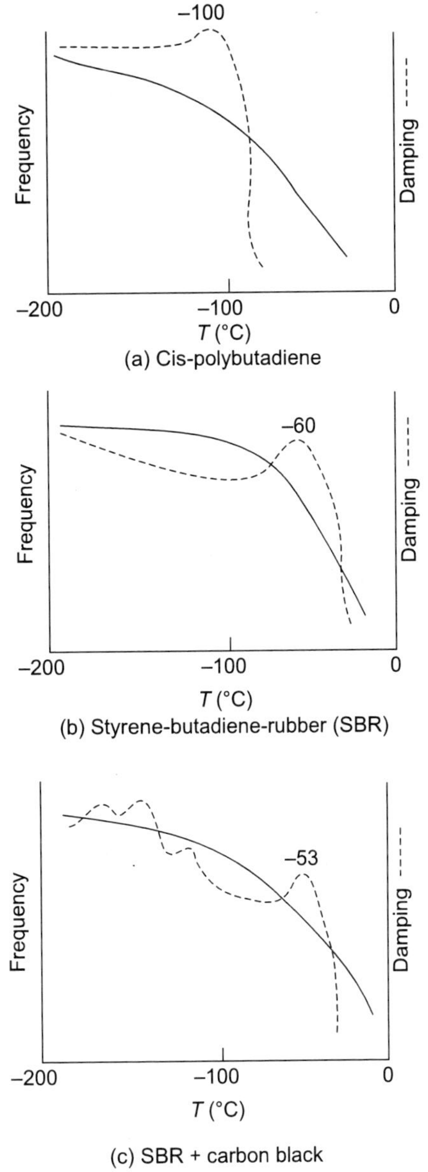

Fig. 6.26 DMA curve of some rubber samples (Ref. 2)

In a study, Kaiserberger et al. have compared the performance of TMA, DMA and DSC for determining the viscoelastic properties of polymers and conclude that DMA has superior sensitivity in detecting phase transition of second and high orders. Internal package stress is also critical information when selecting a material. The combination of the DMA and TMA techniques can provide quantitative information on bulk material internal stress. DMA measures the viscoelastic properties of a material and it provides the modulus value of the material at different temperatures. As material goes through thermal transitions, the modulus value changes allowing the analyst to easily point out the thermal transitions like T_g, crystallization or melting.

DMA gives information on the glass transition, change in crystallinity, the occurrence of cross-linking and also shows features of the polymer chains. The information has been used in various practical areas such as studies of vibration dissipation, impact resistance and noise abatement.

DMA is used to study the effect of ageing on paper. The sample was subjected to accelerated ageing at 100°C and 50% humidity and DMA curve of original and aged samples were recorded. DMA technique was found very sensitive to record second order transition, e.g., frozen solution of sucrose showed two glass transitions at around ~32°C, which suggests that as the solution is frozen tiny localized inclusions are formed, where the concentration differs from the bulk concentration. These inclusions have higher T_g than bulk solution. DSC was not sensitive to record these transitions. In conclusion, the viscoelastic properties recorded by DMA were observed superior to TMA and DSC for detecting phase transition of second and higher orders.

The DMA has wide range of applications in pharmacy research and development. The information helps the scientists to design appropriate properties in the medicine. The Bashaiwoldu et al. [11] has used dynamic mechanical analysis (DMA) to determine mechanical properties of coated paracetamol tablet. The film coatings are applied to pellets for controlled release, taste masking, improved stability, elegance, and mechanical integrity. The tablet was designed to vary its mechanical properties by the incorporation of lactose, and glyceryl monostearate (GMS). DMA was performed on single pellets to determine the storage modulus and phase angle of the coated pellets. This work demonstrated that the coating film affected the mechanical properties of the pellets differently depending on the properties of the core pellets. Analysis of variance established a significant increase in the strength of the soft GMS- or glycerol-containing pellets with coating, while the effect of the coating material was not significant with respect to the elastic modulus, storage modulus, and phase angle of such pellet. The mechanical properties of the pellets were determined from the force-displacement curve obtained during diametral compression test. Coating the pellets with a different weight-by-weight percentage of ethyl cellulose film affected the surface tensile strength of the pellets, but to different extent based on the formulations of the pellets (Fig. 6.27). From the linear regression of the results, it was found that the values of the surface tensile strength of GMS or glycerol-containing pellets increased by 20.4 and 29.5%, respectively, while those of lactose-containing pellets or those of pellets produced with ethanol reduced by 7.4 and 13.5%, respectively, when they were coated to 20% weight gain.

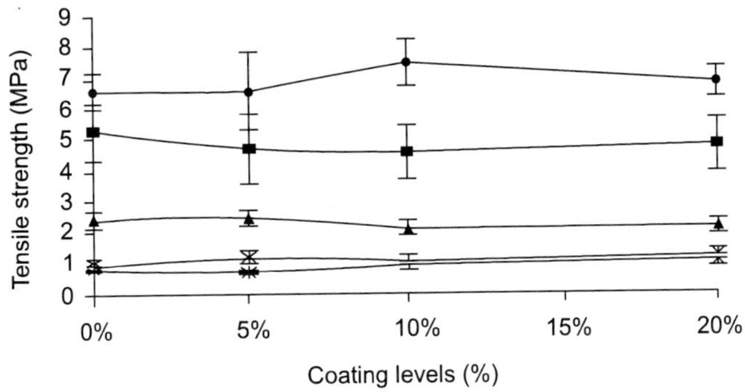

Fig. 6.27 Tensile strength measurement of coated paracetamol tablet with (◆) Microcrysatalline cellulose (MCC); (■) lactose; (▲) ethanol; (×) glyceryl monostearate (GMS); (✱) glycerol

The researcher also concluded the dependency of strength of the pellet, elasticity and deformability and storage modulus of the pellet.

Another important information required for the medicine is its glass transition temperature (T_g). The solid state characterisation of novel pharmaceuticals and polymeric materials frequently involves the measurement of a glass transition temperature (T_g). When the amount of amorphous content is low, DSC encounters difficulties in detecting T_g. DMA's high sensitivity towards the glass transition. It indicates that DMA is an interesting alternative for the detection and measurement of T_g for powdered samples when other techniques fail. Thus, a new DMA sample holder designed to measure glass transition temperatures associated with powders and to determine the experimental parameters that will provide accurate and repeatable values of T_g. Generally, T_g is measured by different techniques such as DTA, DSC as well as DMA. However, DMA is reported as, more sensitive than other methods. The T_g of hydroxypropyl methyl cellulose (HPMC) has been investigated by Mahlin et al. [12] at oscillating strains (1 Hz, 10 Hz and 100 Hz) whilst heated at 3°C/min. The storage and loss modulus signals showed a large reduction in the mechanical strength above 150°C, which was attributed to a glass transition. Thus, DMA to measure the T_g of pharmaceutical powders and a set of parameters were determined that gave an accurate and repeatable value for the T_g of a medicine.

One important use of DMA is characterization of composites and blends. Mixing of different polymers has revealed a new realm of technically important materials. Their properties can be altered by varying the composition of the polymer blends. A blend of two components is classified miscible, thermodynamically, if the Gibbs free energy of mixing is less than zero and the second derivative of the Gibbs free energy of mixing is zero or positive. Experimental evidence of miscibility is often found when a single and a sharp glass transition temperature, T_g, is observed in between the T_g's of the individual components. PVC is well known for its efficiency to form miscible systems with various low or high molecular weight polymers which act as plasticizers. Miscible blends of PVC include its blends with nitrile butadiene rubber (NBR), chlorinated polyethylene (CPE) and epoxidized natural rubber (ENR). PVC: NBR blend is perhaps the oldest commercial blend. The tan δ curves for NBR, PVC: NBR blend and PVC are shown in Fig. 6.28 [13]. The tan δ curve of NBR shows a peak 6.28 at 15°C due to the α transition arising from the segmental motion. This corresponds to the T_g of nitrile rubber. The T_g of the PVC is at 90°C, as shown by a tan δ peak at that temperature in the curve. The tan δ curve for the blend PVC: NBR lies in between (24°C), the two homopolymers indicating miscibility. The tan δ curve for the blend is compared with that of the homopolymers. The tan δ curve shows a peak at –8°C which corresponds to the T_g of that polymer.

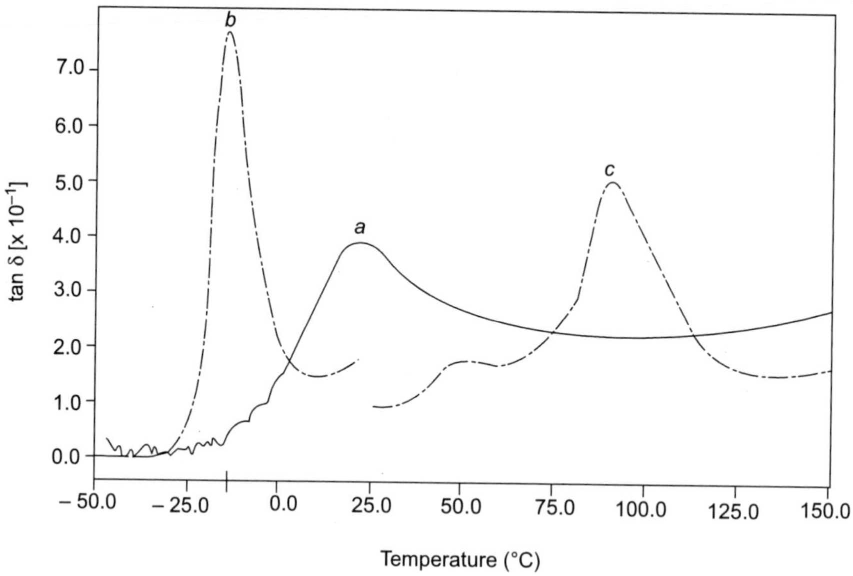

Fig. 6.28 tan δ vs temperature for (a) NBR (b) PVC/NBR and (c) PVC. (Graph adapted from Ref. 13)

Similarly, the technique has promised a wide spectrum of application in material science and technology.

EXERCISE

1. Discuss the principle and significance of thermal mechanical analyzer.
2. What is linear variable differential transformer?
3. Discuss the applications of thermal dilatometer (TD).
4. Discuss the applications of TMA in characterization of wax.
5. Explain the dumping curve with examples.
6. Describe the different probes for TMA analysis with their comparisons.

7. Explain the following abbreviations:
   (a) TD
   (b) MT DMA
   (c) Tg
   (d) MT-DSC
   (e) LVDT
8. Write notes on the following:
   (a) Shear stress
   (b) Liquid dilatometry
   (c) Optoelectronic transducer
   (d) Static force thermomechanometry.
9. Compare the principles of DMA and MT DMA.
10. Explain the thermal mechanical properties of kaolin.
11. Describe the specific features of liquid dilatometer.
12. Compare the following:
    (a) Tensile stress and shear stress
    (b) Glass transition temperature and softening temperature
    (c) Expansion and penetration measurements
    (d) Complex modulus and storage modulus
13. Describe DMA along with its application in characterization of polymers.

REFERENCES

1. Wandlandt, W. W. (1986). Thermal Analysis. John Wiley & Sons, Inc., New York, USA.
2. Brown, M. E. (2001). Introduction to Thermal Analysis: Techniques and Applications. 2nd Edition, Kluwer Academic Publisher, London.
3. Garn, P. D. (1965). Thermo Analytical Methods of Investigations. Academic Press, New York.
4. Willard, H. H., Merrit Jr., L. L., Dean, J. A. and Settle Jr., F. A. (1986). Instrumental Method of Analysis. Wadsworth Publishing Company, USA.
5. Gunzzler, H. and Williams, A. (2001). Handbook of Analytical Techniques. Wiley–VCH, Weinheim, Vol-2.
6. Ewing, G. W. (1990). Analytical Instrumentation Handbook. Marcel Dekker Inc, New York.
7. Meyer, R. A. (2000). Encyclopaedia of Analytical Chemistry. John Wiley & Sons Ltd., Vol 15.
8. Zarrellia, M., Partridge, I. K. and D' Amore, A. (2006). Warpage induced in bi-material specimens: Coefficient of thermal expansion, chemical shrinkage and viscoelastic modulus evolution during cure. Composites: Part-A. 37, 565-570.

9. Chen, H. -L. and Porter, R. S. (1994). Observation of two-stage crystallization of poly (ether-ether ketone) by thermal mechanical analysis. *Thermochim. Acta.* 243, 109-113.
10. Ehrenstein, G. W., Riedel, G. and Trawiel, P. (2004). Thermal Analysis of Plastics: Theory and Practice. Hanser Publisher, Munich.
11. Bashaiwoldu, A. B., Podczeck, F. and Newton, J. M. (2004). Application of dynamic mechanical analysis (DMA) to the determination of the mechanical properties of coated pellets. *Int. J. Pharm.* 274(1-2), 53-63.
12. Mahlin, D., Wood, J., Hawkins, N., Mahey, J. and Royall, P. G. (2009). A novel powder sample holder for the determination of glass transition temperatures by DMA. *Int. J. Pharm.* 371, 120-125.
13. Perera, M. C. S., Ishiaku, U. S. and Ishak, Z. A. M. (2001). Characterisation of PVC/NBR and PVC/ENR50 binary blends and PVC/ENR50/NBR ternary blends by DMA and solid-state NMR. *Eur. Polym. J.* 37, 167-178.

7

Thermo-optometry

7.1 INTRODUCTION

Thermo-optometry refers to the monitoring of optical properties of a material as a function of temperature and is also called thermo-optical analysis [1-4]. The technique can be categorised as:

(a) Thermomicroscopy
(b) Thermophotometry
(c) Thermorefractometry
(d) Thermoluminescence
(e) Microthermal analysis

The technique not only provides important properties of materials but also adds latest amplitude to the other techniques, e.g., scanning tunneling microscope and atomic force microscope, which have led to a new field of micro-thermal analysis. The basic properties derived from thermo-optometry are listed in Table 7.1.

TABLE 7.1 Properties of the techniques of thermo-optometry

Techniques	Properties
Thermomicroscopy	Surface visual appearance
Thermoluminescence	Emitted radiation
Thermophotometry	Intensity of total reflected or transmitted radiation
Thermo-spectrophotometry	Reflected or transmitted radiation of specific wavelength

7.2 THERMOMICROSCOPY

The technique measures the surface appearance, the most obvious property of a material, i.e., its outward appearance, which is often not examined or monitored during the heating process and yet can provide reliable first-hand evidence of the process occurring with the function of temperature while the other analytical techniques give results which have to be associated by the inference with the thermal event and some process such as sintering, decrepitation and creeping are only detectable by direct observation. The schematic diagram is given in Fig. 7.1, consisting of a microscope with hot and cold stage, a sample holder, gaseous atmosphere control unit including vacuum pump, light source and a system for recording and processing visual observations, as well as the temperature of the sample.

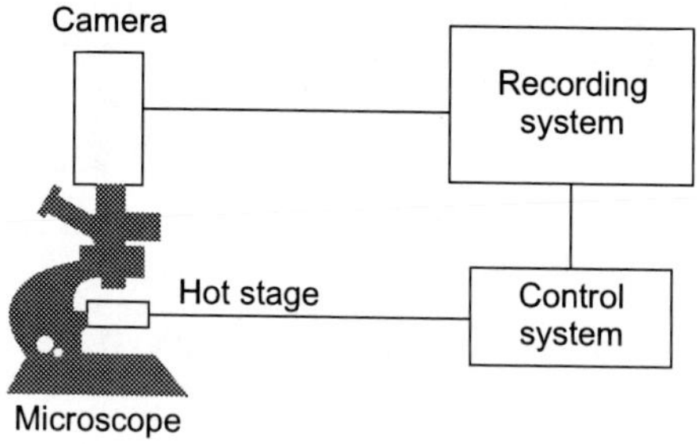

Fig. 7.1 A representative diagram of thermomicroscopy

The main requirement for thermomicroscopy is a suitable heat-resistant objective lens with adequate focal length, controlled temperature stages for microscopes readily available and can cover a wide temperature range from –180°C to 3000°C, allowing the study of varied problems, especially phase study. Some stages are designed to allow combination with the other measurement techniques such as DSC or TG. The stages have to be as thin as possible to enable them to fit between microscope's objective lens and the microscope table. Provision has to be made for heating by the illuminating source, or a heater filter may be necessary. The movement of the sample holder is essential so that different regions of sample can be examined. The sapphire sample glass pan is recommended and standard photographic equipment or video camera can be used for reading observations. Further, different types of software can be used for image analysis to get more analytical information. Sawyer and Grubb have published a review on the use of thermomicroscopy in the study of polymers. Where overlapping transitions, such as melting and degradation or crystallization can lead to ambiguities of interpretation. Kuhnert–Brandstallar has written a classical book on use of thermomicroscopy in pharmaceutics.

7.3 THERMOPHOTOMETRY

This technique is based on the measurement of the intensity of the reflected or transmitted light by the sample. The block diagram of the instrument shown in Fig. 7.2 consists of (i) hot stage – to heat the sample, (ii) photometer – to measure the intensity of reflected light, (iii) camera – to record the visual photographs, (iv) recording system – simple recorder or computer and (v) control system. If the samples are viewed, they transmit light in that arising from the rotation of the plane of the polarization of light, caused by any change in crystalline structure of the sample. Melting or formation of an isotropic structure results in complete extinction of the light.

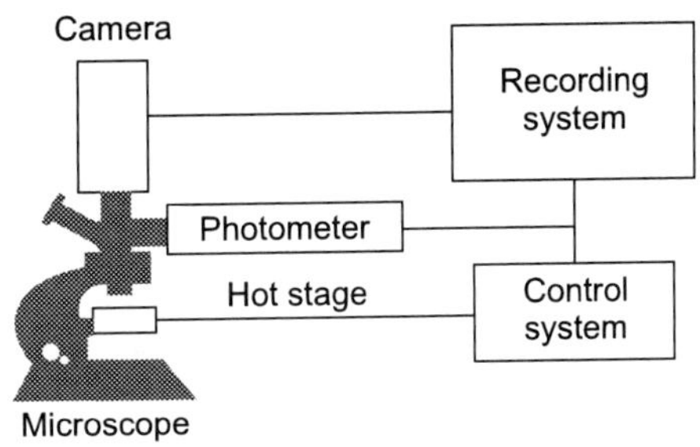

Fig. 7.2 Thermophotometry

The thermophotometry curve of KNO_3: (a) transmitted light and (b) reflected light is shown in Fig. 7.3 as an example.

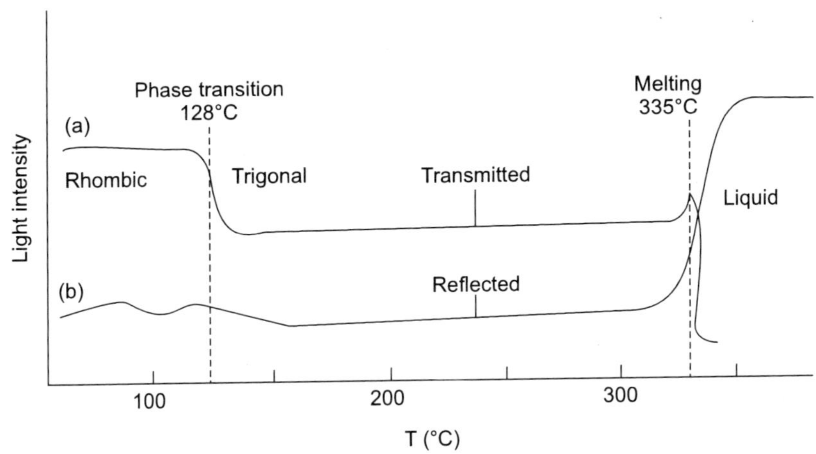

Fig. 7.3 Thermophotometry curve of KNO_3: (a) transmitted light, (b) reflected light

7.4 THERMOLUMINESCENCE

The principle of this technique involved in the measurement of light, being re-emitted (secondary emission) by the sample as function of heating. **Luminescence** generally arises from the annealing of defects in the solid sample as the temperature is raised. The defects are present in the sample on account of its previous thermal history, or through the irradiation of the sample. TL has been frequently used in radiation dosimetry, archeological dating, geological activity and evolution of catalysts, etc. A self-explanatory block diagram of thermoluminescence is shown in Fig. 7.4.

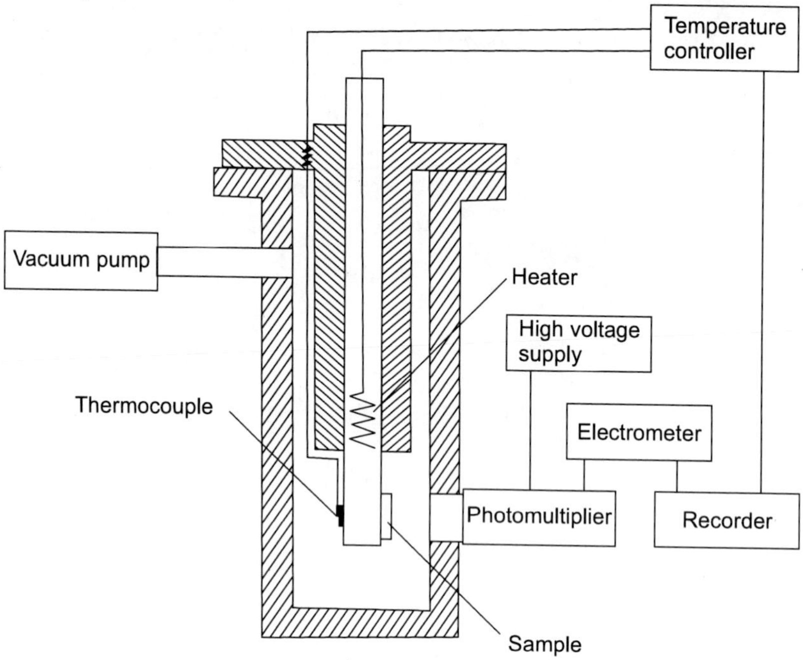

Fig. 7.4 Representative diagram of thermoluminescence

The basic components of TL are: (i) temperature controller, (ii) heater, (iii) photomultiplier, (iv) electrometer, (v) recorder, (vi) thermocouple, and (vii) vacuum pump. Various modifications have also been recommended for making it more useful. TL measurement could be made under isothermal or non-isothermal (dynamic) temperature conditions. Sometimes arrangement for the measurement under reduced pressure is also described.

The use of TL in authentication and characterization of ancient or medieval records has been done by Mriliani et al. [5]. The characterization is based on the change in TL intensity and wavelength of materials such as the pottery. Some materials when heated in air or oxygen exhibit low level light emission called chemiluminesecnce (oxygen luminescence). The intensity of light is proportional to the concentration of oxygen in contact with the sample surface and the presence of stabilizer decreases the intensity. Such measurements can provide valuable insight into the oxidative degradation of polymers and many other organic compounds. A typical result of TL curve is shown in Fig. 7.5. The vigorous redox reaction between the oxidizing group and reducing ligands in coordination compound may also be accompanied by light emission.

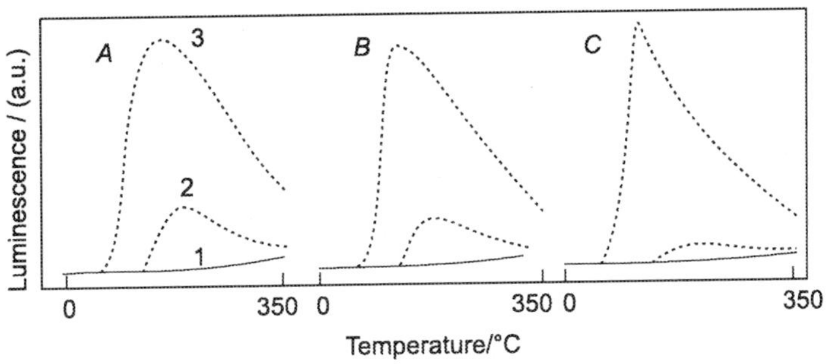

Fig. 7.5 Some representative thermoluminescence curves

Thermoluminescence was demonstrated simultaneously with coupled differential scanning calorimeter and differential thermal analysis by E.P. Manche in 1982 [6]. Modification has been made by employing two-channel potentiometer recorder by Perkin-Elmer Instrument in its dedicated model. One channel measure is the thermoluminescence signal while the other channel measures enthalpy change. The sample and reference materials are heated and the radiance curve is recorded with a polarimeter. The measured area represents the relative thermoluminescence intensity. The TL and DSC curve of KNO_3 in lithium fluoride (LiF) is shown in Fig. 7.6.

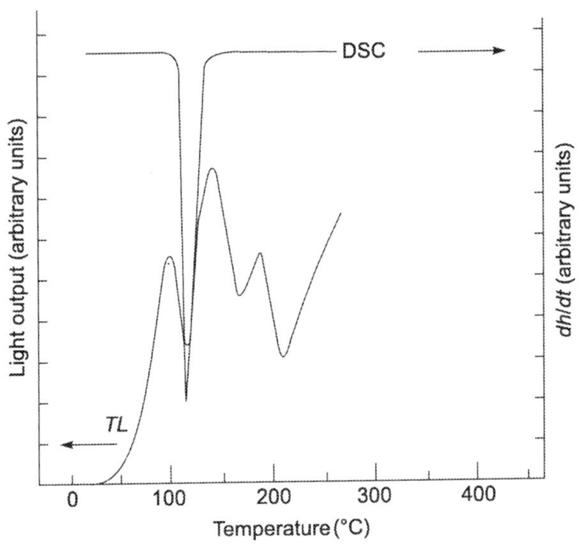

Fig. 7.6 TL and DSC curve of KNO_3 in LiF (Graph adapted from Ref. 3)

Further, with simultaneous technique [7-8], it was possible to correlate the dissociation reaction enthalpy effect with the light emission process. Figure 7.7 shows simultaneous LE-DTA curves of $[Co(NH_3)_6](NO_3)_3$. The result infers that decomposition is as follows:

$$12\,[Co(NH_3)_6](NO_3)_3 \rightarrow 4CO_3O_4 + 18NH_3 + 73H_2O + 4NH_4NO_3 + 39N_2$$

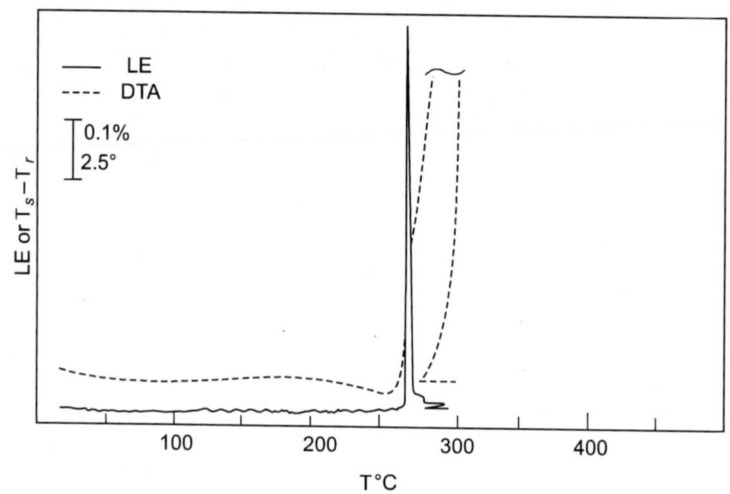

Fig. 7.7 LE-DTA curve of $[Co(NH_3)_6](NO_3)_3$

7.5 THERMOMICROSCOPE COMBINED WITH OTHER TECHNIQUES

The TM and TG-DSC techniques are combined for better application of the technique. The schematic diagram of simultaneous microscope with TG and DSC is shown in Figs. 7.8 and 7.9. Efforts have been made to illustrate the principle by comparing the methods with conventional DSC in Fig. 7.9. Some other techniques may also be combined with TM such as XRD, FT-IR, etc.

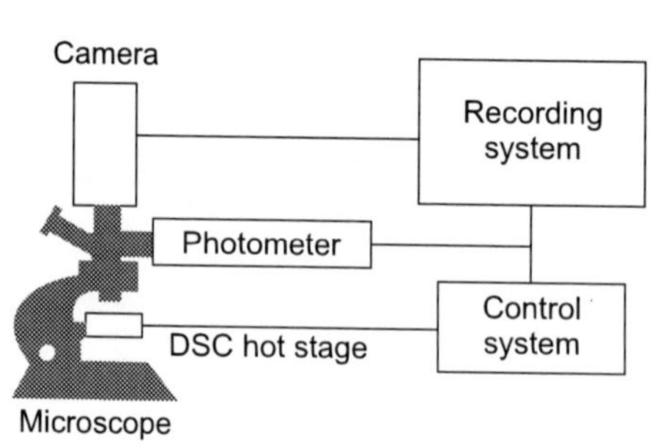

Fig. 7.8 TM combined with DSC

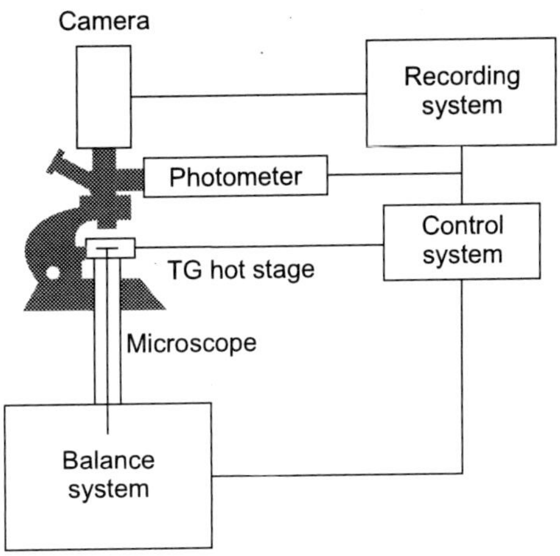

Fig. 7.9 TM combined with TGA

7.6 MICROTHERMAL ANALYSIS

7.6.1 Introduction

Microthermal analysis (μ,TA) is a combination of thermal analysis and atomic force microscopy. This instrument is basically an atomic force microscope (AFM) fitted with a tip that acts as a temperature sensor as well as heat source. The configuration permits thermal characterization of the surface region in addition to the localized high resolution topography imaging ability of conventional AFM. The first application was made by meticulously gluing a tiny shard of diamond onto one end of a tiny strip of gold foil. Further, in 1985 Gerd Binnig and Christoph Gerber used this technique to examine insulating surfaces.

The technique is used to analyze a specimen or regions of specimen as small as 2^2 μm. The sensor is very fine pointed platinum wire mounted on the tip of an atomic force microscope probe (Refer Fig. 4.27). The sample or sample region is heated at a linear rate by passing a current through the wire and the resistance of wire is used simultaneously to measure the sample temperature. In another development, pulsed force mode (PFM) was invented based on a standard AFM setup with a sinusoidal modulation of the *z*-axis piezo. The tip starts above the surface of the sample, approaches, makes contact, indents then is pulled away from the surface. This makes possible a force-distance curve measurement as shown in Fig. 7.11.

Micro-thermal analysis combines the imaging facility of scanning probe microscopy with the ability to characterize, with high spatial resolution, the thermal behaviour of materials [9-10]. The material may be visualized according to its surface topography and also its relative thermal conductivity. Areas of interest may then be selected and localized thermal analysis (TMA and modulated temperature DTA) performed. Applications of this new technique to study semiconductors, polymer blends and biological specimens are described.

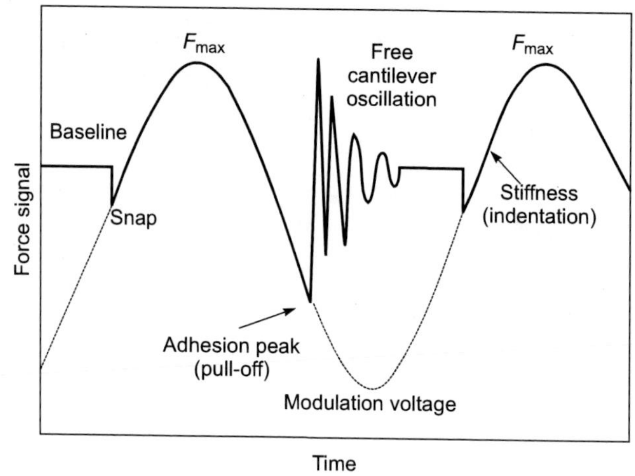

Fig. 7.10 An illustration of force-distance curves

The forces experiencing a probe approaching a surface are illustrated in Fig. 7.10. Initially, there is a weak repulsive force until the probe "jumps" to contact with the surface layer. Further movement towards the surface results in bending of the cantilever due to the stiffness of the surface. Force distance curves are shown for a blend of polystyrene (PS) and poly(methyl methacrylate) (PMMA) as shown in Fig. 7.11. The inset shows a non-contact AFM image of the surface taken at 110°C. The PS forms domains within the PMMA matrix. This can be confirmed by the force distance measurement that shows that the PMMA behaves elastically (it is still below its T_g) whereas the PS behaves in a viscous manner.

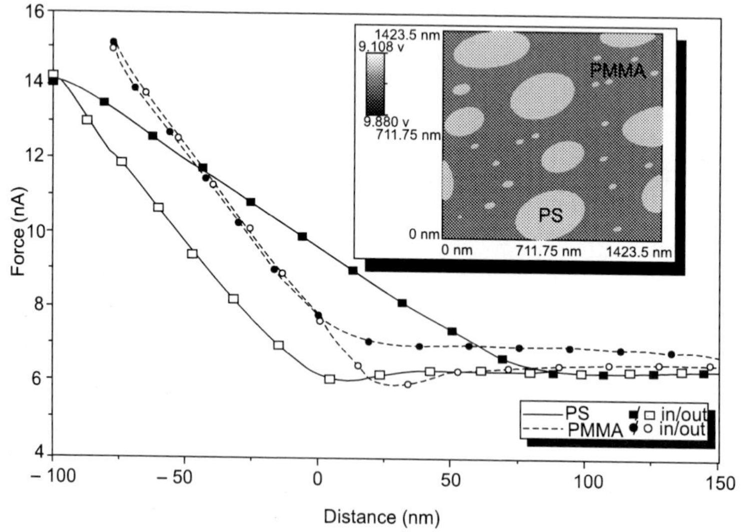

Fig. 7.11 Force distance curves of a blend of polystyrene (PS) and poly(methyl methacrylate) (PMMA)

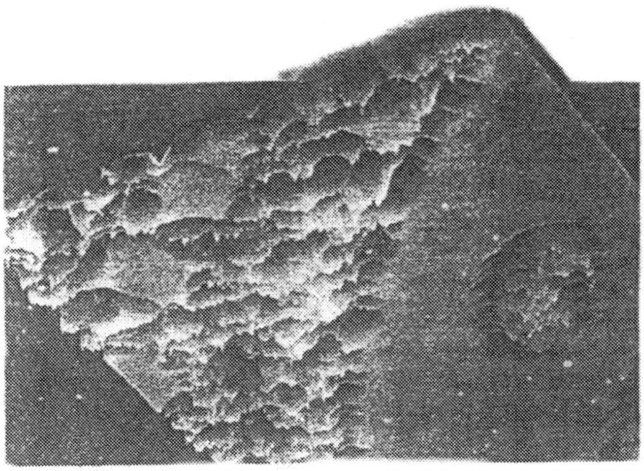

Fig. 7.12 Thermal conductivity image of a polybutadiene/polyvinyl chloride blend

Because of the small thermal mass of the sensor, the heating rate up to 25°C per second can be used. The imaging capabilities of the AFM are used to map the surface area of interest and to select regions for subsequent thermal analysis. The use of a controlled temperature stage allows the entire sample to be heated or cooled and a thermal probe is then not necessary. Standard AFM contact or non-contact probe can be used to produce very high resolution images.

Micro-calorimetric analysis can also be performed by immersion into liquid nitrogen or liquid helium for better arise the internal surface area of the sample. In μDTA, a comparison of the temperature difference or the difference in power supplied to the sample with respect to the reference is done. If the probe is loaded with a small zone, so that it can penetrate the melted specimen, it explores the possibility as micro-thermal mechanical analysis (μ-TMA). The use of sinusoidal-temperature program has led to modulated μTA techniques.

The μTA arrangement can be used to decompose the selected area of sample by heating the probe to 600-800°C. The evolved gas can be absorbed in a suitable sampling tube placed near the probe and analysed later by coupled complimentary techniques such as GC-Mass. μTA has been used for identification of the material being analysed are multilayer film, pharmaceuticals, polymeric blends and composites.

Temperature calibration in μTA is more difficult than in macro TA because the calibrant comes into direct contact with sensor, so metals like In, Sn and Zn are not suitable. Thus, organic compounds and polymers have been used for this purpose. The sensor is easily cleaned after experiment by heating in air, powder-like organic materials need to be formed in large crystals by crystallization from the melt. Some recommended calibrants are given in Table 7.2.

Generally, a two-point calibration procedure is recommended. A suitable crystal with flat surface can be fixed to the μTA sample mounts using double-sided tape. The μDTA response for polymer is different from that of the pure organic compounds

TABLE 7.2 Temperature calibration of different calibrants

Compounds	MP (°C)
Biphenyl	69.3
Benzyl	94.5
Benzoic acid	122.4
Biphenyl acetic acid	147.3
Anisic acid	183.3
2-Chloroanthroquinone	209.6

because the organic melt tends to shrink away from the hot probe and lose contact. However, the less mobile polymer material remains in contact. Thus, for polymer studies, calibration by using Nylon 6, Nylon 66 or polyethylene is recommended. The use of polymer film simplifies the calibration procedure. The precision of the temperature measurement depends on the heating rate. Higher heating rates, 600-1500°C min^{-1} gave better precision than lower heating rates, 120-480°C min^{-1}.

By controlling the temperature of the tip of the μTA probe, the heat flow from the tip to the surface of the sample can be used to record the thermal transport properties of the surface. Such measurements have important applications in microelectronics, cell biology, polymer science, etc. However, calibration is necessarily carried out by using hard materials with known thermal conductivity. The following steps are involved in μTA analysis:

(i) Prepare a sample.
(ii) Heat and calibrate the sensor.
(iii) Position the plane parallel to the sample under the sensor.
(iv) Place the sensor on the area of interest.
(v) Record the thermal conductivity and/or topography image.
(vi) Select the measuring points for thermal analysis.
(vii) Set a suitable measuring program (temperature, frequency, load).

7.6.2 Factors affecting the Results

7.6.2.1 Sample Preparation

The sample analysed on μTA must meet a number of conditions: its roughness should not exceed 10 μm; the sample is polished in the same way as the sample to be examined under microscope. If the study is not aimed at establishing the thermal history, it may be more appropriate to prepare a film from the melt. Furthermore, the sample must placed parallel inside the instrument. If it does not meet this specification by itself, a grinding press and modelling clay may help. Once all the conditions of measuring the sample have been met, the sample is fixed by magnetic support and immobilized on the sample bench. In general "a smooth plane parallel sample is required".

7.6.2.2 *Generating Surface Image*

In the preparation for the measurement, the laser beam needs to be aligned with the centre of the detector. The tip may be carefully positioned over the sample surface. The designed scan area is selected and then the tip is automatically placed on the sample. The surface is scanned over a maximum area of 100 × 100 μm.

The following types of modulation may be required:

(a) Topography: of the scanned area.
(b) Sensors: a continuous visualization of the internal sensor emphasizing topography feature.
(c) Conductivity: a thermal conductivity image generated from the powder needed to keep at desired temperature. An example of image is shown in (Refer Fig. 4.27).
(d) Modulation temperature (Amplitude): visualization based on the response signal for a specified temperature modulated amplitude.
(e) Modulation temperature (Phase): visualization based on the phase shift between the specified and attained temperature modulation.
(f) Z Piezo: the voltage [V] of the Z piezo position is visualized to emphasize topography features.

7.6.2.3 *Choice of Measuring Points*

Choice of measuring points depends on the objective. With a multilayer sample, for example, points in every layer are chosen, at an interface they should be in each surface and boundary zone. As the tip usually penetrates the sample during the measurement and leaves a small crater, the point should be at least 10 μm apart.

Between the individual measuring points, a baseline is measured in air and subtracted from the subsequent experimental curve. When the melt residue clings to the tip, it must be burnt off through heating.

Load

In μTA, a mirror attached to the cantilever reflects the laser beam towards the detector every time the top is deflected. The mirror system and the sensor can be adjusted via screw (load) to make the laser impinge on the centre of the detector. When the tip automatically makes contact on the sample surface, the resultant deflection on the detector can be preset. Because it is possible to adjust the reflection of the laser beam. When the contact has been made and position of the sample predetermined, this affords an indirect means of varying the contact force of the sensor. On the sample, however scope of varying the contact force is very restricted by the great flexibility of the cantilever. Therefore, it is vital to strike a compromise between ready identification of penetration into the sample during softening and ready removal again. If the tip penetrates so far, it may pull some amount of the sample and may even damage.

Measuring Program

In μTA, the starting temperature should be low enough to clearly detect the effect to be measured. The end temperature should be far enough above the measuring effect and below the decomposition temperature. The heating rate in μTA is higher than DSC and TMA because the sample area is very small. However, a higher rate may sometimes mask small and lower visualization effects from beneath the surface. The change in rates shows shifts in melting point and glass transition of materials.

7.6.3 Evaluation

Owing to the very small contact between sensor tip and the sample, it is essential that the experimental result is reproduced several times (at least three). Otherwise the effect cannot be considered significant. The onset temperature of the effect concerned is usually employed in the evaluation of various curves.

EXERCISE

1. Explain thermal optical analysis.
2. List the different components of a thermal microscope.
3. Compare thermo photometry and thermomicroscopy.
4. Describe the instrumentation of thermoluminescence in brief.
5. Explain the main applications of thermoluminescence.
6. Discuss the principle and applications of microthermal analysis.
7. Compare the TG-DTA and TL curves of a compound.
8. Explain the factors affecting microthermal analysis.
9. Describe the principle and applications of atomic force microscope.
10. Write notes on the following:
    (a) Radiation dosimetry
    (b) Electrometer
    (c) Localized thermal analysis
    (d) Force distance curves
    (e) Calibrant
11. Describe the application of TL in archeological dating with examples.
12. Explain the following:
    (a) Use of vacuum in thermoluminescence
    (b) TL is a supplement to differential scanning calorimeter for thermal characterization of a material.

(c) Significance of hyphenated techniques in thermomicroscopy.
(d) Spatial resolution in microthermal analysis.
(e) Force distance curves of a blend of polystyrene.

REFERENCES

1. Brown, M. E. and Kemp, R. B. (1998). Handbook of Thermal Analysis and Calorimetry. P. K. Gallagher (Ed.). New York: Elsevier.
2. Sawyer, L. C. and Grubb, D. T. (1987). Polymer Microscope. Chapman and Hall, New York.
3. Brown, M. E. (2001). Introduction to Thermal Analysis. Kluwer Academic Publisher, London.
4. Gallagher, P. K. (1997). Thermal Characterization of Polymeric Materials, (Ed. Turi, E. A.). Academic Press, New York.
5. Mriliani, C., Forini, N., Morrer, A., Romani, A. and Spigarelli, S. (1998). Authentication and characterisation of pottery sherds from Apricena (FG). *Thermochimica Acta*. 321, 191-195.
6. Manche, E. P. and Carroll, B. (1982). Simultaneous thermoluminescence and differential scanning calorimetry. *Anal. Chem*. 54, 1236-1237.
7. Nuzzio, D. B. (1982). A new instrument for thermoluminescence research. *Thermochim. Acta*. 52, 245-254.
8. a. Wandlandt, W. W. (1980). The thermophotometry of some coordination compounds. *Thermochim. Acta*. 35, 247-253; b. Wandlandt, W. W. (1980). Simultaneous light emission—Differential thermal analysis of selected coordination compounds. *Thermochimica Acta*. 39, 313-320.
9. Price, D., Reading, M. and Lever, T. (1999). Applications of microthermal analysis. *J. Therm. Anal. Calorim*. 56, 673-679.
10. Wandlandt, W. W. (1980). The State-of-the-Art of Thermal Analysis. Eds. Merris, O., Rook, H. L. and Garn, P. D. NBS Spec. Tech. Publ. S 80.

8

Complimentary Techniques

8.1 INTRODUCTION

Thermal analytical techniques are based on the measurement of an exclusive change in properties while material is subjected to program heating. Sometimes observed thermal transitions are not reflected in measured properties. Thus, the measurement is unable to reflect to all the transitions occurring in the material. Hence, there is an increasing preference for the application of coupled techniques [1]. to get all the information related to the materials. The operation of coupled technique provides an analyst a more comprehensive picture of the sample which is being investigated. For example, from DTA, an analyst will be able to identify the various endothermic and exothermic reactions that are occurring as the sample undergoes the prescribed thermal treatment. However, it would be extremely useful if the analyst could also determine if the reaction corresponded to either physical or chemical changes. A thermogravimetry curve would show some physical and chemical changes of a sample as it is heated. Hence, the coupling of these two techniques would lead to a more complete profile for the material that is being analyzed. Very often coupled techniques, e.g., TG/DTA are also referred to as 'hyphenated techniques.

When two analytical techniques are coupled and employed on any given sample within a single run, the term 'simultaneous' is used to refer to this procedure. Now it is also important to understand the advantages and disadvantages associated with the use of simultaneous techniques. The concise discussion on various thermal analytical techniques has already been done in the previous chapter. The present chapter is dedicated to the methods used to detect evolved gases after thermal treatment. The methods used to detect and identify these gases include mass spectrometry (MS), Fourier transform infrared spectroscopy (FTIR), and gas chromatography (GC). The ultimate purpose of developing simultaneous techniques is to devise advanced systems which are capable of incorporating various conventional and thermal analytical techniques within a single system.

8.2 SIMULTANEOUS TG-DTA UNIT

The demand of parallel results stimulates the combining of TG/DTG/DTA or DSC techniques. These units typically comprise thermobalance that monitors the mass change within a sample and the temperature difference between the sample and reference cell as the sample is subjected to a controlled temperature regime. The pioneering TG/DTA system involves the use of a derivatograph to allow for the simultaneous collection of TG/DTA data. A diagram of TG-DTA/DSC coupled instrument model STA PT1010, Linseis International, Germany is shown in Fig. 8.1. The success of the simultaneous unit spurred the emergence of improved by manufacturer, e.g., Setram simultaneous TG/DSC, TA instruments SDT 2960 (TG/DTA) and the Netzsch STA 409 (TG/DSC) system.

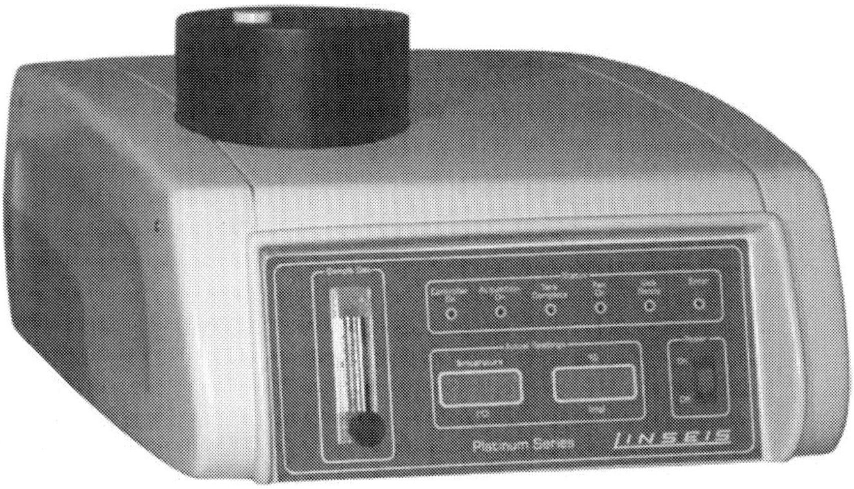

Fig. 8.1 TG-DTA/DSC coupled instrument (STA PT1010, Linseis International, Germany)

TG/DTA unit design is generally based on thermobalance that is equipped with thermocouple leads. The first involves placing the sample and reference cells on one common vertical platform. The second arrangement entails separate sample and reference cell. In this setup, there are two thermocouple leads, and each of them is attached to the base of either sample cell or the reference cell. The modern TG/DTA system should consist of the microbalance with attached thermocouple lead; there should also be a computer station linked to the system. Generally, a direct current (DC) amplifier is also included in the system to enhance the DTA signal. Lastly, there should be a recorder to collect the TG and DTA signals.

In newer model, some more modifications have been incorporated. For example, the TA instrument 2960 includes a horizontal beam balance setup to counteract the effect of purge gas flow into the chamber unit. Setram TG/DSC 111 involves the use of symmetrical dual; furnace is linked with the calvet microcalorimeters, while the Netzsch STA 449 (TG/DSC) incorporates the use of a heat flux panel within the unit. The model allows attachment of other techniques such as IR, GC, as shown in Fig. 8.2.

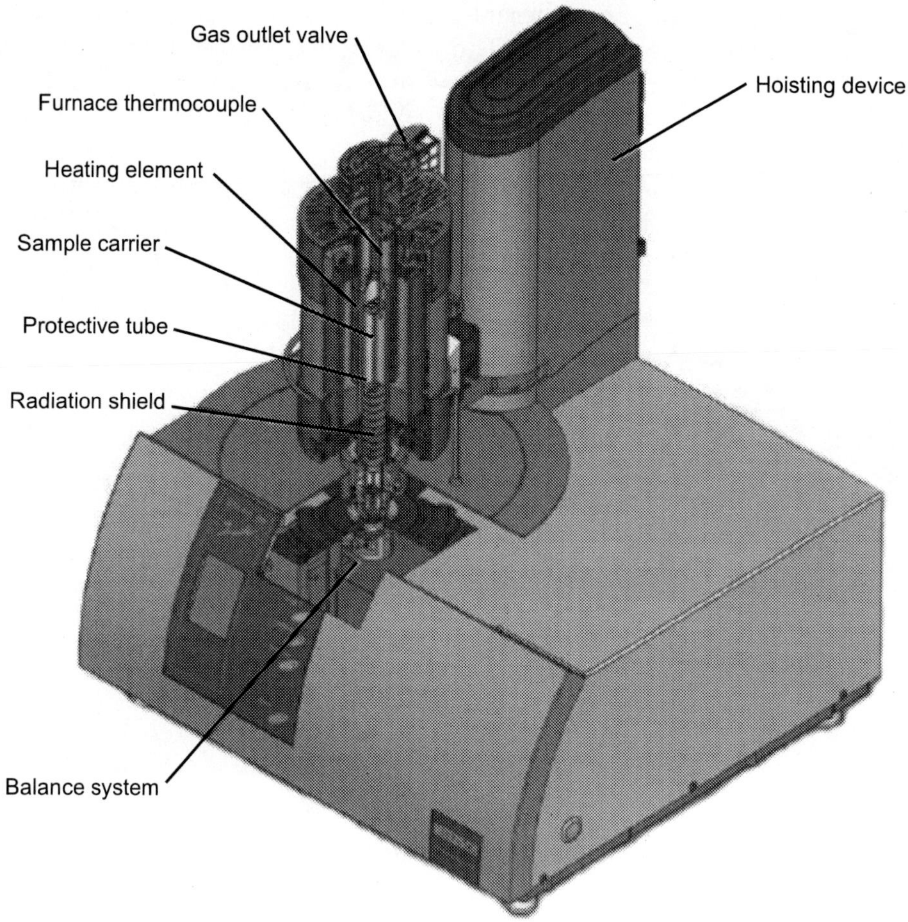

Fig. 8.2 Netzsch STA 449 (TG/DTA/DSC). (Reprinted with permission)

8.3 LIMITATIONS AND ADVANTAGES OF SIMULTANEOUS TG-DTA UNIT

A fundamental design limitation of the TG/DTA technique is that the sensitivity of both the TG and DTA methods is somewhat reduced when coupled. In practice, small sample size and fast heating rates are the recommended parameters if one desires to achieve well resolved DTA curve. This is necessary to ensure rapid development and loss of temperature gradient as the sample is heated. Conversely, in order to enhance the output from TG experiments, large sample size and lower heating rates are advocated. Thus, a compromise has to be made to optimize both signals from the TG/DTA system.

Another disadvantage of TG/DTA equipment is the tendency for sample temperature to deviate from the program temperature over the course of the thermal events. This deviation may have noticeable influence on the measurements involving the rate of weight loss for the sample under investigation. The DTA signal obtained from DTA is slightly different from TG/DTA. The signal obtained from conventional DTA is from crumpled crucible, while in TG/DTA, it is generated from open crucible. The event of the sample undergoing sublimation is inevitable and has a marked effect.

There are several advantages for advocating simultaneous TG/DTA techniques. The obvious incentive is time and cost efficiency from administrative point of view. However, the technical aspect is if separate techniques were used to analyze non-representative sample materials, a direct comparison would be difficult but by using simultaneous techniques, this problem would be eliminated.

Another advantage of simultaneous techniques is that the analyst may ensure that the thermal environment and sample temperature are identical for collecting the data set. In addition, the use of simultaneous techniques also allows for regulation of the ambient temperature and the flow rate condition of an experimental setup. It is often assumed that the atmospheric environment immediately in contact with the sample of two different techniques remains unchanged. In practice, this may differ by virtue of the discrepancies in the geometry of the sample container and the furnace enclosing the sample.

8.4 EVOLVED GAS ANALYSIS AND ASSOCIATED ANALYTICAL TECHNIQUES

The importance of the simultaneous techniques such as TG/DTA leads to the coupling of other conventional analytical techniques with a TG/ DTA system such as MS, FTIR, GC, NMR, XRD, etc. The main goals of coupling these techniques are to enhance the specificity of a single system and to eliminate the problems that are associated with trying to compare data sets that were collected under different experimental conditions. These techniques provide excellent analysis of affluent materials. Some of these techniques are discussed in the following sections.

8.5 MASS SPECTROMETRY

8.5.1 Introduction

The advent of the word *spectrograph* was used in 1884 in "*International Scientific Vocabulary*". In early days, a mass spectroscope configuration was used in early instruments when it was desired that the effects of adjustments be quickly observed. Once the instrument was properly adjusted, a photographic plate was inserted and exposed. The term mass spectroscope continued to be used even though the direct illumination of a phosphor screen was replaced by indirect measurements with an oscilloscope. Mass spectrometry is often abbreviated as *mass-spec* or simply as *MS*. Thomson has also noted that a *mass spectroscope* is similar to a *mass spectrograph* except that the beam of ions is directed onto a phosphor screen. The suffix 'scope' here denotes the direct viewing of the spectra (range) of masses.

Modern techniques of mass spectrometry [2-5] were devised by Arthur Jeffrey Dempster and F.W. Aston in 1918 and 1919, respectively. They won the 1922 Nobel Prize in Chemistry for their work in mass spectrometry. In 1989, half of the Nobel Prize in Physics was awarded to Hans Dehmelt and Wolfgang Paul for the development of the ion trap technique in the 1950s and 1960s. In 2002, the Nobel Prize in Chemistry was awarded to John Bennett Fenn for the development of electrospray ionization (ESI) and Koichi Tanaka for the development of soft laser desorption (SLD) in 1987. However earlier, matrix-assisted laser desorption/ionization (MALDI), was developed by Franz Hillenkamp and Michael Karas; this technique has been widely used for protein analysis. These inventions reflect the utility of the technique.

Mass spectrometry identifies the chemical composition of a compound or sample based on the mass-to-charge ratio of charged particles. In this technique the sample undergoes chemical fragmentation, thereby forming charged particles (ions). The ratio of charge-to-mass of the particles is calculated by passing them through electric and magnetic fields in a mass spectrometer shown in schematic diagram Fig. 8.3. The design of a mass spectrometer has three essential components: an *ion source*, which transforms the molecules in a sample into ionized fragments; a *mass analyzer*, which sorts the ions by their masses by applying electric and magnetic fields; and a *detector*, which measures the value of some indicator quantity and thus provides data for calculating the abundance of each ion fragment present. The technique has both qualitative and quantitative uses, such as identifying unknown compounds; determining the isotopic composition of elements in a compound; determining the structure of a compound by observing its fragmentation; quantifying the amount of a compound in a sample; studying the fundamentals of gas phase ion chemistry (the chemistry of ions and neutrals in a vacuum); and determining other physical, chemical, or biological properties of compounds.

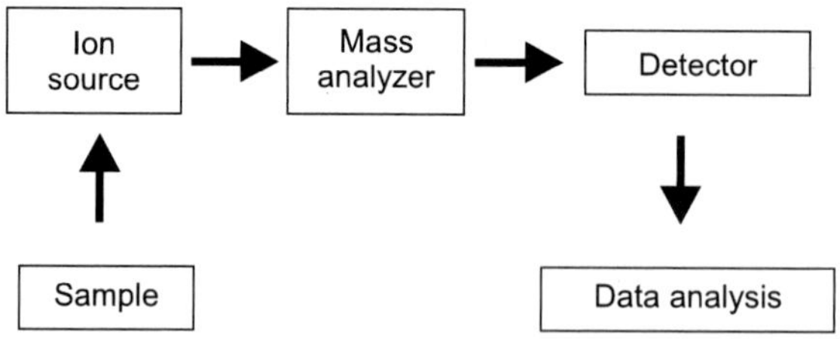

Fig. 8.3 Schematic representation of mass spectrometer

8.5.2 Instrumentation

The technique is based on the measurement of e/m ratio of a charge species generated after fragmentation of analysed species. Different types of measuring techniques have been developed and are in commercial use. Figure 8.4 shows working diagram of mass spectroscopy.

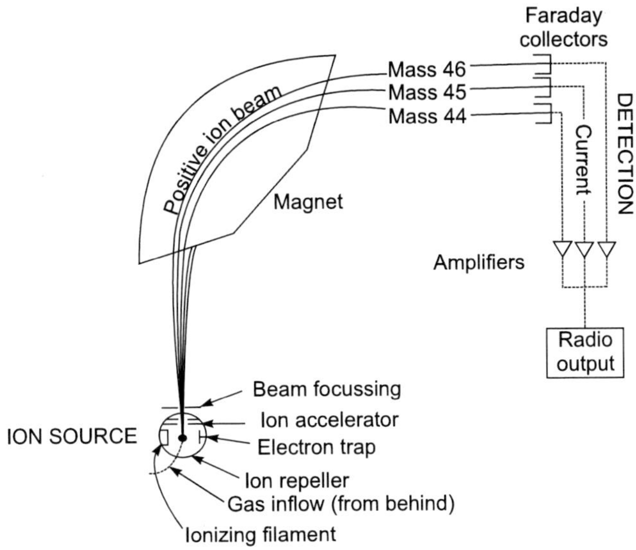

Fig. 8.4 Schematic diagram of a simple mass spectrometer

The instrument consists of the following parts:

1. Ion source

The ion source is the part of the mass spectrometer that ionizes the material under analysis (analyte). It is the integrated form of many components as shown in Fig. 8.4. The ions are then transported by magnetic or electric fields to the mass analyzer. Techniques for ionization have been key for determining what types of samples can be analyzed by mass spectrometry. Some of these are:

(a) Electron and chemical ionization are used for gases and vapours.

(b) Two techniques often used with liquid and solid biological samples include electrospray ionization (invented by John Fenn) and matrix-assisted laser desorption/ionization (MALDI, developed by K. Tanaka and M. Karas and F. Hillenkamp) separately.

(c) Inductively coupled plasma (ICP) sources are used primarily for metal analysis on a wide array of sample types.

(d) Glow discharge, field desorption (FD), fast atom bombardment (FAB), thermospray, desorption/ionization on silicon (DIOS), direct analysis in real time (DART), atmospheric pressure chemical ionization (APCI), secondary ion mass spectrometry (SIMS), spark ionization and thermal ionization (TIMS) are other tools for ionization.

(e) Ion attachment ionization is a new soft ionization technique that allows for fragmentation free analysis.

2. Mass analyzer technologies

The effectiveness of mass spectroscopy depends on the mass analyzer technology applied. Some of them are as follows:

(a) **Time-of-flight:** The time-of-flight (ToF) analyzer uses an electric field to accelerate the ions through the same potential, and then measures the time they take to reach the detector. If the particles have the same charge, the kinetic energies will be identical, and their velocities depend only on their masses. Lighter ions will reach the detector first.

(b) **Quadrupole:** Quadrupole mass analyzers use oscillating electrical fields to selectively stabilize or destabilize ions passing through a radio frequency (RF) or quadrupole field. A quadrupole mass analyzer acts as a mass selective filter and is closely related to the quadrupole ion trap, particularly the linear quadrupole ion trap except that it operates without trapping the ions and is for that reason referred to as a transmission quadrupole. A common variation of the quadrupole is the triple quadrupole.

(c) **Quadrupole ion trap:** The quadrupole ion trap works on the same physical principle as the quadrupole mass analyzer, but the ions are trapped and sequentially ejected. Ions are created and trapped in a mainly quadrupole RF potential and separated by m/Q, non-destructively.

(d) **Fourier transform ion cyclotron resonance:** Fourier transform ion cyclotron resonance MS measures mass by detecting the image current produced by ions cyclotroning in the presence of a magnetic field. Instead of measuring the deflection of ions with a detector such as an electron multiplier, the ions are injected into a penning trap (a static electric/magnetic ion trap) where they effectively form part of a circuit.

3. Detector

The final element of a mass spectrometer is the detector. It records either the charge induced or the current produced when an ion passes by or hits a surface. In a scanning instrument, the signal produced in the detector during the course of the scan versus where the instrument is in the scan (at what m/Q) will produce a mass spectrum, a record of ions as a function of m/Q.

Typically, some type of electron multiplier is used, though other detectors including Faraday cups and ion-to-photon detectors are also used. Because the number of ions leaving the mass analyzer at a particular instant is typically quite small, considerable amplification is often necessary to get a signal. Generally, micro-channel plate detectors are commonly used in modern commercial instruments. In FTMS, the detector consists of a pair of metal surfaces within the mass analyzer/ion trap region in which the ions pass near oscillate.

8.5.3 Interfacing of TG-MS

The critical requirement to interface a mass spectrometer to TG is by a capillary in a common instrument. The details of combination can be understood with the help of Figs. 8.5 and 8.6.

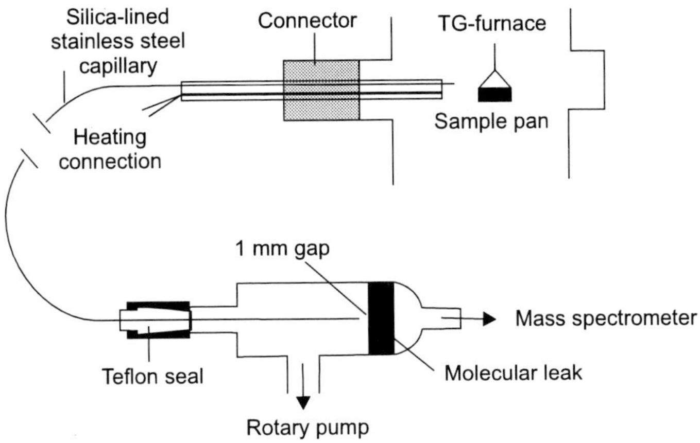

Fig. 8.5 Interfacing of TA and MS instrument (Adapted from www.tgperkin.com)

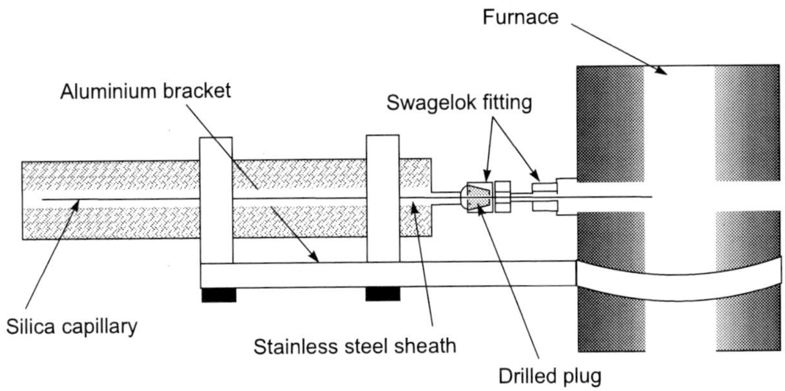

Fig. 8.6 TG-MS capillary interface (Adapted from www.tgperkin.com)

The additional benefit of attaching MS spectra with TG is additional information for the interpretation of the reactions in the TGA results can be inferred, the sensitivity of the method for the analysis of gaseous reaction products can be achieved. The exact control of the furnace atmosphere before starting and during the experiment can be made and the location of air leaks around the furnace will be achieved.

8.5.4 Applications

(a) Interpretation of mass spectra [6]: The mass spectra used to derive precise structure of a molecule is deciphered through a set of fragment masses; the interpretation of mass spectra requires combined use of various techniques. Usually, the first strategy for identifying an unknown compound is to compare its experimental mass spectrum against a library of mass spectra or software assisted interpretation of mass spectra are performed. Software taking advantage of this idea has been developed for both small molecules and complex molecules for analysis.

Another way of interpreting mass spectra involves spectra with accurate mass. A mass-to-charge ratio value (m/z) with only integer precision can represent an immense number of theoretically possible ion structures. More precise mass figures significantly reduce the number of candidate molecular formulas, albeit each can still represent large number of structurally diverse compounds. A computer algorithm called formula generator calculates all molecular formulas that theoretically fit a given mass with specified tolerance and are currently in use.

A recent technique for structure elucidation in mass spectrometry, called precursor ion fingerprinting identifies individual pieces of structural information by conducting a search of the tandem spectra of the molecule under investigation against a library of the product-ion spectra of structurally characterized precursor ions.

(b) Identification of effluent chemical species: We already have discussed in details the thermal decomposition mechanism which occurs only because of elimination of a volatile product due to thermal decomposition in Chapter 3. A case of calcium oxalate is discussed for decomposition in three steps but exclusively which chemical species is coming out can be only confirmed by TG-MS spectra as shown in Fig. 8.7. The spectra is showing three peaks for 18, 28 and 44 m/e. They are reasonably concluded because of the elimination of H_2O, CO and CO_2 chemical species. The intensity of these peaks once again supports the percentage contribution of these gases.

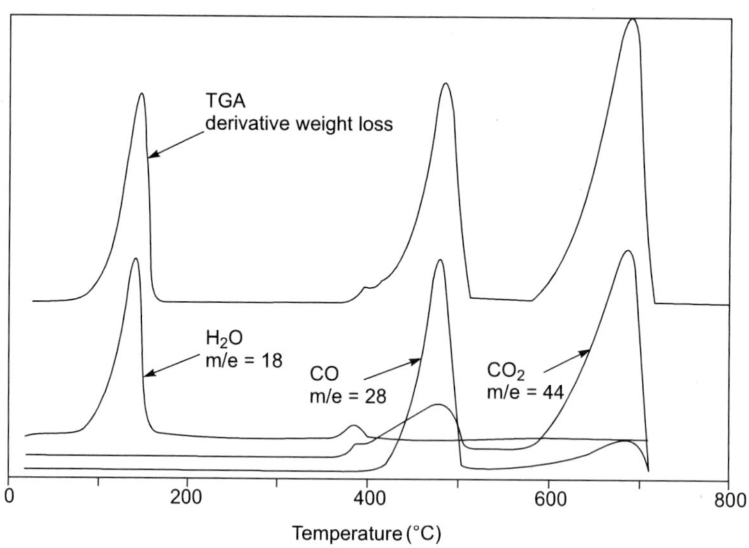

Fig. 8.7 TG-MS spectra of hydrated calcium oxalate (Adapted from free materials available on website of TA instrument)

(c) Determination of polymer composition: The conclusive composition can be determined by using TG-MS analysis. The TG-MS result of ethylene vinyl acetate copolymer is recorded in Figs. 8.8(a) and (b). The first peak observed for 60 amu refers to elimination of acetic acid with its exact quantitative elimination 14% and 40%, while the second peak above 400°C is for removal of hydrocarbon for 56 amu. The result concludes exact composition of copolymerized monomers.

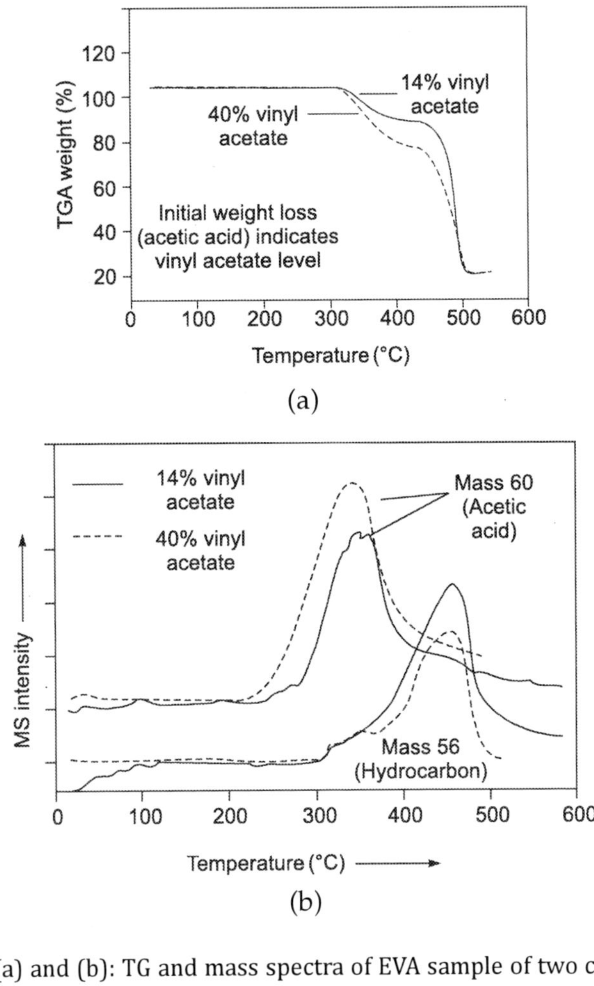

Fig. 8.8 (a) and (b): TG and mass spectra of EVA sample of two compositions

(d) Smoke generation in flame retardant polymers: The burning of polymers generates smoke in different amounts. This can be estimated by weight loss but which chemical species is produced during burning can only be concluded by couple TG-MS spectra. Figure 8.9 is TG-MS spectra of flame-retardant polymers containing MoO_3 as an additive. The MS spectra shows peaks for 78 amu because benzene is in lower intensity if it is doped with molybdenum oxide while high intense peaks in case of pure PVC. The result also confirms it being effective flame retardant.

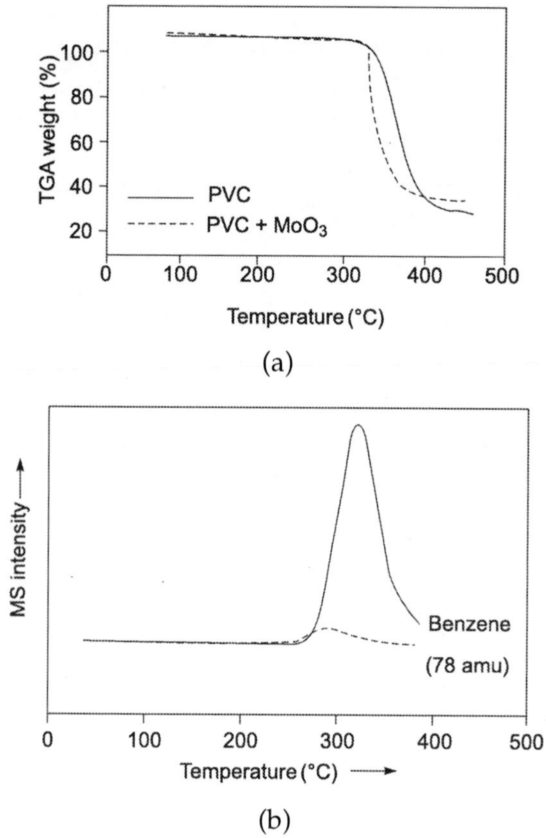

Fig. 8.9 (a) and (b): Spectra of flame retardant polymers containing MoO_3 as an additive (adapted from TA instrument's website)

The totally dedicated MS spectrometry can be used for many advanced applications other than TG-MS.

TG-MS coupled technique can also be used for establishing degradation behaviour such as the presence of azobenzenic moieties in the chain induces a liquid crystalline behaviour, but, due to the high values of the transition temperature, some precautions during the thermal characterization are necessary. In the case of azobenzenic samples, the degradation reactions begin, as a function of the chemical structure. A degradation mechanism is based on chain transfer reactions. The chain flexibility influences the thermal degradation mechanism; in the case of rigid polymers, the chain transfer reactions are less probable. For the flexible chains, the thermal stability is not essentially influenced by the copolymerization ratio of the monomers. Some important areas for the use of TG- MS studies are: trace gas analysis, pharmacokinetics, protein characterization, space exploration, respired gas monitor, etc.

However, the effectiveness of TG-MS also needs to be evaluated and a procedure for the evaluation of the performance of a TG/MS system is proposed by M. Statheropoulos et al. [7]. This procedure includes tests for monitoring the mass-flow stability, the gas transfer delay, the evolved-gas condensation effect and the interferences of the coupling system on TGA and specifications. The results of such evaluation procedure can be used for monitoring the performance of a specific TG/MS system and, thus, can be a basis for design improvements. In addition, they may be used for comparison of the results of different TG/MS systems, or for inter-laboratory comparisons.

8.6 TG-INFRARED SPECTROSCOPY (TG-IR)

8.6.1 Introduction

The coupling of IR with thermal analysis unit to perform EGA is another well established method for analysis of effluent gaseous mixture. The main features require the utilization of infrared instrument. A brief introduction is discussed herewith.

Infrared spectroscopy (IR spectroscopy) is the subset of spectroscopy that deals with the infrared region of the electromagnetic spectrum [8-10]. It covers a range of techniques, the most common being a form of absorption spectroscopy. It can be used to identify compounds or investigate sample composition.

The infrared portion of the electromagnetic spectrum is divided into three regions; the near-, mid- and far-infrared, named for their relation to the visible spectrum. The far-infrared, approximately 400–10 cm^{-1} (1000–30 μm), lying adjacent to the microwave region, has low energy and may be used for rotational spectroscopy. The mid-infrared, approximately 4000–400 cm^{-1} (30–1.4 μm) may be used to study the fundamental vibrations and associated rotational-vibrational structure. The higher energy near-IR, approximately 14000–4000 cm^{-1} (1.4–0.8 μm) can excite overtone or harmonic vibrations. The names and classification of these sub-regions are merely conventions. They are neither strict divisions nor based on exact molecular or electromagnetic properties.

Infrared spectroscopy exploits the fact that molecules have specific frequencies at which they rotate or vibrate corresponding to discrete energy levels (vibrational modes). These resonant frequencies are determined by the shape of the molecular potential energy surfaces, the masses of the atoms and, by the associated vibronic coupling. In order for a vibrational mode in a molecule to be IR active, it must be associated with changes in the permanent dipole, particularly, in the Born-Oppenheimer approximations, i.e., when the molecular Hamiltonian corresponding to the electronic ground state can be approximated by a harmonic oscillator in the neighbourhood of the equilibrium molecular geometry, the resonant frequencies are determined by the normal modes corresponding to the molecular electronic ground state potential energy surface. Nevertheless, the resonant frequencies can be in the first approach related to the strength of the bond, and the mass of the atoms at either end of it. Thus, the frequency of the vibrations can be associated with a particular bond type. Summary of absorption of bonds in molecules is given in Fig. 8.10. Wave numbers are listed in cm^{-1}.

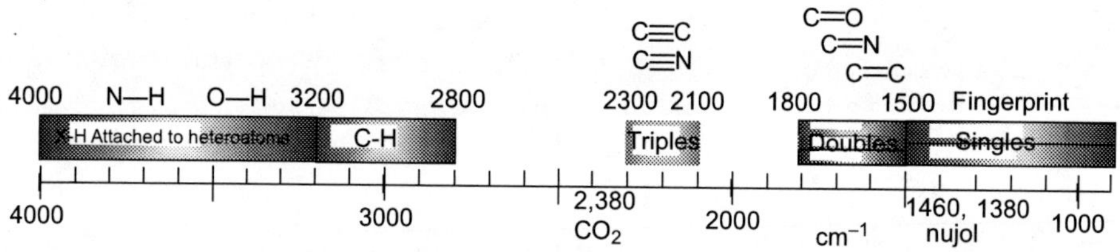

Fig. 8.10 Summary of IR bands

The simple diatomic molecules have only one bond, which may stretch but more complex molecules have many bonds, and vibrations can be conjugated, leading to infrared absorptions at characteristic frequencies that may be related to chemical groups. For example, the atoms in a CH_2 group, commonly found in organic compounds can vibrate in six different ways: symmetrical and antisymmetrical stretching, scissoring, rocking, wagging and twisting.

8.6.2 Instrumentation

The infrared spectrum of a sample is collected by passing a beam of infrared light through the sample. Examination of the transmitted light reveals how much energy was absorbed at each wavelength. This can be done with a monochromatic beam, which changes in wavelength over time, or by using a Fourier transform instrument to measure all wavelengths at once. From this, a transmittance or absorbance spectrum can be produced, showing which IR wavelengths the sample absorbs. Analysis of these absorption characteristics reveals details about the molecular structure of the sample. The block diagram of the instrument is shown in Fig. 8.11 with coupled TG indicating three blocks: (a) IR spectroscopy, (b) Interface and (c) TG instrument.

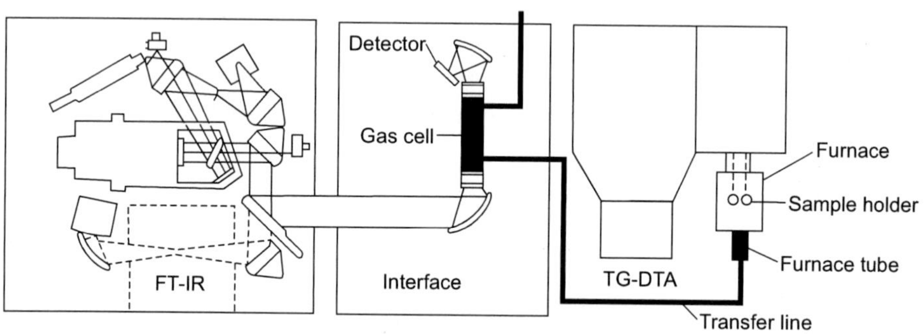

Fig. 8.11 Schematic diagram of TG-DTA coupled FT-IR technique

This technique works almost exclusively on samples with covalent bonds. Simple spectra are obtained from samples with few IR active bonds and high levels of purity. More complex molecular structures lead to more absorption bands and more complex spectra. The technique has been used for the characterization of very complex mixtures.

In general, the number of modes of vibrations are either 3N-5 or 3N-6 for linear and nonlinear molecules. However, these frequencies of vibration are related as per Eq. (1)

$$\nu = \frac{1}{2\pi}\sqrt{\frac{f}{\mu}} \text{ (Explain the term)} \quad (1)$$

where, ν - Frequency
F - Force constant
μ - Reduced mass

There are several advantages to this method of TG-IR including high specificity and relatively fast data acquisition times. In addition, this is also an ideal technique for the analysis of samples which have proven to be difficult to study the mass spectrometer. Some typical examples of such materials are carbon monoxide, carbon dioxide and H_2O.

In a typical IR apparatus, a beam of infrared light is produced and split into two separate beams. One is passed through the sample, the other through a reference. The two signals are then compared and a printout is obtained.

8.6.3 Applications

Infrared spectroscopy is widely used in both research and industry as a simple and reliable technique for measurement, quality control and dynamic measurement. It is of special use in forensic analysis in both criminal and civil cases, enabling identification of polymer degradation for example. IR is perhaps the most widely used method of applied spectroscopy. The instruments are now small, and can be transported, even for use in field trials. With advanced technology in computer filtering and manipulation of the results, samples in solution can now be measured accurately (water produces a broad absorbance across the range of interest, and thus renders the spectra unreadable without this computer treatment). Some instruments will also automatically tell you what substance is being measured from a store of thousands of reference spectra held in storage.

By measuring at a specific frequency over time, changes in the character or quantity of a particular bond can be measured. This is especially useful in measuring the degree of polymerization in polymer sample. Modern research instruments can take infrared measurements across the whole range of interest as frequently as 32 times a second. This can be done whilst simultaneous measurements are made using other techniques. This makes the observations of chemical reactions and processes quicker and more accurate.

Techniques have been developed to assess the quality of natural products such as tea leaves using infrared spectroscopy. This will mean that highly trained experts (also called 'noses') can be used more sparingly, at significant cost saving.

Infrared spectroscopy has been highly successful for applications in both organic and inorganic chemistry. Infrared spectroscopy has also been successfully utilized in the field of semiconductor microelectronics. For example, infrared spectroscopy can be applied to semiconductors like silicon, gallium arsenide, gallium nitride, zinc selenide, amorphous silicon, silicon nitride, etc.

TG-IR analysis of layers of a cable is used in the automotive industry. The cable is composed of five concentric layers disposed in the following sequence: rubber, fabric, rubber, fabric, rubber. As the polymeric material of the cable is carbon black-filled, its identification by infrared spectroscopy was not applicable. Thus TG-IR technique can be an optimized technique, as it allows the identification of a material through its decomposition products. Such analysis is completely free from interferences due to the presence of carbon.

During the TG analysis process, spectra of the gas released by the sample are acquired at a sampling frequency of one spectrum every 4 seconds, and with a spectral resolution of 4 cm^{-1}. Three main weight losses are seen in TG-IR spectra of cable: the first around 15th minute, the second around 24th minute, and the third around 29th minute, together with the weight loss in TG, three TG-IR peaks are detected as shown in Fig. 8.12(a). These peaks were observed because the sample degrades with releasing gases. The details of the acquired spectra then proceeds to

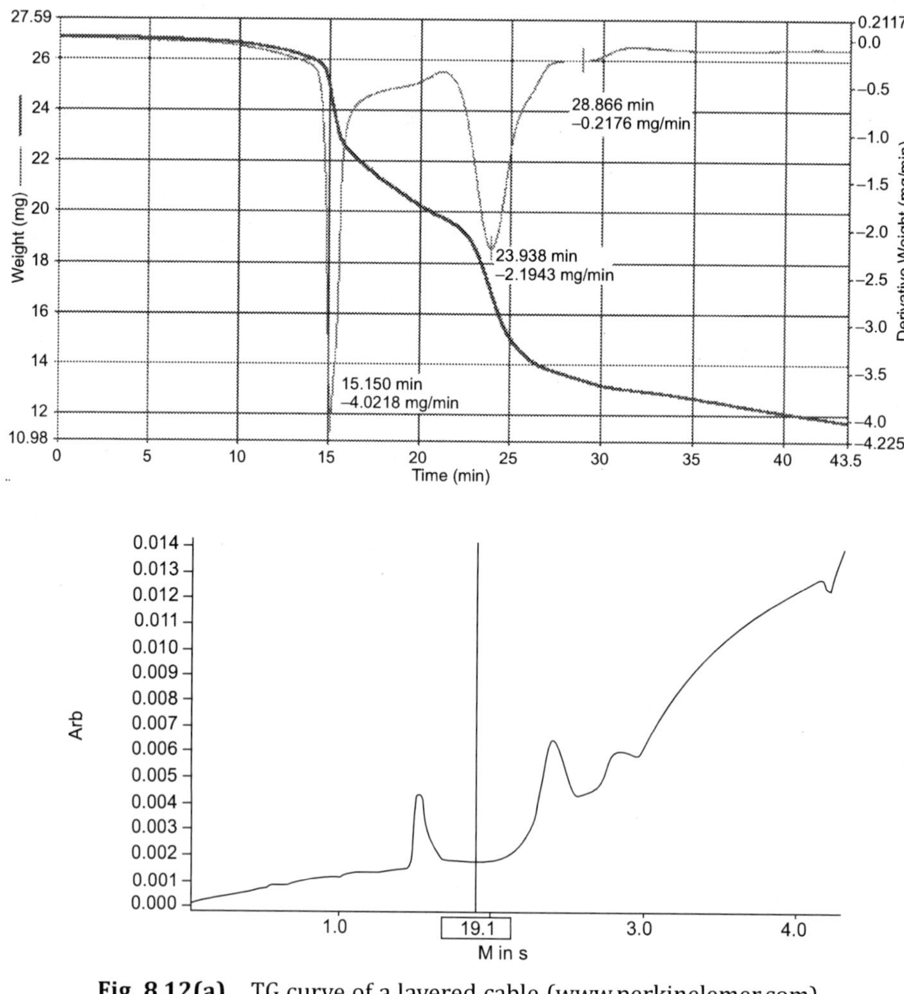

Fig. 8.12(a) TG curve of a layered cable (www.perkinelemer.com)

the identification of the released gases. Ten minutes after start-up of the analysis (~T = 200°C), there is a slight decline in the curve; from this point onwards the spectra shows the stretching band of the $-CH_2$ and $-CH_3$ groups. The limited amount of CO_2 generation starts and reaches a maximum at 15th minute. The carbon dioxide peak probably derives from the CO_2 loss of the vinyl-acetic group of the polymer. Further, the peak related to the organic part could originate from the evaporation of an additive or an early decomposition of the polymeric matrix. The possibility of hypothesis would also appear to be supported by the fact that immediately after, a marked generation of HCl occurs. In Fig. 8.12(b) the intensity profiles related to hydrochloric acid and A peak are compared; shows the HCl spectrum superimposed on the A peak as it appears before data processing. Fig. 8.12(b) shows the HCl band after the subtraction of peak A: Hydrochloric acid emission during the analysis confirms the presence of chlorine in the sample. Let us now move to the analysis of the gases released during the second important weight loss, occurring around 24th minute (about 500°C). At this temperature, the polymer decomposes markedly, giving off different gases due to chain fragmentation: the presence of methane, ethylene and medium-short alkyl chain fragments can be noted. Obviously, the spectra of these

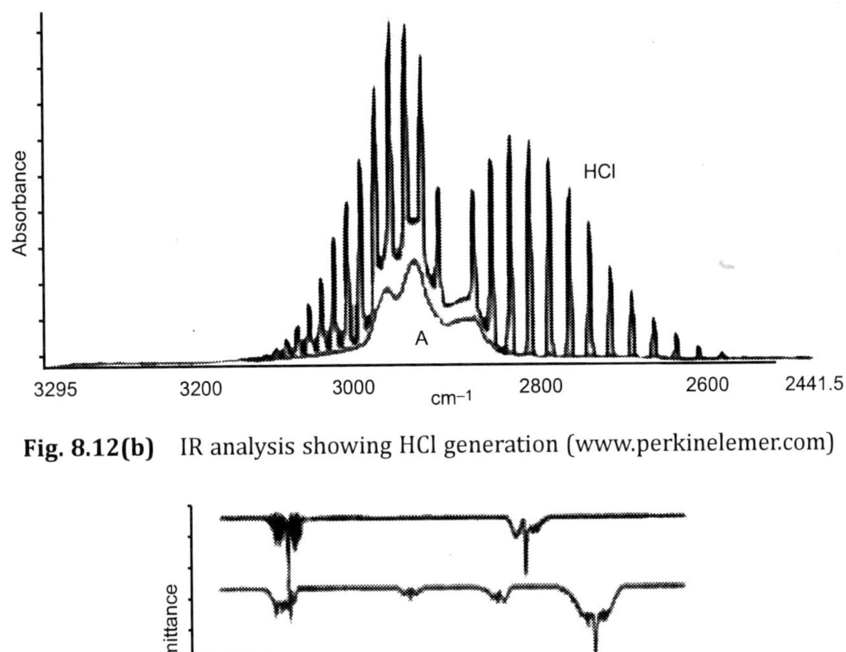

Fig. 8.12(b) IR analysis showing HCl generation (www.perkinelemer.com)

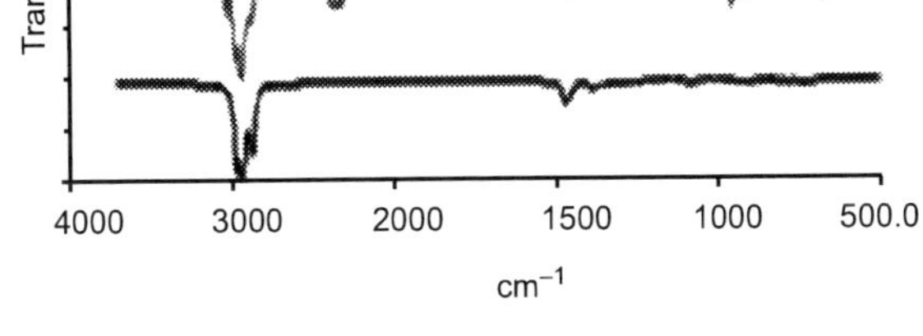

Fig. 8.12(c) IR analysis showing degradation of cable (www.perkinelemer.com)

gases are partially overlapping. Figure 8.12(c) shows the spectra of the spectra atlas, and the spectrum recorded at 24.4th minute, showing the profiles relative to each of the gases: Once the generation of these gases is over, from 27 minutes onwards and until the end of the analysis, emission of CO_2 and traces of CO can be detected.

Another application of TG- IR for analysis of thermal degradation of polyesters is simultaneous TG-DTA/FT-IR analysis. This technique gave spectroscopic and weight loss information about the thermal degradation process of engineering polyesters; poly(ethylene terephthalate) (PET) and poly(butylene terephthalate) (PBT) investigated by Kinoshita et al. [11]. The evolved gases from PET were benzoic acid, carbon dioxide and carbon monoxide, while those from PBT were terephthalic acid esters and benzoic acid. Figure 8.13 shows TG-DTA curves of PET (a) and PBT (b). A single weight loss due to thermal decomposition was observed for both polyesters in the experimental temperature range. The weight loss started at ~360°C and approached a constant value at ~550°C for PET. The weight loss of PBT started at lower temperature (~320°C and stopped at ~450°C). The residues at ~800°C were about 10% and 5% of the starting materials for PET and PBT, respectively. FT-IR spectra of evolved gases at various temperatures from PET and PBT were shown in Figs. 8.14 and 8.15 respectively. The difference between both spectra was observed at around 3000 cm^{-1}, 2350 cm^{-1}, 1700 cm^{-1}, 1300-1000 cm^{-1} and 700 cm^{-1}. Since the absorption band at 2350 cm^{-1} was due to carbon dioxide, PET produced a much larger amount of carbon dioxide during thermal decomposition than PBT. The spectra of PBT showed the separated double absorption peaks at 1760 and 1740 cm^{-1}, while the spectra of PET showed single peak at 1760 cm^{-1}. The absorption band at 1740 cm^{-1} was assigned to aromatic esters, therefore, the evolved gases from PBT contained a much larger amount of aromatic esters than those from PET.

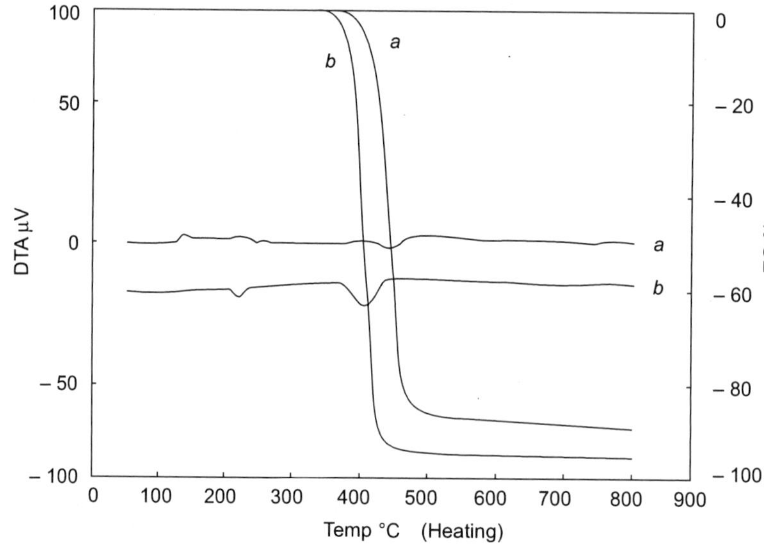

Fig. 8.13 TG-DTA graph of (a) PET and (b) PBT

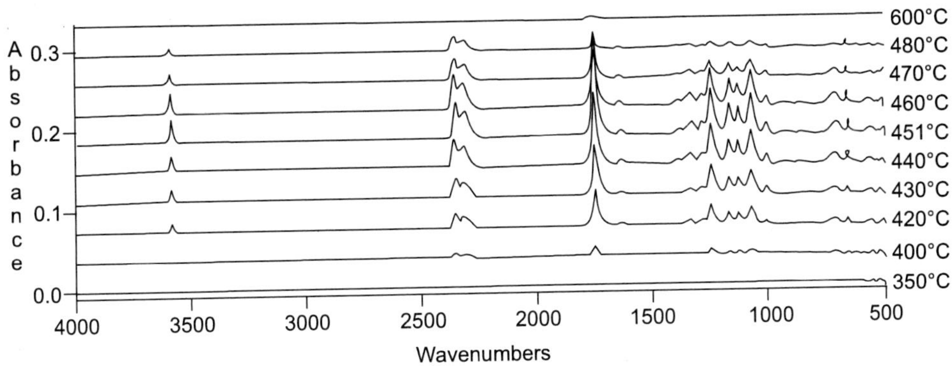

Fig. 8.14 FT-IR spectra of various gases evolved for PET at various temperatures

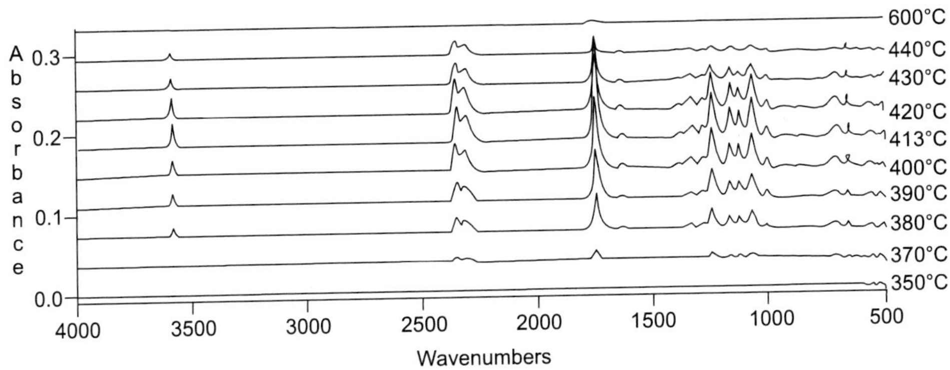

Fig. 8.15 FT-IR spectra of various gases evolved for PBT at various temperatures

This coupled technique is also used frequently in analysis of archaeological samples and local clays such as bricks, terracotta. Thermogravimetric (TG) and differential (TG–DTG) by Paama et al. [12] analysis and Fourier transform infrared spectrometry (FTIR) methods can be used for the study of characteristic reactions, associated with the course of the heating. The TG–DTG curves were obtained in the temperature range 25–900°C. The thermal curves of clays were recorded in the temperature range 25–900°C in the air and nitrogen atmosphere as shown in Fig. 8.16. The weight loss peak around 100°C is induced by hygroscopic water (i.e., physically absorbed water), whereas those appearing at 200–350°C are attributed to the bound water. The dehydration of hydromagnesite can be identified at 250–280°C and $Mg(OH)_2$ at 350–420°C. $Ca(OH)_2$ dehydration is detected between 400 to 520°C. Magnesium carbonate decomposes in the 450–520°C range, and calcium carbonate is associated with the 700–900°C range. The infrared spectra of clay samples show hydraulic water peaks at 1520 and 3746 cm^{-1}, and CO_2 peaks at 670 and 2358 cm^{-1}. The quantitative estimation of evolved gas can also be done by comparing the intensity as shown in Fig. 8.17. Thus, TG–IR thermograms of thermal decomposition of clays show elimination of water and carbon dioxide, the concentration of carbon monoxide and sulphur dioxide was on the low level.

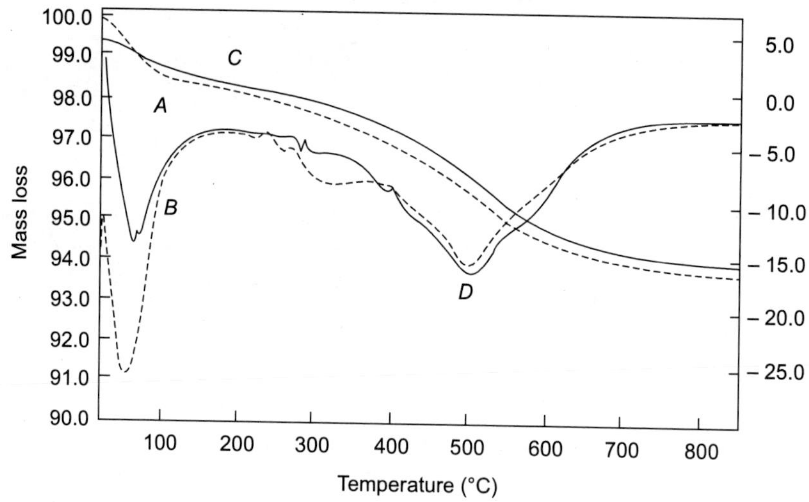

Fig. 8.16 TG–DTG analysis of clay in the dynamic air and nitrogen atmosphere: (A) TG curve in air; (B) DTG curve in air; (C) TG curve in nitrogen; and (D) DTG curve in nitrogen

The efficiency of the purification procedures was followed by thermogravimetry (TG), IR spectroscopy. The application can be done by evaluating the IR frequencies for concluding the ion-ion interactions. Similarly, a series of applications can be made by coupling the techniques for understanding the decomposition process more clearly for pure and hybrid materials.

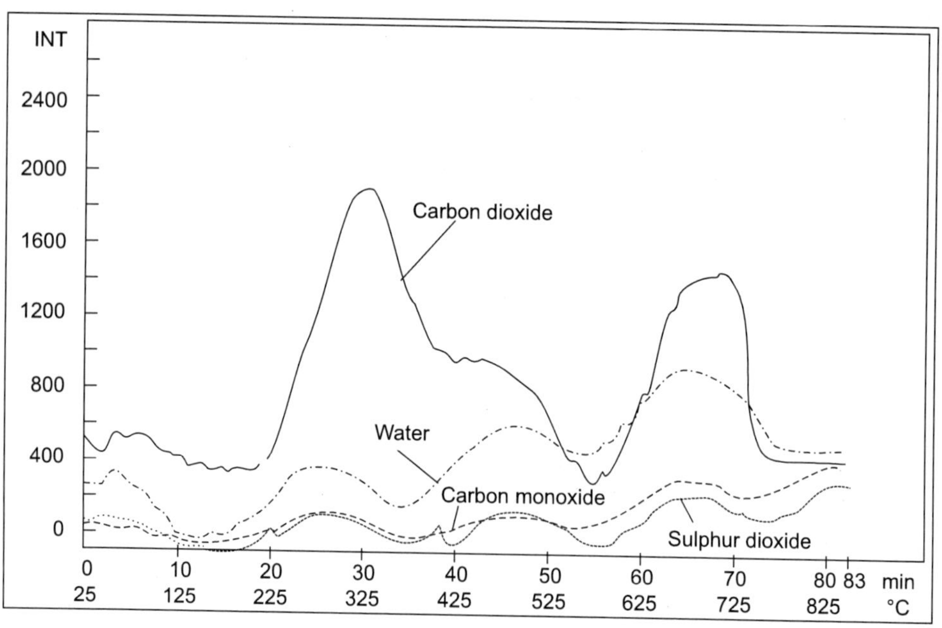

Fig. 8.17 TG–IR thermograms of thermal decomposition of clay sample

8.7 TG-GAS CHROMATOGRAPHY (TG-GC)

8.7.1 Introduction

This is a powerful analytical tool and can be used to determine the composition of evolved gas during pyrolysis [9, 13-14]. It is fairly a simple way to attach a gas chromatography to the thermal analyzer instrument. The GC can be combined simply to the pyrolysis chamber by means of a gas sampling valve. The valve can be operated manually or automatically. Sometimes GC can be combined with mass spectrometer also for better identification of gaseous product.

Gas chromatography, specifically gas-liquid chromatography, involves a sample being vapourized and injected onto the head of the chromatographic column. The sample is transported through the column by the flow of inert, gaseous mobile phase. The column itself contains a liquid stationary phase which is adsorbed onto the surface of an inert solid. A schematic diagram of TG-GC is presented in Fig. 8.18.

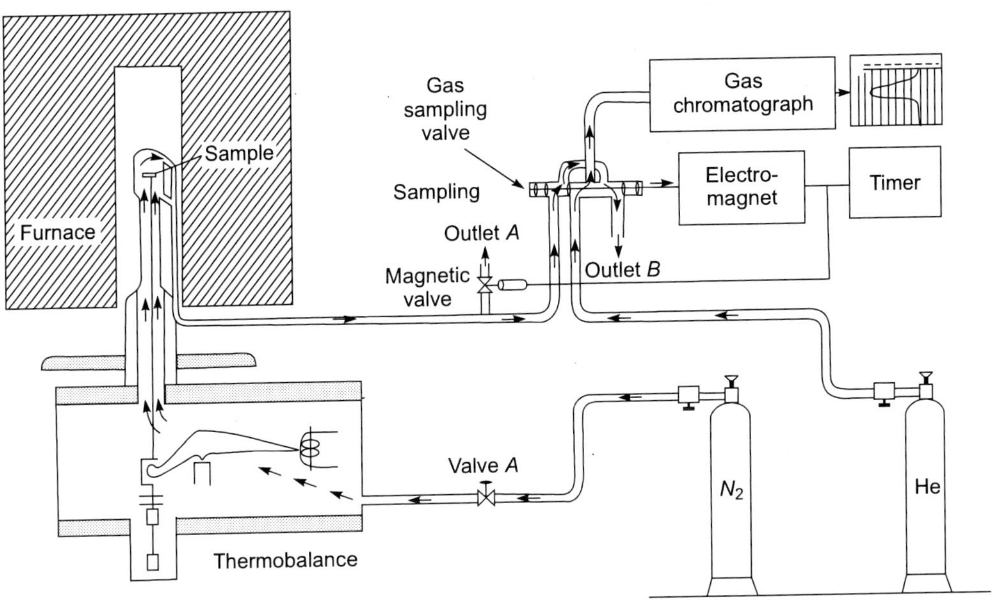

Fig. 8.18 Schematic diagram of TG-GC instrument (Adapted from Ref. 8)

In a GC analysis, a known volume of gaseous or liquid analyte is injected into the "entrance" (head) of the column. As the carrier gas sweeps the analyte molecules through the column, this motion is inhibited by the adsorption of the analyte molecules either onto the column walls or onto packing materials in the column. The rate at which the molecules progress along the column depends on the strength of adsorption, which in turn, depends on the type of molecule and on the stationary phase materials. Since each type of molecule has a different rate of progression. Thus, various components of the analyte mixture are separated as they progress along the column and reach the end of the column at different times (retention time). A detector is used to monitor the outlet stream from the column. Thus, the time at which each component reaches the outlet and the amount of that component can be determined. The substances are identified (qualitatively) by the order in which they emerge (elute) from the column and by the retention time of the analyte in the column.

In this technique, gaseous product from pyrolysis chamber passes through a pair of sampling valve. These valves operate at programmed intervals to withdraw the sample of the gas and pass through the column and detector. The temperature furnace, flow rate of gas and sampling time are essentially required to control during analysis. Generally, different types of column are used, e.g., fluoropak, silica gel, molecular sieve, etc.

8.7.2 Instrumentation

The basic components of gas chromatographs are described below:

(a) Carrier gas: The carrier gas must be chemically inert; commonly used gases include nitrogen, helium, argon and carbon dioxide. The choice of carrier gas often depends upon the type of detector used. The carrier gas system also contains a molecular sieve to remove water and other impurities.

(b) Sample injection port: The introduction of sample inside the GC is made by this part with the emphasis for optimum column efficiency. The sample should not be too large, and should be introduced onto the column as a "plug" of vapour. Slow injection of large samples causes band broadening and loss of resolution. The most common injection method is where a microsyringe is used to inject the sample through a rubber septum into a flash vaporizer port at the head of the column. The temperature of the sample port is usually about 50°C higher than the boiling point of the least volatile component of the sample.

For packed columns, sample size ranges from tenths of a microlitre up to 20 microlitres. Capillary columns, on the other hand, need much less sample, typically around 10^{-3} μL.

The injector can be used in one of two modes; split or splitless. The injector contains a heated chamber containing a glass liner into which the sample is injected through the septum. The carrier gas enters the chamber and can leave by three routes (when the injector is in split mode). The sample vaporises to form a mixture of carrier gas, vaporised solvent and vaporised solutes. A proportion of this mixture passes onto the column, but most of it exits through the split outlet. The septum purge outlet prevents septum bleed components from entering the column.

(c) Columns: The most important part of GC is the **column, which is responsible for the separation of individual components.** There are two general types of columns, packed and capillary. Packed columns contain a finely divided, inert, solid support material (commonly based on diatomaceous earth) coated with liquid stationary phase. Most packed columns are 1.5–10 m in length and have an internal diameter of 2–4 mm.

Capillary columns have an internal diameter of a few tenths of a millimeter. They can be one of two types; wall-coated open tubular (WCOT) or support-coated open tubular (SCOT). Wall coated columns consist of a capillary tube whose walls are coated with liquid stationary phase. In support-coated columns, the inner wall of the capillary is lined with a thin layer of support material such as diatomaceous earth, onto which the stationary phase has been adsorbed. SCOT columns are generally less efficient than WCOT columns. Both types of capillary column are more efficient than packed columns.

(d) Column temperature controller: For precise work, column temperature must be controlled within tenths of a degree. The optimum column temperature is dependant upon the boiling point of the sample. As a rule of thumb, a temperature slightly above the average boiling point of the sample results in an elution time of 2–30 minutes. Minimal temperatures give good resolution, but increased elution times. If a sample has a wide boiling range, then temperature programming can be useful. The column temperature is increased (either continuously or in steps) as separation proceeds.

(e) Detector: This is the identifying units of GC and there are many detectors used in gas chromatography. Different detectors will give different types of selectivity. A non-selective detector responds to all compounds except the carrier gas, a selective detector responds to a range of compounds with a common physical or chemical property and a specific detector responds to a single chemical compound. Detectors can also be grouped into concentration-dependant detectors and mass flow dependent detectors. The signal from a concentration-dependent detector is related to the concentration of the solute in the detector, and does not usually destroy the sample dilution of with make-up gas will lower the detector's response. Mass flow dependent detectors usually destroy the sample, and the signal is related to the rate at which the solute molecules enter the detector. The response of a mass flow dependent detector is unaffected by make-up gas. A tabular summary of common GC detectors is as follows in Table 8.1.

The FID is a useful general detector for the analysis of organic compounds; it has high sensitivity, a large linear response range, and low noise. It is also robust and easy to use, but unfortunately, it destroys the sample where the effluent from the column is mixed with hydrogen and air, and ignited. Organic compounds burning in the flame produce ions and electrons which can conduct electricity through the flame. A large electrical potential is applied at the burner tip, and a collector electrode is located above the flame. The current resulting from the pyrolysis of any organic compound is measured. FIDs are mass sensitive rather than concentration sensitive; this gives the advantage that changes in mobile phase flow rate do not affect the detector's response.

TABLE 8.1 Types of detectors and their properties

Detector	Type	Support gases	Selectivity	Delectability	Dynamic range
Flame ionization (FID)	Mass flow	Hydrogen and air	Most organic compounds	100 pg	10^7
Thermal conductivity (TCD)	Concentration	Reference	Universal	1 ng	10^7
Electron capture (ECD)	Concentration	Make-up	Halides, nitrates, nitriles, peroxides, anhydrides, organometallics	50 fg	10^5
Nitrogen-phosphorus	Mass flow	Hydrogen and air	Nitrogen, phosphorus	10 pg	10^6
Flame photometric (FPD)	Mass flow	Hydrogen and air, possibly oxygen	Sulphur, phosphorus, tin, boron, arsenic, germanium, selenium, chromium	100 pg	10^3
Photo-ionization (PID)	Concentration	Make-up	Aliphatics, aromatics, ketones, esters, aldehydes, amines, heterocyclics, organosulphurs, some organometallics	2 pg	10^7
Hall electrolytic conductivity	Mass flow	Hydrogen, oxygen	Halide, nitrogen, nitrosamine, sulphur	–	–

(f) Auto samplers: Auto samplers provide the means to introduce automatically a sample into the inlets. Manual insertion of the sample is possible but is no longer common. However, automatic insertion provides better reproducibility and time optimization.

Different kinds of auto-samplers exist. Auto samplers can be classified in relation to sample capacity or to analysis.

8.7.3 Applications

Qualitative analysis: The chromatographic data is presented as a graph of detector response (y-axis) against retention time (x-axis). This provides a spectrum of peaks for a sample representing the analytes present in a sample eluting from the column at different times. Retention time can be used to identify analytes if the method conditions are constant. Also, the pattern of peaks will be constant for a sample under constant conditions and can identify complex mixtures of analytes. However, the GC is connected to a mass spectrometer or similar detector that is capable of identifying the analytes represented by the peaks.

Quantitative analysis: The area under a peak is proportional to the amount (concentration) of the analyte present. Thus, by calculating the area of the peak using the mathematical function of integration, the concentration of an analyte in the original sample can be determined. Concentration can also be calculated using a calibration curve created by finding the response for a series of concentrations of the analyte, or by determining the relative response factor of an analyte with standard concentration. The relative response factor is the expected ratio of an analyte to an internal standard (or external standard) and is calculated by finding the response of a known amount of analyte and a constant amount of internal standard (a chemical added to the sample at a constant concentration, with a distinct retention time to the analyte).

Presently, in modern GC systems, computer software is used to draw and integrate peaks, and match MS spectra to library spectra stored in the instrument.

A typical result of TG-GC analysis of $CaC_2O_4{\cdot}H_2O$ is illustrated in Fig. 8.19.

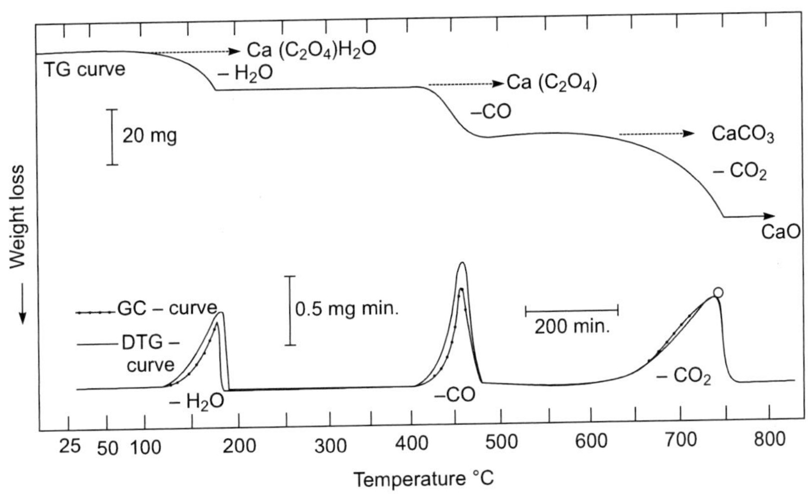

Fig. 8.19 TG-DTG-GC result of $CaC_2O_4{\cdot}H_2O$

During the decomposition of hydrated calcium oxalate, H_2O, CO_2 and CO evolved in inert atmosphere. However, CO is further oxidized to carbon dioxide if decomposed in ambient or oxidising atmosphere.

Vast literature is available for deciding the suitable set of analysis condition for the given sample and its expected result.

The thermal decomposition of polyimide/polytetrafluoroethylene/graphite (PI/PTFE/GR) resin can be efficiently studied by this technique (Fig. 8.20)[15]. The technique improves resolution and enhances the accuracy of identification and quantification.

Generally, thermal decomposition of PI/PTFE/GR resin occurred from the PTFE portion at first, and subsequently the PI portion. In an intermediate temperature region, both decompositions proceeded simultaneously. With conventional TG since the sample is exposed to higher temperature than that required for decomposition to each component, analysis of the evolved gases is further complicated. However, in high resolution TG-GC/MS, since the decomposition components on clearly separated steps are expressed as a simple profile, analytical treatment becomes easy and improvement of accuracy in identification and quantification is accomplished.

The TG-GC coupled techniques can be also used to analyze Polyethylene-Poly (vinyl acetate) copolymer system. Acetic acid gas is evolved rapidly and quantitatively during the initial period of thermal decomposition in an inert atmosphere [16]. The combination of GC and TG adds another feature of separating and identifying the substances evolved during the various weight loss stages, thus further establishing the decomposition mechanism can be established. The TG and GC scans for pyrolysis of a low-density polyethylene, a poly(vinyl acetate) homopolymer, and an ethylene-vinyl acetate copolymer containing 30% by weight of vinyl acetate shown in Fig. 8.21. The weight loss step 1 liberates mainly acetic acid with a retention time of 3.2 minutes for either poly (vinyl acetate) or ethylene-vinyl acetate copolymer. Through expanded weight scale, we can measure the amount of polymer coating as small as 0.01 of the glass weight from its weight loss at elevated temperatures. The nature of the coating can then be identified by the GC scans of the volatilized products. The two-step weight loss pattern and the decomposition temperature range are reminiscent of those of poly (vinyl acetate). Indeed, the pyrogram of weight loss step 1 shows acetic acid, and that of step 2 very similar to that observed earlier for poly(vinyl acetate). The presence of other peaks suggests other components in the sample.

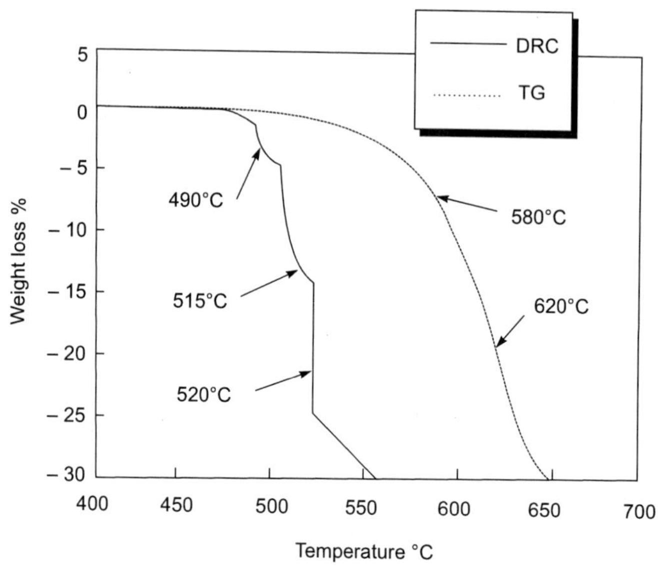

Fig. 8.20 TG and TG-GC result of polyimide/polytetrafluoroethylene/graphite (PI/PTFE/GR) (Adapted from Ref. 15)

Similarly, various combinations of modified polymers, e.g., blend, composite, copolymers, polymeric mixture can be effectively studied by TG-GC coupled technique.

A major disadvantage of technique, when applied to the sample such as polymer, which degrades to produce a variety of fragments simultaneously is the lack of separation of product before analysis. The other techniques [17] such as NMR, XRD are also successfully used for analyzing the materials in combination of thermal analyser.

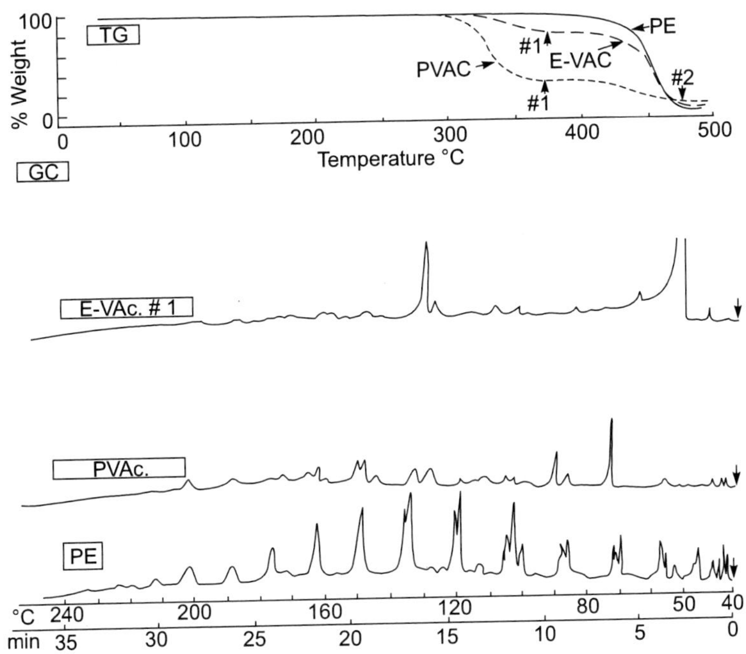

Fig. 8.21 TG- GC curve of polymer blend (Adapted from Ref. 16)

EXERCISE

1. Explain the need of integrating various techniques.
2. Compare time-of-flight and quadrupole.
3. What are the ion sources? List the different ion sources.
4. List the different components of a mass spectrometer.
5. Describe the simultaneous TG-DTA unit with its advantages.
6. Write a note on application of mass spectra in estimation of flame retardancy.
7. List the significance of complementary techniques or hyphenated techniques with examples.
8. Explain the supplementary behaviour of mass spectroscopy to thermo-gravimetric analysis.

9. Compare ESI and MALDI mass spectrometers with examples.
10. How can one identify the effluent species of thermal analysis?
11. Determine the percentage compositions of ethylene vinyl acetates in the given spectra.

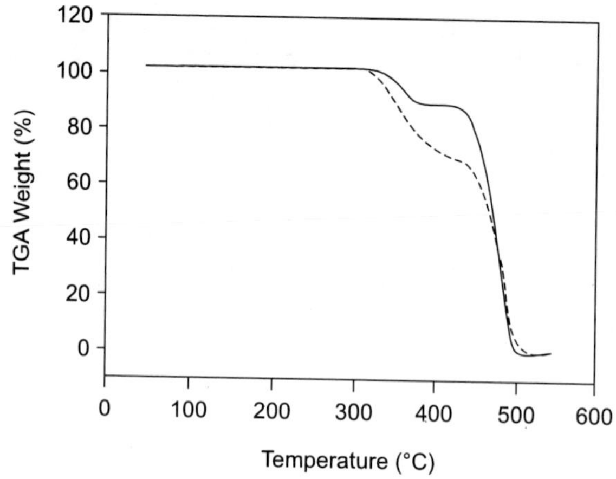

12. Write notes on the following:
    (a) EGA
    (b) Flame-retardant polymers
    (c) Quadrupole ion trap
    (d) Thermal ionization
13. Answer the following questions:
    (a) Discuss the advantages of combining gas chromatograph with mass spectrometer.
    (b) DTA and DSC are supplement to thermal gravimetric analysis. Explain.
    (c) What are the advantages of microcalorimeter over calorimeter?
    (d) Describe the limitations and advantages of simultaneous TG-DTA technique.
14. Elaborate the following abbreviations:
    (a) MALDI (b) ESI
    (c) FD (d) FAB
    (e) SIMS (f) TOF
    (g) EGA

REFERENCES

1. Meyers, R. A. (Eds.), (2000). Encyclopedia of Analytical Chemistry: Applications, Theory, and Instrumentation. John Wiley & Sons, Chichester, UK.
2. Sparkman, O. D. (2000). Mass spectrometry desk reference. *J. Am. Soc. Mass Spectrom.* 12(11), 1144.

3. Downard, K. M. (2007). Francis William Aston: The man behind the mass spectrograph. *Eur. J. Mass Spectrom.* 13, 177-190.
4. Bruins, A. P. (1991). Mass spectrometry with ion sources operating at atmospheric pressure. *Mass Spectrom. Rev.* 10, 53-77.
5. Wollnik, H. (1993). Time-of-flight mass analyzers. *Mass Spectrom. Rev.* 12, 89-114.
6. McLafferty, F. W. and Tureček, F. (1993). Interpretation of Mass Spectra. University Science Books, Sausalito, California.
7. Statheropoulos, M., Kyriakou, S. and Tzamtzis, N. (1998). Performance evaluation of a TG/MS system. *Thermochim. Acta.* 322, 167-173.
8. Wandlandt, W. W. (1986). Thermal Analysis. John Wiley & Sons, Inc., New York, USA.
9. Willard, H. H., Merrit Jr, L. L., Dean, J. A. and Settle Jr., F. A. (1988). Instrumental Method of Analysis. 7th Edition, Wadsworth Publishing Company, Florence, KY, USA.
10. Lau, W. S. (1999). Infrared characterization for microelectronics. World Scientific Private Co., Singapore.
11. Kinoshita, R., Teramoto, Y., Nakanoand, T. and Yoshida, H. (1992). Thermal degradation of polyesters by simultaneous TG-DTA/FT-IR analysis. *J. Therm. Anal.* 38, 1891-1900.
12. Paama, L., Pitkãnen, I. and Perämäki, P. (2000). Analysis of archaeological samples and local clays using ICP-AES, TG–DTG and FTIR techniques. *Talanta.* 51, 349-357.
13. Hachenburg, H. (1973). Industrial Gas Chromatographic Trace Analysis. Heyden, London.
14. Thompson, B. (1977). Fundamental of Gas Analysis and Gas Chromatography. Varian, Palo, Ltd.
15. Arii, T., Senda, T. and Fujii, N. (1995). A combined thermogravimetric-gas chromatographic/mass spectrometric analysis (TG-GC/MS) using a high resolution TG technique. *Thermochim. Acta.* 267, 209-221.
16. Chiu, J. (1968). Polymer characterization by coupled thermogravimetry-gas chromatography. *Anal. Chem.* 40, 1516-1520.
17. Brown, M. E. (2001). Introduction to Thermal Analysis. Kluwer Academic Publisher, London.

9

Experiments Based on Thermal Analysis

Introduction

The earlier discussed thermal techniques are widely used by scientists to evaluate the various properties of materials to explore their appropriate applications and standardize the inter-laboratory results. So far students' training is concerned, laboratory experiments start very late in PG level or mostly in the form of theory classes. However, a scientifically trained person starts serving industry and R&D labs after graduation. Thus, keeping the significance of techniques in mind, it is essential to train students through experimental assignments. Many efforts have been made in this direction: (a) differential thermal analysis of potassium perchlorate is illustrated from the point of view of the characterization of this compound rather than the details of its decomposition by Gordon [1]. Similarly, all the techniques [2-3] were employed for evaluation of chemical, mechanical, electrical, optical, and magnetic behaviour. This chapter is designed with some introductory experiments, which may be performed or demonstrated in a common laboratory.

EXPERIMENT 1

Aim: Investigation of thermal decomposition of calcium oxalate by means of TG analysis.

Principle

This experiment is based on a TG curve and supplemented with DTA [4]. It is understood that most of the compounds undergo various physical and chemical changes (thermal decomposition) on heating. These changes are always associated with enthalpy change and some of them followed by weight change due to evolution of gaseous products which have already been discussed in previous chapters. Here, only a thermoanalytical curve is used to explain the formation of intermediates and end products. The mass change is directly related to the formation of various chemical species.

Apparatus and Chemicals

Analytical balance, derivatograph or TG-DTA instrument, calcium oxalate monohydrate, aluminium oxide, drying equipment (oven).

Procedure

The following steps are required to complete the exercise:

1. Calibrate the TG-DTA instrument for mass, temperature and baseline as discussed in Chapter 2.
2. Clean the sample crucible with required solvent and remove it by heating at ~120°C in an oven for 30 minutes.
3. Put ~10.0 mg of calcium oxalate monohydtrate and Al_2O_3 in the sample and reference pans, respectively.
4. Set the analysis parameter at heating rate 5°C per minute, and temperature range ambient to 1000°C.
5. The flow of N_2 /air will be 20 mL per minute in case of ambient and 100 mL per minute for N_2 atmosphere.
6. Record the TG curve.
7. The TG graph is recorded (Fig. 9.1) in N_2 atmosphere.

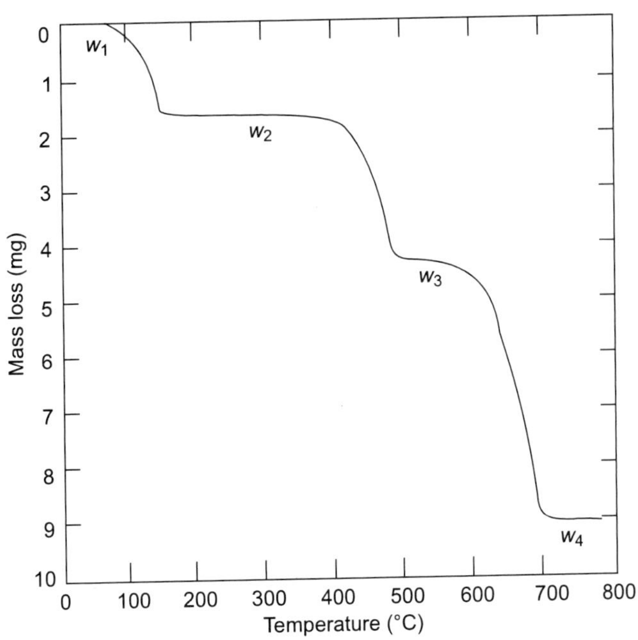

Fig. 9.1 TG curve of monohydrated calcium oxalate

Analysis

Qualitative

In the case of investigation of calcium oxalate, the conclusion has been made by comparing the condition of analysis (ambient or N_2 atmosphere). The present case is an example of analysis in the ambient atmosphere, where the TG shows decomposition in three steps. It can be concluded by:

$$CaC_2O_4 \cdot H_2O \xrightarrow{\text{max. } 250°C} CaC_2O_4 + H_2O \uparrow$$

$$CaC_2O_4 \xrightarrow{\text{max. } 470°C} CaCO_3 + CO \uparrow$$

$$CaCO_3 \xrightarrow{\text{max. } 880°C} CaO + CO_2 \uparrow$$

Now if we compare the decomposition pattern and temperature with reported values. It is quite similar to earlier reported graphs. So one can conclude that the thermogram is for calcium oxalate. Further, the purity can be evaluated like:

Quantitative

After analyzing the individual steps, one can interpret the TG curve for quantitative (stoichiometric) calculation on the basis of mass change in these temperature intervals and with the original weight of the sample. Calculation can be done as follows:

1. Amount of CaC_2O_4 (theoretical: 87.6%)

$$x_1 = \frac{W_{(II)}}{W_{(I)}} \times 100\% \; CaC_2O_4$$

2. Amount of $CaCO_3$ (theoretical: 68.5%)

$$x_2 = \frac{W_{(III)}}{W_{(I)}} \times 100\% \; CaCO_3$$

3. Amount of CaO (theoretical: 38.4%)

$$x_3 = \frac{W_{(IV)}}{W_{(I)}} \times 100\% \; CaO$$

4. Amount of H_2O (theoretical: 12.33%)

$$x_4 = \frac{W_{(I)} - W_{(II)}}{W_{(I)}} \times 100\% \; H_2O$$

5. Amount of CO (theoretical: 19.2%)

$$x_5 = \frac{W_{(II)} - W_{(III)}}{W_{(I)}} \times 100\% \; CO_2$$

6. Amount of CO_2 (theoretical: 30.1%)

$$x_6 = \frac{W_{(III)} - W_{(IV)}}{W_{(I)}} \times 100\%\ CO_2$$

where $W_{(I)}$ is weight of the original sample, $W_{(II)}$ weight of the sample after first decomposition, $W_{(III)}$ weight of the sample after the second decomposition and W_{IV} is the weight end product process as labelled in the curve.

The information can be supported by DTA, since peak area of DTA curves are also proportional to the amount of heat absorbed or evolved during the transformation. Therefore, we can also determine the approximate value (with the accuracy of 5 to 10%) of transformation of materials if we calibrate the instrument under the identical conditions.

EXPERIMENT 2

Aim: The study of the TG (thermogravimetry), DTG (differential thermogravimetry), DTA (differential thermal analysis), TGT (thermo-gas-titrimetry), DTGT (differential thermo-gas-titrimetry) and T (temperature) curve of tetra amine cupric sulphate with thermo-gas titrimeter.

Principle

The derivatogram of tetra amine cupric sulphate is shown in Fig. 9.2. The first DTG peak is doublet, whereas a single peak can be observed in the DTGT curve. Thus, it is clear that change in weight is a composite process, this means, hydrated water and ammonia escape simultaneously. It can be measured by thermo-gas-tritimetry with a scheme such as determination of ammonia content may be carried out by acid base titration with a glass and saturated calomel electrode and with 0.1 N aq. HCl solution as titrant. While adsorbent solution is distilled water containing 3% ethyl alcohol whose pH is adjusted to 5.0 with hydrochloric acid. The reaction is energy compensatory thus can be studied by this scheme, wherever complimentary techniques such as GC, MS are not available.

Apparatus and Chemicals

Thermo-gas titrimeter, pH meter, glass electrode, saturated calomel electrode, tetra amine cupric sulphate, distilled water, ethanol, and hydrochloric acid.

Procedure

The instrument is put into the operation and adsorbent solution (~200 mL), whose pH was 5.0, then poured into the absorber prior to starting the titration. If the pH of solution deviates by 0.1 pH from adjusted value after absorption of decomposed product. The delivery of the titrant starts as long as the pH of 5.0 is retained again. The apparatus records the required titrant as a function of time.

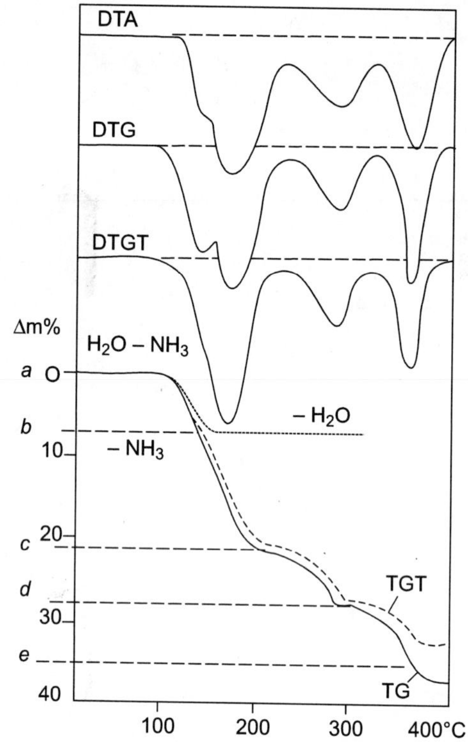

Fig. 9.2 DTA, DTG, DTGT, TGT and TG curves of tetra amine cupric sulpahte (a = $Cu(NH_3)_4SO_4 \cdot H_2O$, b = $Cu(NH_3)_4SO_4$, c = $Cu(NH_3)_2SO_4$, d = $CuNH_3SO_4$ and e = $CuSO_4$)

Evaluation

Calculate the amount of the decomposition product titrated on the basis of TGT curve in milligram unit and determine the composition of various phases formed in the course of the decomposition. Draw the TG (% mass), DTA (change in temperature, ΔT), DTG (rate of change of mass, dm/dt), DTGT (rate of change of mass, dm/dt), TGT (absolute mass by thermo gas titrimetry, m) curves plotted against the temperature.

EXPERIMENT 3

Aim: Investigation of a liquid crystalline (mesomorphous) transition (DSC).

Principle

The inter-conversion of transitions in materials is associated with absorption or emission of heat depending on the difference of their matrix energy. Generally, solid crystalline substances which show anisotropy in their properties are transformed into an isotropic liquid at their melting points. This reveals that a material of a high degree of orientation is transformed into one of low orientation.

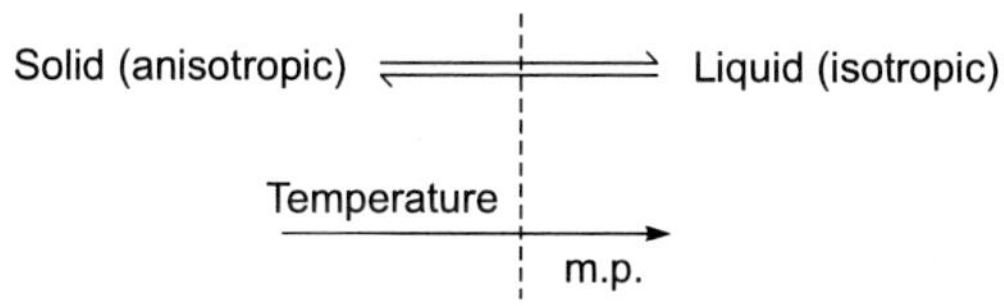

However, in the case of number of organic compounds between these two degrees of orientation there exits a third one, the so-called mesomorphous or liquid crystalline state reached from anisotropic state through two points of transition. The transition appears different through mesomorphous phase.

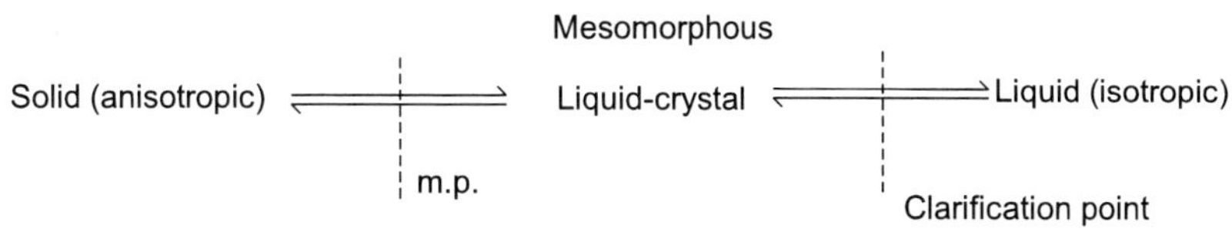

The characteristics of a substance in the mesomorphous state is that in some of its properties, e.g., mechanical it behaves as liquid (isotropy), whereas in others, e.g., optical and electrical ones it shows anisotropy.

By means of the differential scanning calorimeter, it is possible to determine the characteristic temperature of these transitions along with the value of the energy requirement of mesomorphous → isotropic transition. By recording a cooling curve and the curve of repeated heating, even the monotropic and enantiotropic nature of these transitions can be observed.

Apparatus and Chemicals

Cholesterol capronate, DSC apparatus, microbalance, and solvents

Procedure

The study of mesomorphic transition involves the following steps:

1. Weigh 4-5 mg of cholesterol capronate into the sample holder and put it into the measuring cell and close the sample sealer.
2. The temperature programmer is set to a heating rate of 5°C per minute and in the required temperature analysis range.
3. The Dewar flask is filled with ice to maintain low temperature.
4. Adjust the essential parameters such as zero point of the temperature scale and start the temperature program and recording.
5. Simultaneously with this measurement, observe the behaviour of cholesterol capronate upon heating and cooling and repeated heating under microscope equipped with a hot stage.
6. Make note on the observations and determine the temperature values corresponding to the transformations.

Analysis

Observation is used to determine the individual transition temperature from DSC curves. On the basis of DSC curve observations revealed, the following transition scheme shows the direction of the transition with arrow noting the corresponding transition temperature.

S_1 (solid phase 1) → L (liquid phase)

S_2 (solid phase 2) → M (mesomorphic phase) → L (liquid phase)

This further decides whether the transition is monotropic or enantiotropic and successively, the nature of solid phase type 1 or 2.

EXPERIMENT 4

Aim: Evaluation of purity and determination of melting point (DSC).

Principle

Determination of purity by DSC is based on the shift in the physical transition such as depression in melting point of the material. This is related to the mole fraction of impurity. The impurity concentration holds the following equation

$$\Delta T = T_{mp} - T = \frac{RT_{mp}^2}{\Delta H_{s1}} X \tag{1}$$

where, T_{mp} is the melting point of pure substance, T is the melting point of the sample, ΔH_{s1} is the heat of fusion of the pure substance, R is the gas constant (1.987 cal/mol°C) and X is the mole fraction of impurity. However, difficulties persist since in practice, neither melting point nor enthalpy (ΔH) are known with accuracy. In spite of this fact, it can be used to determine purity of the sample with a single record of DSC because DSC curve is used to measure ΔT, ΔH_s, and T_{mp}.

Let us consider the cryoscopic constant in light of Eq. (1):

$$T = T_{mp} - KX$$

$$K = \frac{RT_{mp}^2}{\Delta H_{s1}} \tag{2}$$

The actual melting point (T) of the sample can be measured, but T_{mp}, ΔH_{s_1} and X are unknown. If a fraction of the weighted sample, i.e., an r-th fraction of the weighing is made to melt, the impurity will be enriched in this molten fraction and its mole fraction will increase to X/r. Consequently, the melting point, according to previous equation, will be

$$T = T_{mp} - \frac{KX}{r} \tag{3}$$

where r, Melt fraction $= \dfrac{\text{Melt (mg)}}{\text{Weight (mg)}}$

Knowing the melting point (T_r) values pertaining to the melt fraction (r) values, and plotting the latter against $1/r$, a straight line is obtained. Whose point of intersection with ordinate gives T_{mp} values without the necessity of having the pure substance. Further, if the heat of fusion has been determined by another DSC measurement, the mole fraction of the impurity can be calculated from the slope(s) of the straight line.

$$s = K \cdot X = \frac{RT_{mp}^2}{\Delta H} X \tag{4}$$

Accordingly, the heat of fusion of the pure material is substituted by that of the sample. The only equation that remains now is the following: How is it possible to obtain the function?

$$T_4 = \frac{1}{r}$$

Figure 9.3, DSC curve is used: r, baseline of apparatus, range of evaluation.

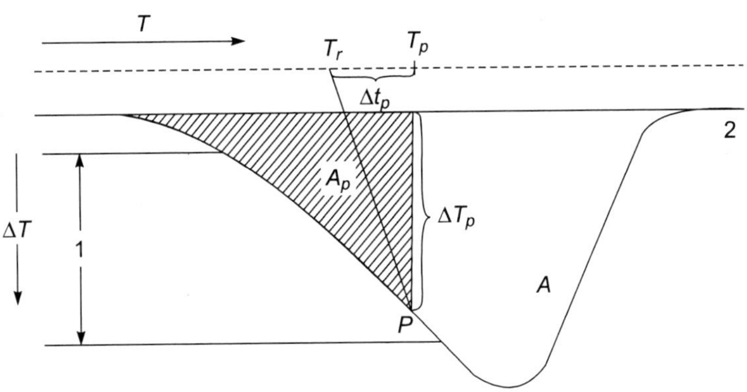

Fig. 9.3 Graphical representation of DSC curve using purity measurement: 1 is range of evaluation and 2 is baseline of apparatus

Evaluation can be carried out on the downward branch of DSC curve in the $0.1\langle\dfrac{A_p}{A}\rangle 0.5$ range at four to five points. During fusion, the amount of the material molten to the point P is proportional to the amount of heat input and that, in turn, to the area under that point, i.e., A_p and consequently, the melt fraction can be calculated by the following:

$$r_p = \frac{A_p}{A} \tag{5}$$

where r_p is the melt fraction in point P, A_p is the area under the curve until point P, and A is the total area under the curve.

Here temperature is recorded by a thermocouple placed on the reference side: this follows the temperature program and the value read in the point P as T_p.

However the temperature (T_r) of the sample of the melt fraction r_p is lower than those values by ΔT_p, if the difference and the sample and reference temperatures and consequently T_r, can be calculated from T_p by simple calculation as:

$$T_r = T_p - \Delta T_p \tag{6}$$

Thus corresponding $T_r - 1/r$ value pairs are plotted in Fig. 9.4.

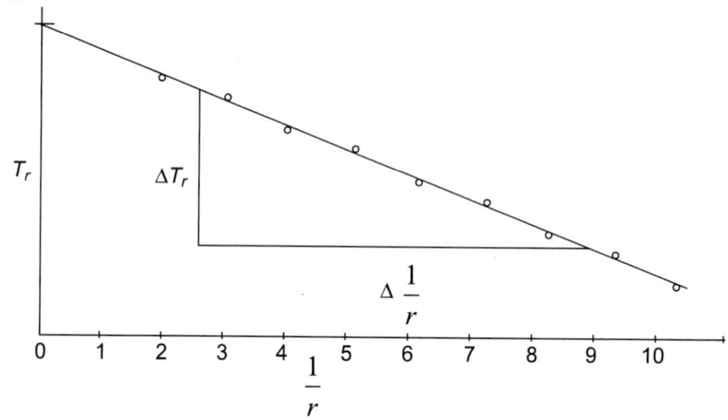

Fig. 9.4 T_r versus $\frac{1}{r}$

The diagram enables one to read the T_{mp} value and can be used to calculate the slope(s) as:

$$s = \frac{\Delta T_r}{\Delta \frac{1}{r}} \tag{7}$$

and accordingly, the degree of purity or the impurity content of the material can be calculated from Eq. (4).

Purity calculation by DSC has several advantages and can be compared with other techniques for simplicity, speed and general applications. Further, DSC also yields two more thermodynamic parameters: T_{mp} and ΔH. It is an advantage of the technique but drawback in some cases for adequate accuracy of about 1-2%. The case of purity of the order of 95% but error in the mole fraction of impurity may be as high as 20-25%. The theoretical calculations are valid for highly diluted solutions or melts. The range of applicability of the technique can be narrowed by the following conditions of calculation:

1. The components (sample and impurity) produce a eutectic composition, they do not form a mixed crystal.
2. Impurities form an ideal solution with melt of the sample.
3. The heat of fusion is independent of temperature and the concentration of the impurity.

Apparatus and Chemicals

DSC, microbalance, phenacetine, p-aminobenzoic acid or benzamide.

Procedure

The experiment is performed in two steps:

1. The heat of fusion of the sample is determined in the course of first heating cycle experiment.
2. In the second cycle, heating is plotted against temperature and in the vicinity of melting point. The sensitivity of the instrument should be fixed as per requirement.

Calculation

Evaluate the heat of fusion from the first record in calories per mole. Determine the corresponding T_r and $1/r$ values at five points n from the second record and draw curve between T_r and $1/r$. Carry out the calculation with T_{mp} and H_{s1} values taken from literature and compare the mole fraction value obtained. Further, explain the the difference in the measured and given impurity level.

EXPERIMENT 5

Aim: Determination of active chlorine in chlorinating materials.

Apparatus and Chemicals

Thermometric apparatus, pipette, bleaching powder, pharmaceutical benzene sulfachloramide sodium (neomagnol), KI, HCl, hypo, and saturated chlorine water.

Principle

The details of this method are discussed in Chapter, "Thermometric Titration". The basic principle is that "every chemical transition is accompanied by changes in enthalpy and temperature under adiabatic condition", which is also directly related to limiting concentration of reacting reagent.

The amount of heat formed of a reaction system,

$$Q = n\Delta H \tag{1}$$

where, n is the mole number of reacting component and ΔH is the molar enthalpy change. The Q is fundamentally related to Δt as

$$Q = k\Delta T \tag{2}$$

where, k is the heat capacity of the system, thus from Eqs. (1) and (2) get

$$\Delta T = n\frac{\Delta H}{K} \tag{3}$$

Since ΔH is constant at constant pressure and in a small temperature interval, if the heat capacity of the system (volume of test solution, its specific heat, etc.) is always the same, the value of k is constant; so, under adiabatic conditions

$$\Delta T = k'n \tag{4}$$

Thus, the observed change in temperature is proportional to the number of moles of the component investigated.

Preparation of Calibration Curve

Three parallel experiments should be carried out in each case with different amounts of chlorine water. In the first experiment 2 cc, 5 cc, 10 cc and 20 cc, respectively, of chlorine water are added to the distilled water in the beaker of apparatus. The required amount of distilled water at room temperature added was completed exactly to 200 cc.

Fill the immersion pipette with the reagent, clamp it into the measuring head, and dip into the solution. Start stirring, wait for three minutes until thermal equilibrium is established. Adjust the apparatus at appropriate sensitivity, then add the reagent to the solution and read the deflection of the instrument.

Since chlorine is volatile in nature, thus determine the amount of the chlorine present in the reaction volume during the reaction by the measurement of the equivalent amount of elementary iodine liberated from the reagent, then iodine is titrated with sodium thiosulphate (hypo) solution (a reaction with energy compensatory nature). The deflection is plotted as the function of the chlorine in milligram determined in the 200 cc test solution. This graph serves the purpose of calibration curve for unknown reactions.

Determination of Active Chlorine in Bleaching Powder

Add known amount of bleaching powder (0.1 to 0.2 gm) in 200 mL water. After stirring the solution for 3 minutes neutralize it by adding drops of HCl, finally carry out the measurement in the way described above.

Determination of Active Chlorine in Neomagnol

Break the tablet into pieces and use 0.1 to 0.2 gm for one measurement, after weighing the sample put it into the titrating beaker containing 200 cc of water. While the solution is stirred for 3 minutes usually, the tablet dissolves and also thermal equilibrium will be established. Carry out the measurement as described above.

Note

Generally, 17% of deviation can be observed with respect to calibration curve taken because Neomagnol tablet is accompanied by 8% binding material causing side reactions. Therefore, if required a new calibration line must be prepared on the basis of titrating the iodine liberated in the course of the process with sodium thiosulphate solution.

EXPERIMENT 6

Aim: Evaluate the thermochemical transition of a chemical reagent.

Principle

Thermal equilibrium between sample solution and the reagent is established speedily, when the concentrated reagent solutions are mixed with more dilute sample solution. The mixture warms up or cools down on account of hydration. To avoid this, indifferent additives are admixed to the reagent until thermal compensation is maintained. A case for example, concentrated hydrofluoric acid (HF) is applied as reagent, the solution is warmed up, and let the loss in heat be Q_1. However, if the urea is dissolved in HF, a loose complex compound is formed which dissociates upon dilution. It takes up heat quantity of Q_2. By applying the urea in appropriate proportion, $Q_1 + Q_2 = 0$. This can be an illustration for understanding other chemical transitions. It was found empirically that for the purpose 150 gm urea is to be dissolved in 1000 cc of 40% HF solution.

Apparatus and Chemicals

Beckmann thermometer, graduated cylinder, burette, thermometric apparatus, pipette, HCl, HF, urea, and mercuric chloride.

Procedure

10 cc of $HgCl_2$ solution is added to 60 cc of HCl solution and the volume is made to 200 cc. The temperature of the mixture is adjusted to room temperature. The mixture is poured in the beaker of thermometric apparatus. 5 cc of urea containing HF is poured into sample solution.The sample is stirred and the reagent equalizes. Switch

the apparatus and adjust the galvanometer in the appropriate scale. The HF is poured into solution to be tested. If temperature of the solution is increases upon the addition of reagent, some more urea is dissolved in the HF, whereas if the temperature is decreased, a few c.c HF is added and the deflection of galvanometer is repeatedly checked. The procedure is continued till deflection is observed.

EXPERIMENT 7

Aim: Determination of the Fe_2O_3 and SiO_2 content of concrete.

Principle

Ferric ions in the concrete sample dissolved in HCl are reduced to ferrous ions with sodium polysulphide. The excess of sulphide is removed from the solution with mercuric chloride. Further, the ferrous ions are oxidized to ferric ions with hydrogen peroxide, this reaction causes, increase in temperature. It is proportional to the iron oxide content and can be measured. Further, HF is added to the solution which forms a hexafluoro complex with silicic acid along with change in temperature. The change in temperature brought about by the heat of reaction is proportional to the SiO_2 content.

Apparatus and Chemicals

Thermometric apparatus, pipette, magnetic stirrer, Beckmann thermometer, graduated cylinder, burette, hydrochloric acid, sodium polysulphide solution, mercuric chloride, H_2O_2, HF, urea, distilled water, and H_2SO_4.

Procedure

Weigh ~2.00 gm pulverized concrete sample in a 500 mL dry glass beaker. 100 cc of distilled water is added and the mixture is vigorously stirred with a magnetic stirrer in order to prevent sedimentation of the concrete. With continuous stirring, 60 cc HCl is added and stirred for further few minutes (~5 min). Saturated sodium polysulphide solution is dropped to the mixture as long as it becomes colourless (approx. 1 cc). Further 10 cc of saturated $HgCl_2$ is added and the volume of solution is made to 200 cc. The temperature of the solution is adjusted to room temperature and it is poured into thermally insulated flask. The pipettes are filled: one used for HF with 5 cc and other with 1-2 cc H_2O_2. The pipettes are immersed in solution and magnetic stirrer is used to equalise the temperature.

Switch on the apparatus. The reaction comes to end in about 20 to 30 sec, and observed in the maximum deflection. The value read from the scale gives SiO_2 content if it is compared with standard deflection for SiO_2. Similarly, for the determination of Fe_2O_3, a separate stunt is calibrated and circuit switched on, the galvanometer is repeatedly zeroed, and H_2O_2 is brought into the sample solution. The maximum of the deflection indicating the completeness of the reaction is recorded.

Evaluation

Calculate the Fe_2O_3 and SiO_2 content in the sample in per cent unit.

EXPERIMENT 8

Aim: Determination of crystallinity in polymer sample (DSC).

Principle

The relative amorphous and crystalline nature of a polymeric sample determines its utility for various technical applications which can be measured by different techniques, e.g., XRD, DTA, DSC, etc. However, the thermal analysis provides information with better accuracy. In general, the heat of fusion of a solid material is related to the percentage of crystallinity. Thus, the percentage crystallinity of a polymer is equal to the ratio of heat observed and standard heat of fusion of the same polymeric sample with known crystallinity.

Percentage Crystallinity

The crystallinity of a sample can be calculated by using enthalpy of a reaction observed in melting process employing the formula,

$$\text{Crystallinity of specimen sample} = \Delta H_m/\Delta \text{H°m} \times 100$$

ΔH_m experimental heat of fusion, **ΔH°m** literature value for % crystalline material heat fusion. However, the initial reaction and final reaction temperatures should be measured carefully.

$$\%\text{DOC} = \frac{\Delta H_m + \Delta H_c}{\Delta H_f} \times 100 \tag{1}$$

DOC is degree of crystallinity, $\Delta H_m + \Delta H_c$ is heat of fusion observed in analyzed sample, ΔH_f is heat of fusion observed in identical polymer sample with 100% crystallinity.

Apparatus and Chemicals

DSC, microbalance, polyphenylene sulphide (PPS), and Al_2O_3.

Procedure

The procedure is completed in two steps (heating cycles):

In the course of the first heating cycle, the heat of fusion of the sample is determined as described in Chapter 4.

In the second cycle heating is plotted against temperature and in the vicinity of the melting point, the sensitivity of the recorder is increased. The observed typical result is shown in Fig. 9.5. The area of both peaks indicates the heat of fusion and heat of crystallization.

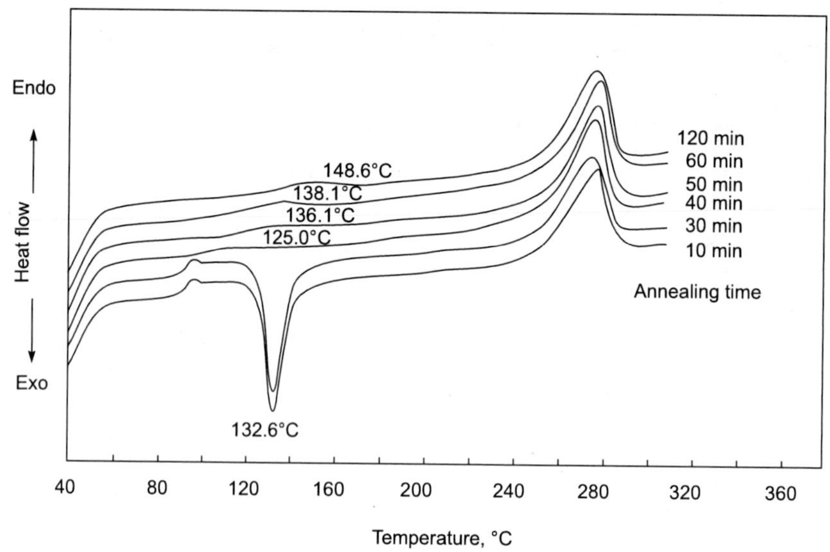

Fig. 9.5 DSC curves of a composite annealed at different temperatures (Adapted from Ref. 5)

Evaluation

Calculate the heat of fusion in calories per mole from the second record because crystallization is a second order change. Carry out calculations also with the equation (1) and the value of Hf_2 quoted for PPS are about basis of a value of 500 J g^{-1} taken from the literature, while the observed one is 10-12 cal g^{-1} (or 41.80-50.16 J g^{-1}). The values for other polymers may be taken form literature (some are listed in the Appendix). Explain the difference between the measured and the reported values.

Results

The calculated crystallinity is 22.44%. Which may vary for a polymeric sample annealed at different temperature and time.

EXPERIMENT 9

Aim: Determination of transition temperature (T_g) and heat of transition.

Apparatus and Chemicals

DSC apparatus, microbalance, potassium nitrate, and indium.

Principle

The basis of the principle described in Chapter 4, differential scanning calorimeter enables the determination of various types of transition temperature and heat of transition. For this purpose, first of all the apparatus is to be calibrated, and the value of the constant K has to be established. The standard materials are choosen such that their transition temperature lies in the required calibration temperature range. The transition temperature, $\beta \rightarrow \alpha$ transition in KNO_3 is on the basis of literature data taken, 127.7°C and the heat of transition is 12.05 cal g^{-1}. Calibration against metallic indium also seems to be adequate with melting point 156.6°C and heat of fusion 6.79 cal g^{-1}. Weigh the potassium nitrate and indium samples in sample holder. In order to determine the transition temperature, in the course of the first heating cycle recorded in the thermogram. After cooling, the second heating cycle is performed and recording is made as plotted against time. The evaluating procedure is explained below.

Evaluation of Parameters [6-7]

Glass Transition (T_g): The thermal analysis curves are used to evaluate the glass transition from a range of curves observed between **glassy-to-rubbery** phase. The evaluation of transition temperature from DTA or DSC curve is an important job. The various reported standard methods are: (i) Midpoint temperature method; (ii) Inflection method; (iii) Half-width method, and (iv) Comparing area method. A graphical illustration has been made in Fig. 9.7 to explain the methods along with comparison in Table 9.1.

TABLE 9.1 Evaluation methods

Methods	Features
Midpoint temperature method	Reliable method but unsteady baseline frequently makes errors.
Inflection method	The discrepancies with software, since they are based on different algorithms.
Half-width method	Yields precise value but superposed transitions hamper the results.
Comparing area method	The method is of little use but theoretically correct.

The T_g of polycarbonate, evaluated based on Fig. 9.6 by the inflection method, is 150 ± 0.6°C but half-width method yields 170 ± 1.50°C as shown in Fig. 9.7 by half width and area comparing methods. The obtained value for same polymer differs by used method and also on the basis of instrument used, e.g., DSC: heat flux or power compensated.

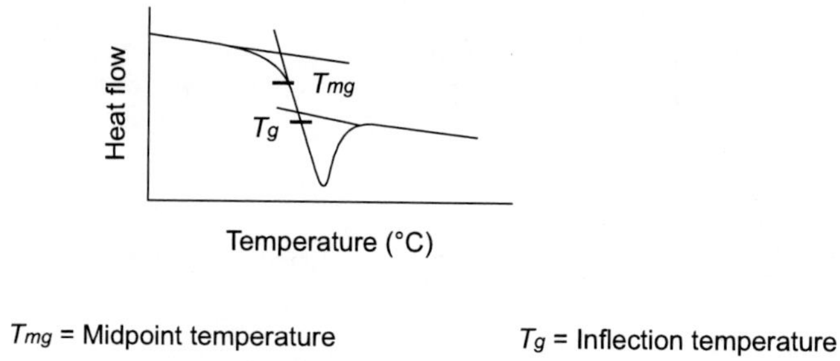

Fig. 9.6 Evaluation of T_g by inflection and half-width methods

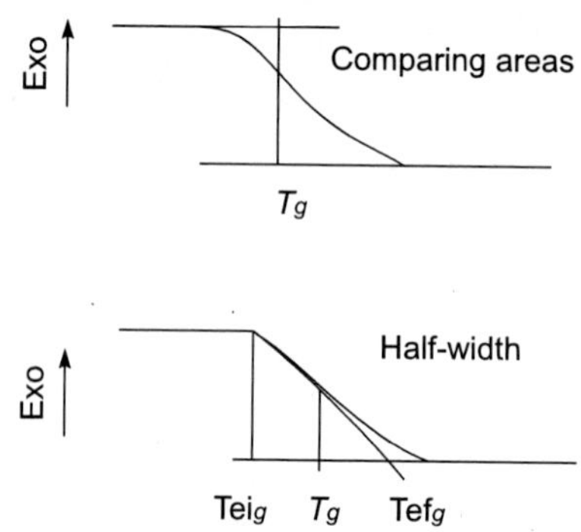

Fig. 9.7 Evaluating the glass transition by half-width and area comparing methods

Enthalpy (Heat of Reaction): The enthalpy of a reaction change is based on the area in a curve. It is calculated from the initial sample mass and the area between the curve. During the estimation, the following considerations should be taken.

(a) Baseline: Consideration of baseline is very important, a minor change can cause significant errors. The enthalpy, ΔH, calculation generally takes baseline as a straight line by joining before and after transition temperatures. The curved line is also taken sometimes. The determined value significantly changes on selected baseline. If the curve is not readily shaped, errors up to 30% may be observed. The illustration for consideration of baseline is made in Fig. 9.8.

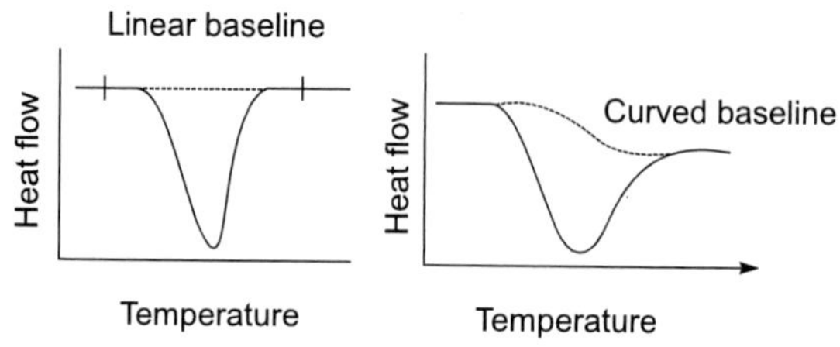

Fig. 9.8 Schematic diagram of different types of baselines used in DSC curve

(b) Limits of measurement: The temperature at which the tangent constructs the baseline, more significantly it covers the temperature in which the reaction is completed. This greatly affects the results concluded. Different types of limits and baselines are shown in Fig. 9.9. This mostly depends on the individuals and discretion of the operator.

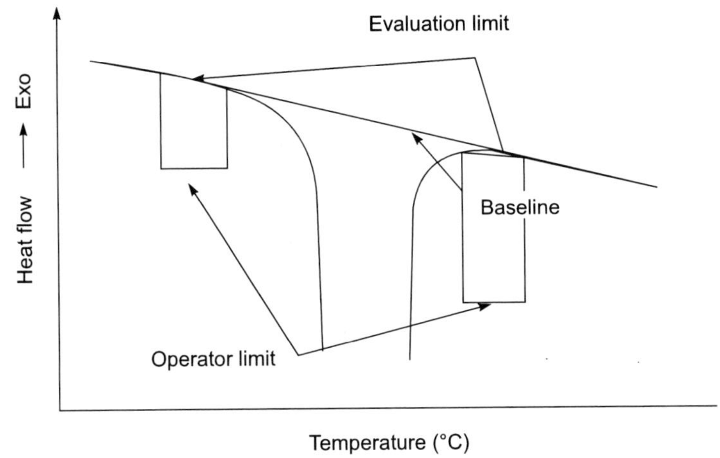

Fig. 9.9 Representation of baseline limits used in DSC curve

But, whenever a comparable measurement is required it is advisable to use the same types of baselines and evaluation limits after close inspection.

(c) Estimation of specific heat capacity

The specific heat capacity (C_p) can be determined of an unknown material at constant pressure by performing three consecutive DSC measurements:

(i) with two empty pans,
(ii) a pan filled with reference and other empty, and
(iii) a pan filled with sample and other empty.

The specific heat capacity is calculated by the difference between heat flux and between empty pan and the sample by employing Eq. (1)

$$C_p = Q(\text{sample}) - Q(\text{empty pan})/V \cdot m \tag{1}$$

where V is volume and m is mass.

However, the instrument should be calibrated by a standard material such as sapphire in the range of 0-200°C. The reported value changes from 0.718 to 1.016 J/g°C.

(d) Reaction temperature

The concept is explained with an example of melting point; the illustrative graph is shown in Fig. 9.10. The process of evaluation of melting point differs from crystalline materials to glassy or partially amorphous material like plastics. If the material is crystalline in nature, melting starts at the onset temperature (T_{im}) and continues until the entire specimen has melted at the melting peak temperature, T_{pm}. Then the curve returns to the baseline. This is where temperature compensation occurs between the specimen and reference materials. The melting point reported is either extrapolated temperature, T_{eim} or the melting peak temperature, T_{pm}. But if the material is plastic this analogy need to modify. Since at temperature T_i, the less crystallites begins to melt, T_{pm} the most crystallites melts but at T_{fm} all the crystallites have been melted or destroyed. Thus, the temperature (T_{fm}) is called experimental temperature or melting temperature (T_m).

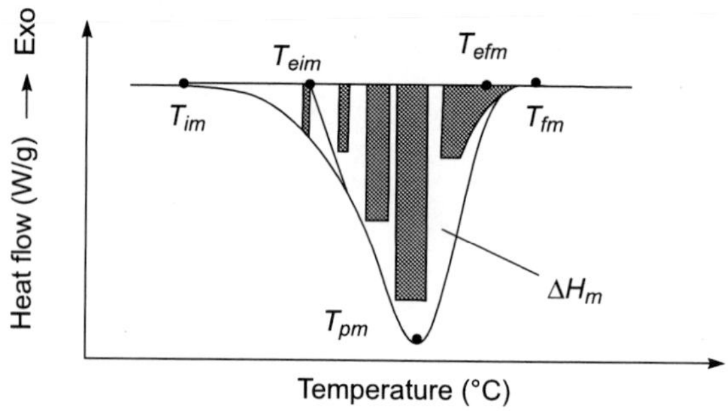

Fig. 9.10 Representation of evaluation of melting point or reaction temperature

EXPERIMENT 10

Aim: Calibration and determination of glass transition temperature by thermo-mechanical analyser (TMA).

Calibration of TMA

TMA instrument must be calibrated in terms of length and temperature if the load is applied electromagnetically. It should also be calibrated with the aid of known weights if the instrument is in constant use. However, a new calibration must be performed every time when a probe changes. Further, calibration and the test must be performed using same probe and applied load, heating rate and temperature range.

In general, calibration made for : Calibration parameter = Experimental parameters.

Calibration of Length

Length is calibrated mostly with a metallic specimen that has known reversible coefficient of linear thermal expansion. Aluminium specimen is frequently used. The calibration factor,

$$F = \alpha(\text{literature})/\alpha(\text{experimental})$$

Calibration of Temperature

The temperature is calibrated with a pure metal that has known reproducible melting point which lies in the temperature range of interest.

Measurement of T_g

Principle

Many methods have been discussed earlier to estimate crucial thermal properties by lab scientists to industry personnel with different ranges of sensitivity. Generally, at glass transition temperature (T_g), many physical properties are associated to amorphous domain to semicrystalline nature. These properties vary in several orders prior and after T_g. These properties are used to evaluate the temperature, one of them is linear thermal expansion explored in TMA for T_g assessment.

Procedure

(i) Calibrate the instrument as per the procedure described.

(ii) Record the expansion or penetration curve above the glass transition depending on the expansion, probe area, applied load and temperature dependent modulus of elasticity.

Evaluation of T_g

A representative TMA curve is shown in Fig. 9.11. The curve shows the linear expansion coefficient of expansion of a plastic in T_g range. The T_g is evaluated from the expansion curve by determining the point, where the extrapolated line before and after glass transition intersects, is marked T_g. However, the intersection of the extrapolated line on penetration is called softening temperature (T_s).

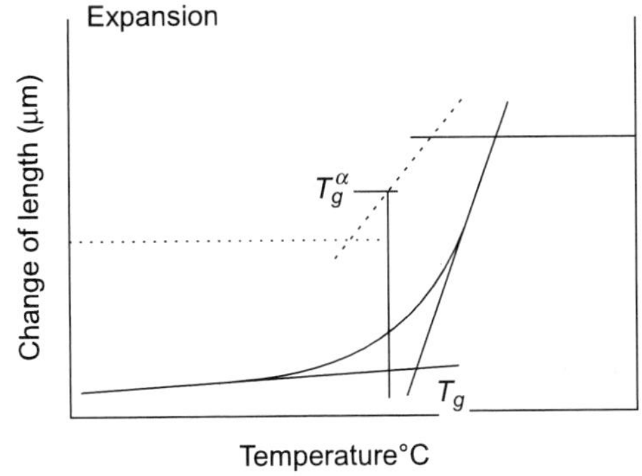

Fig. 9.11 Assessment of T_g by expansion curve: T_g is glass transition from linear expansion; T_g^{α} is glass transition from α

where α can be measured by slope of expansion curve and represented with dotted line.

EXPERIMENT 11

Aim: Estimation of carbon black content in a polymer sample.

Principle

In plastic industry, carbon black is used as a black pigment to generate protection against UV radiation, enhancing the heat stability and also imparting electrical conductivity.

The estimation of carbon black is based on the separation of carbon black and polymer by employing the different purge gases. In general, polymers are pyrolysed between 400 and 600°C in an inert atmosphere. However, the carbon black remains stable but can be oxidatively decomposed in oxygen atmosphere. The method is explained herewith a representative polymer polypropylene as an individual case.

Evaluation

A representative case of polypropylene containing carbon black is shown in Fig. 9.12. The curve reveals that PP pyrolysed at approx. 420°C, then after switch over is made from $N_2 \rightarrow O_2$ and subsequent weight loss appears for approximately 32%. This confirms the presence of 32% carbon black in PP sample due to loss in weight in oxygen atmosphere.

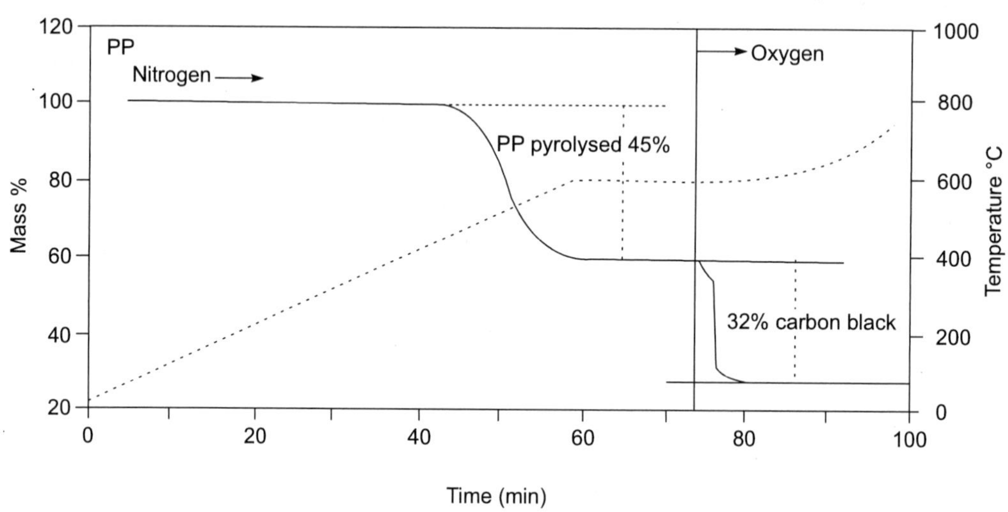

Fig. 9.12 OIT curve of PP containing carbon black

REFERENCES

1. Gordon, S. and Campbell, C. (1955). Differential thermal analysis of inorganic compounds. *Anal. Chem.* 27, 1102-1109.
2. Paulik, J., Wendlandt, W. W. and Paulik, F. (1981). Comprehensive Analytical Chemistry: Thermal Analysis. Simultaneous Thermoanalytical Examinations by Means of the Derivatograph. (Vol. 12), Elsevier, Amsterdam.
3. Wunderlich, B. (1990). Thermal Analysis. Academic Press, San Diego, CA.
4. Purgor, E. (1994). A Practical Guide to Instrumental Analysis. CRC Press, New York, USA.
5. Lee, T. H., Boey, F. Y. C. and Khor, K. A. (1995). On the determination of polymer crystallinity for a thermoplastic PPS composite by thermal analysis. *Compos. Sci. Technol.* 53, 259-214.
6. Gallagher, P. K. (1974). Thermal Analysis: Comparative Studies on Materials. Edited by Kambe, H. and Garn, P. D. John Wiley & Sons, New York.
7. Ehrenstein, G. W., Riedel, G. and Trawiel, P. (2004). Thermal Analysis of Plastics: Theory and Practice. Hanser Publisher, Munich.

Appendices

A. References

A1

(a) Journals dedicated to thermal analysis

(i) *The Journal of Thermal Analysis and Calorimetry,* Springer.

(ii) *Thermochimica Acta,* Elsevier

(b) Journals that mainly cover the techniques

(i) *Analytical Chemistry,* American Chemical Society

(ii) *Talanta,* Elsevier

(iii) *Analytica Chemica Acta*

(iv) *Analyst,* Royal Society

(v) *Journal of Polymers- Analysis and characterization,* Taylor and Francis

A2 Journal – *Analytical Chemistry*

The *Analytical Chemistry* publishes review of update in the field of thermal analytical techniques after the interval of every two years including detail references: *Analytical Chemistry,* 82(2010)4936, 80(2008)27 R, 78(2006)27 R, 76(2004)27 R, 74(2002)27 R, 72(2000)27 R, 70(1998)27 R, 68(1996)63 R, 66(1994)17 R, 64(1992)147 R, 62(1990)44 R.

A3 Reference Books (these have particular historical values)

(a) Mackenzie, R.C. (1984). A History of Thermal Analysis. *Thermochim Acta.* 73: 251-267.

(b) Wendlandt, W.W. (1986). The Development of Thermal Analysis Instrumentation (1955-1985). *Thermochim Acta.* 100: 1-22.

A4 Other Reference Books

1. Brown, M.E. (2001). Introduction to Thermal Analysis. London: Kluwer Academic Publishers.
2. Dean, J.A. (1995). The Analytical Chemistry Handbook. New York: McGraw Hill, Inc.

3. Ehrenstein, G.W., Riedel, G. & Trawiel, P. (2004). Thermal Analysis of Plastics: Theory and Practice. Munich: Hanser.
4. Ewing, G.W. (1990). Analytical Instrumentation Handbook. New York: Marcel Dekker Inc.
5. Gunzzler, H. &Williams, A. (2001). Hand Book of Analytical Techniques. VCH, Weinheim: Wiley. Vol. 2.
6. Hatakeyama, T. & Hatakeyama, H. (2004). Thermal Properties of Green Polymers and Composites. Dodrecht: Kluwer Academic.
7. Höhne, G.W.H., Hemminger, W.H., Flammersheim, H-J. (2003). Differential Scanning Calorimetry. Springer.
8. Menard, von K. (1999). Dynamic Mechanical Analysis: A Practical Introduction. Florida: CRC Press.
9. Meyer, R.A. (2000). Encyclopedia of Analytical Chemistry. John Wiley & Sons Ltd. Vol. 15.
10. Skoog, D.A., James Holler, F. & Nieman, T. (1998). Principles of Instrumental Analysis. Belmont: Brooks/Cole.
11. Sorai, M. (2004). Handbook of Calorimeter and Thermal Analysis. Hoboken, NJ: John Wiley.
12. Turi, E.A. (1997). Thermal Characterization of Polymeric Materials. Two-Volume Set. U.S.: Academic Press Inc.
13. Wunderlich, B. (2005). Thermal Analysis of Polymeric Materials. Berlin: Springer.

B. Societies

1. Indian Thermal Analysis Society
   Analytical Chemistry Division
   Bhabha Atomic Research Centre, Mumbai- 400 085.
   Tel: 91-22-25590645
   Email: arenu@barc.gov.in
2. Canadian Thermal Analysis Society (CTAS)
   CANMET Materials Technology Laboratory
   Natural Resources Canada
   568 Booth Street, Ottawa, ON K1A 0G1
   Tel: 613-947-7058
   Fax: 613-992-8735
   Email: Dmartine@NRCan.gc.ca
3. North American Thermal Analysis Society
   LOIS HALL
   Institute for Combustion Science and Environmental Technology
   2413 Nashville Rd. Bowling Green, KY 42101

Western Kentucky University
Tel: 270-745-2220
Fax: 270-745-2221
Email: natas@wku.edu

4. International Confederation of Thermal Analysis and Calorimetry (ICTAC)
SONY Institute of Higher Education, Department of Informatics and Media Technology
Atsugi, Kanagawa 243-8501
Japan
Fax: +81-46-250-8936
Tel: +81-46-247-3131
Email: ozao@ei.shohoku.ac.jp
5. Dutch Society for Thermal Analysis and Calorimetry, (Thermische Analyse Werkgroep, Netherlands, TAWN)
Dr. P.J. van Ekeren (President)
TNO Defense Security and Safety
Department of Energetic Materials
P.O. Box 45, 2280 AA Rijswijk
Tel: +31 (0)15 2843280
Fax: +31 (0)15 2843958
Email: p.j.vanekeren@chem.uu.nl
Website: http://home.wanadoo.nl/tawn/home-en.htm
6. The Japan Society of Calorimetry and Thermal Analysis
Miyazawa Bldg. 601, 1-6-7 Iwamoto-chyo, Chiyoda-ku,
Tokyo 101-0032, Japan
Fax: +81-3-5821-7439
Email: netsu@mbd.nifty.com
7. Canadian Thermal Analysis Society
Subhash C. Mojumdar (President)
University of Toronto
Dept. of Chem. Engineering and Applied Chemistry
200 College Street
Toronto, Canada M5S 3E5
Tel: 416-978-0798
Fax: 416-978-1376
E-mail: subhash.mojumdar@utoronto.ca

C. Manufacturers

1. TA Instruments - Waters LLC
   Corporate Headquarters
   159 Lukens Drive
   New Castle, DE 19720, USA
   Tel: 302-427-4000
   Fax: 302-427-4001
   E-mail: INFO@tainstruments.com
   Website: http://www.tainstruments.com
2. Shimadzu Corporation, Japan
   1, Nishinokyo-Kuwabara-cho, Nakagyo-ku,
   Kyoto 604-8511, Japan
   Tel: 81(75)823-1111
   Website: www.shimadzu.com
3. SETARAM Instrumentation
   7, rue de l'Oratoire
   69300 Caluire, France
   Tel: + 33 (0)4 72 10 25 25
   Fax: +33 (0)4 78 28 63 55
   Website: www.setaram.com
4. Perkin Elemer Ltd
   PerkinElmer
   940 Winter Street
   Waltham, Massachusetts 02451, USA
   Tel: +1 (781) 663-6900
   Website: www.perkinelmer.com
5. Mettler Toledo Inc
   Mettler-Toledo Inc., 1900 Polaris Parkway, Columbus, OH 43240
   Tel: 1614 4384511, 1 614 438 4646,
   Website: mary.finnegan@mt.com
6. NETZSCH- Gerätebau GmbH
   Wittelsbacherstrasse 42
   95100 Selb, Germany
   Tel: (+49) 9287/881-0
   Fax: (+49) 9287/881-505
   E-mail: at@netzsch.com
   Website: www.netzsch-thermal-analysis.com

7. Linseis Inc.
   109 North Gold Drive
   Robbinsville, NJ 08691, USA
   Tel.: +01 (609) 223 2070
   Fax: +01 (609) 223 2074
   Email: info@linseis.com
   Website: www.linseis.net
8. Instrument Specialists, Inc.
   133 E, Main Street, P.O. Box. 280
   Twin Lakes, WI 53181
   Tel: (262) 877-3600
   Fax No. - (262) 877-3699
   Website: www.instrument-specialists.com
   Email: info@instrument-specialists.com
9. Thorn Scientific Services Ltd.
   7 Chantry Close, Ashtead, Surrey, KT21 2QD, UK
   Tel: 01372 802 537
   Fax: 01372 802 818
   Email: steve@thornta.com
   Website: www.thornta.com

D. Regular Academic Events

1. National Symposium & Workshop on Thermal Analysis (THERMANS) Biannual (ITAS)
2. CTAS Workshop and Exhibition (Annual) (CTAS)
3. Annual Conference Exploring the Frontiers of Thermal Analysis (NATAS)
4. International Congresses on Thermal Analysis and Calorimetry (After every four years)
5. Symposium on "Thermische Analyse Dag" (TAD). (Annually in November)
6. Japanese Conference on Calorimetry and Thermal Analysis (JCCTA)(Annually)

E. Standards and Reference Laboratories

1. TGA: Certified Calibration Kit; Certified Temperature Calibration Kit (P/N 952384-901)
   Secondary Temperature Calibration Materials (Nickel, Alumel)
2. Reference Laboratories (NIST, LGC)
   Universal Magnet
   ASTM E1582 test method
   Detailed ISO style calibration instructions

Glossary

Differential thermal analysis (DTA): The technique in which the temperature difference between a substance and a reference material is measured as a function of temperature whilst the substance and reference material are subjected to a controlled temperature programme. The record is the differential thermal or DTA curve; the temperature difference (AT) should be plotted on the ordinate with endothermic reactions downwards and temperature or time on the abscissa increasing from left to right. The term quantitative differential thermal analysis (quantitative DTA) covers the uses of DTA where the equipment is designed to produce quantitative results in terms of energy and/or any other physical parameter.

Differential scanning calorimetry (DSC): The technique in which the difference in energy inputs into a substance and a reference material is measured as a function of temperature whilst the substance and reference material are subjected to a controlled temperature programme. Two modes, power-compensation differential scanning calorimetry (power compensation DSC) and heat-flux differential scanning calorimetry (heat-flux DSC), can be distinguished depending on the method of measurement used.

Dynamic thermomechanometry: The technique in which the dynamic modulus and/or damping of a substance under oscillatory load is measured as a function of temperature whilst the substance is subjected to a controlled temperature programme. Torsional braid analysis is a particular case of dynamic thermomechanometry in which the material is supported on a braid.

Emanation thermal analysis: The technique in which the release of radioactive emanation from a substance is measured as a function of temperature whilst the substance is subjected to a controlled temperature programme.

Evolved gas analysis (EGA): The technique in which the nature and/or amount of volatile product(s) released by a substance is/are measured as a function of temperature whilst the substance is subjected to a controlled temperature programme.

Evolved gas detection (EGD): The technique in which the evolution of gas from a substance is detected as a function of temperature whilst the substance is subjected to a controlled temperature programme.

Heating-curve determination: The technique in which the temperature of a substance is measured as a function of the programmed temperature whilst the substance is subjected to a controlled temperature programme in the heating mode. Sample temperature should be plotted on the ordinate increasing upwards and programmed temperature or time on the abscissa increasing from left to right. Two derivative curves, heating-rate curves (for dT/dt) against (T or t) and inverse heating-rate curves (for dt/dT against T or t), can be obtained.

Isobaric mass-change determination: The technique in which the equilibrium mass of a substance at constant partial pressure of the volatile product(s) is measured as a function of temperature whilst the substance is subjected to a controlled temperature programme. The record is the isobaric mass-change curve; the mass should be plotted on the ordinate decreasing downwards and temperature on the abscissa increasing from left to right.

Thermoacoustometry: The technique in which the characteristics of imposed acoustic waves are measured as a function of temperature after passing through a substance whilst the substance is subjected to a controlled temperature programme.

Thermodilatometry: The technique in which a dimension of a substance under negligible load is measured as a function of temperature while the substance is subjected to a controlled temperature programme. The record is the thermodilatometric curve; the dimension should be plotted on the ordinate increasing upwards and temperature or time on the abscissa increasing from left to right. Linear thermodilatometry and volume thermodilatometry are distinguished on the basis of the dimensions measured.

Thermoelectrometry: The technique in which an electrical characteristic of a substance is measured as a function of temperature whilst the substance is subjected to a controlled temperature programme. The most common measurements are of resistance, conductance or capacitance.

Thermogravimetry (TG): The technique in which the mass of a substance is measured as a function of temperature whilst the substance is subjected to a controlled temperature programme. The record is the thermogravimetric or TG curve; the mass should be plotted on the ordinate decreasing downwards and temperature (T) or time (t) on the abscissa increasing from left to right.

Thermomagnetometry: The technique in which the magnetic susceptibility of a substance is measured as a function of temperature whilst the substance is subjected to a controlled temperature programme.

Thermomechanical analysis (TMA): The technique in which the deformation of a substance under non-oscillatory load is measured as a function of temperature whilst the substance is subjected to a controlled temperature programme. The mode, as determined by the type of stress applied (compression, tension, flexure or torsion), should always be stated.

Thermoparticulate analysis: The technique in which the release of particulate matter from a substance is measured as a function of temperature while the substance is subjected to a controlled temperature programme.

Thermophotometry: The technique in which an optical characteristic of a substance is measured as a function of temperature while the substance is subjected to a controlled temperature programme.

Thermomicroscopy: Measurement of total light, light of specific wavelength(s), refractive index and luminescence lead to thermophotometry, thermospectrometry, thermorefractometry and thermoluminescence, respectively; observation under the microscope leads to Thermomicroscopy.

Thermosonimetry: The technique in which the sound emitted by a substance is measured as a function of temperature whilst the substance is subjected to a controlled temperature programme.

Index